Springer Aerospace Technology

Series Editors

Sergio De Rosa, DII, University of Naples Federico II, Napoli, Italy

Yao Zheng, School of Aeronautics and Astronautics, Zhejiang University, Hangzhou, Zhejiang, China

The series explores the technology and the science related to the aircraft and spacecraft including concept, design, assembly, control and maintenance. The topics cover aircraft, missiles, space vehicles, aircraft engines and propulsion units. The volumes of the series present the fundamentals, the applications and the advances in all the fields related to aerospace engineering, including:

- structural analysis,
- aerodynamics,
- aeroelasticity,
- aeroacoustics,
- flight mechanics and dynamics,
- orbital maneuvers,
- avionics,
- systems design,
- materials technology,
- launch technology,
- payload and satellite technology,
- space industry, medicine and biology.

The series' scope includes monographs, professional books, advanced textbooks, as well as selected contributions from specialized conferences and workshops.

The volumes of the series are single-blind peer-reviewed.

To submit a proposal or request further information, please contact: Mr. Pierpaolo Riva at pierpaolo.riva@springer.com (Europe and Americas) Mr. Mengchu Huang at mengchu.huang@springer.com (China).

The series is indexed in Scopus and Compendex

More information about this series at http://www.springer.com/series/8613

Denis Alexandrovich Akmaykin ·
Eduard Anatolyevich Bolelov ·
Anatoliy Ivanovich Kozlov ·
Boris Valentinovich Lezhankin ·
Alexander Evgenievich Svistunov ·
Yury Grigorievich Shatrakov

Theoretical Foundations of Radar Location and Radio Navigation

Springer

Denis Alexandrovich Akmaykin
Vladivostok, Russia

Eduard Anatolyevich Bolelov
Moscow, Russia

Anatoliy Ivanovich Kozlov
Moscow, Russia

Boris Valentinovich Lezhankin
Irkutsk City, Russia

Alexander Evgenievich Svistunov
Moscow, Russia

Yury Grigorievich Shatrakov
Saint Petersburg, Russia

ISSN 1869-1730 ISSN 1869-1749 (electronic)
Springer Aerospace Technology
ISBN 978-981-33-6516-2 ISBN 978-981-33-6514-8 (eBook)
https://doi.org/10.1007/978-981-33-6514-8

This Springer imprint is published by the registered company Springer Nature Singapore Pte Ltd.
The registered company address is: 152 Beach Road, #21-01/04 Gateway East, Singapore 189721, Singapore

Preface

Radar location and radio navigation today are inherent attributes of modern science and technology, and its fundamental principles find the widest and multiple application even far from radio-engineering areas of human activity such as medicine, biology or meteorology. That is what accounts for a large number of regularly published monographs, books, academic papers and popular publication and contributions dedicated to a given problems of radar, radio navigation and its application.

The rarer event is a publication of academic books, oriented for higher educational institutions, on fundamentals of radar location and radio navigation, and on corresponding equipment. Here, it is worthwhile to mention those Russian scientists by tutorial materials of which the radar location and radio navigation were comprehended by all who is connected with these areas of science and technology. Among them are Profs. P. A. Bakulev, G. B. Belotserkovskiy, P. I. Dudnik, G. S. Kondratenkov, A. G. Saybel, U. G. Sosulin, E. G. Trubitsin, M. I. Finkelshtein.

The recommended to readers study guide, eventually written by students of the listed pleiad of scientists, first of all reflects a specific character of radar location and radio navigation problems and tasks as applied foremost to tasks and needs of air and maritime transport.

The study guide is divided into two parts: theoretical basics of radar location and radio navigation on the one side, and corresponding to its equipment, from the other side. The paper represents both commonly known approaches, stated in earlier published study books and tutorials, and original authorship material obtained during performance of theoretical and experimental works and at handling of functioning radar location and radio navigation equipment.

Book is intended for students of radio-engineering educational program specialization and will be useful to everyone interested in modern problems of radar location and radio navigation and its application.

Moscow, Russia

Anatoliy Ivanovich Kozlov
Professor, Doctor of Physical and Mathematical Sciences

Contents

Abbreviations

A	Antenna
AA	Antenna assembly
AAIM	Autonomous onboard integrity monitoring
AAS	Aircraft attitude sensor
ABAS	Airborne augmentation system
AC	Alternate current
ACR	Airfield control radar
AdC	Adding circuit
ADM	Antenna drive mechanism
ADP	Antenna directional pattern
ADSM	Antenna driving and stabilization mechanism
AFC	Automatic frequency (tuning) control
AFCS	Automatic flight control system
AFCU	Automatic frequency control unit
AFS	Antenna-feeder system
ADC	Analog-to-digital converter
AGC	Automatic gain (adjusting) control
AGI	Above ground installations
AI	Airborne instrument
AIS	Automatic identification system
AM	Amplitude-modulated
Amp	Amplifier
Anlz	Analyzer
APS	Angular position sensor
APV	Approach procedures with vertical guidance
ARB	Automatic radar beacon
ARC	Automatic radio compasses
ARPS	Automatic radar plotting system
AS	Antenna switch
ASCU	Antenna stabilization and control unit
AT	Auto-tracking
ATC	Air traffic control

ATIS	Automatic terminal information service
ATM	Air traffic management
Att.	Attenuator
AV	Aerial vehicle
BCRS	Broadcasting radio stations
BM	Balanced modulator
BRC	Back-reflection coefficient
BSD	Backscattering diagram
C	Compensator
CCD	Coherent change detection
Circ	Circulator
CLO	Coherent local oscillator
CO	Coherent oscillator
COG	Course over ground
COLREG	International Regulations for Preventing Collisions at Sea
CONV	Converter
Cr. Ready	Course readiness
CRS	Coherent radar station
CRT	Cathode-ray tube
CS	Commercial service
CTrm	Coherent transmitter
CU	Control unit
CuMa	Clutter map
DAPP	Controllers of approach points
DAP	Digital autopilot
DC	Direct current
DCI	Distance to course intersection
DCU	Digital control unit
DF	Doppler frequency
DGNSS	Differential global navigation satellite system
DME	Distance measuring equipment
DME-P	Precise distance-measuring equipment
DMU	Device matching unit
DOP	Dilution of precision
DP	Directional pattern
DPC	Dispatching points of the circle
DPLS	Dispatching points of landing system
DPU	Digital processing unit
DSA	Distance of shortest approach
DSDM	Doppler speed and drift angle meters
DSU	Digital synthesis unit
DU	Display unit
DVOR	Doppler very-high-frequency Omnidirectional Range
DWRS	Doppler weather radar system
EBL	Electronic bearing lines

EqBL	Equal bearing lines
EDDL	Equal differences in distance lines
EGNOS	European geostationary navigation overlay service
EM	Electric motor
EMF	Electromotive force
EMW	Electromagnetic waves
ERA	Effective reflective area
ERBL	Electronic range and bearing lines
ESDL	Equal sum in distance lines
ESL	Equal space line
ETA	Estimated time of arrival
EWS	Electromagnetic wave scattering
Ex.S	External systems
FA	Final approach
FAGC	Fast automatic gain control
FAS	Final approach segment
FC	Feeder circuit/path
FDD	Frequency difference detector
FEC	Forward error correction
FM	Frequency modulation
FMB	Frequency multiplier block
FMp	Frequency multiplier
FPAP	Flight-path alignment point
FrD	Frequency divider
GAGAS	Geostationary navigation supplement of the GPS
GB	Gearbox
GBAS	Ground-based augmentation system
GC	Gyrocompass
GCS	Ground control segment
GCU	Gain control unit
GDOP	Geometrical dilution of precision
GEO	Geostationary orbit
GES	Ground earth stations
GIC	GPS integrity channel
GLONASS	Global navigation satellite system
GMS	Ground mission segment
GNSS	Global navigation satellite system
GPA	Glide-path angle
GPB	Glide-path beacon
GPIP	Glide-path intersection point
GPS	Global Positioning System
GR RNS	Goniometric-ranging (rho-theta) RNS
GRNS	Goniometric radio-navigational systems
GSS	Ground sensor stations
GST	Galileo system time

GTRF	Galileo terrestrial reference frame
HA	Heading angle
HFM	High-frequency modulator
HFO	High-frequency oscillator
HNS	Hyperbolic navigation systems
HP	Horizontal plane
HRNS	Hyperbolic (differential-ranging) radio navigation systems
Htd.	Heterodyne
IA	Initial approach
IAGC	Instantaneous automatic gain control
ICAO	International Civil Aviation Association
ICSt	Integral control station
ICS	Integrated control system
IDD	Input distribution device
IERS	International Earth Rotation Service
IFA	Intermediate-frequency amplifier
IFO	Intermediate-frequency oscillator
IFPA	Intermediate-frequency pre-amplifier
IFq	Intermediate frequency
IFx	Intermediate fix
ILS	Instrument landing system
Int.	Integrator
IRf	Interfering reflections
IRNSS	Indian regional navigation satellite system
ITB	International Time Bureau
ITRF	International terrestrial coordinate system
LA	Limiting amplifier
LAL	Local airlines
LB	Locator beacon
LCCS	Local control and corrective system
LDP	Local dispatching points
LFM	Linear frequency modulation
LNA	Low-noise amplifier
LOC	Localizer
LRL	Landing (approach) radar locator
LRRNS	Long-range radio-technical navigation systems
LWC	Liquid water content
M	Modulator
Magn.	Magnetron
MDAP	Main dispatching approach points
MFD	Multi-function display
MLC	Matching loss coefficient
MLS	Microwave landing system
MMR	Multi-mode receiver
MP	Monochrome pulse

MRB	Magnetic radio bearing
MRL	Meteorological radar locator
MS	Master station
MSAS	Space augmentation system
MSK	Minimum shift keying
MT	Message types
MTI	Moving target indicator
MU	Matching unit
Mx	Mixer
NCRS	Non-coherent radar system
NGP	Navigational guide points
NSV	Navigation space vehicles
OCS	Operational control segment
OOW	Officer of the watch
OPC	Over-periodical compensation
PA	Power amplifier
PC	Personal computer
PCM	Pulse-code modulation
PCRS	Pseudo-coherent radar system
PD	Phase detector
PDOP	Position dilution of precision
PFD	Power flow density
PI	Potential indicator
PkD	Peak detector
PLL	Phase-locked-loop frequency control
POB	Persons onboard
PPAA	Passive phased antenna array
PPI	Plan position indicator
PPS	Precise Positioning Service
PRS	Pseudo-random sequence
PS	Phase shifter
PU	Processing unit
PuA	Pulse amplifier
PVOR	Precision VHF omnidirectional range
QZSS	Quasi-zenith satellite system
R	Rectifier
RAIM	Receiver autonomous integrity monitoring
RB	Radio beacon
RBLS	Radio beacon landing system
RBS	Radar beacon sequencer
RBT	Radar beacon transponder
RCS	Radar cross section
Rcv	Receiver
RFA	Radio-frequency amplifier
RfS	Reference stations

RLS	Radar location station
RM	Relative motion
RNP	Radio navigation point
RNS	Radio navigation station
RNY	Runway
RQ	Range request
RRNS	Range-finding radio navigation systems
RRP	Radar reference points
RS	Radar station
RS	Range response
RSP	Runway supervisory points
SART	Search and rescue radar transponder
SBAS	Satellite-based augmentation system
SHF	Super-high-frequency band
SLO	Stable local oscillator
SLRL	Surveillance and landing radar locator
SM	Summator
SOG	Speed over ground
SPS	Standard Positioning Service
SRL	Secondary radar locator
SRL-A	Aerodrome surveillance radar locator
SRL-R	Route surveillance radar locator
SRNS	Satellite radio navigation systems
ST	Selsyn transmitter
STP	Signal transmission path
SWM	Surface-wave magnetron
Synch.	Synchronizing unit
SyP	Synchro pickup
TACAN	Tactical control and navigation system
TAGC	Temporal automatic gain control
TAI	Temps atomique international/international atomic time
TAP	Terminal Access Point/terminal area path
TCG	Time corrected gain
TCH	Threshold crossing height
TCI	Time to course intersection point
TDOP	Time dilution of precision
TDP	Taxiing dispatch points
TDU	Turbulence detection unit
TEC	Total number of electrons
TM	True motion
TP	Target position
T-R	Transmitter/receiver/transceiver
TRB	True radio bearing
TRD	Transmitting device
TRL	Transmitting limiter

Trm	Transmitter
TRU	Transmit/receive unit
TSA	Time of shortest approach
TSS	Traffic separation schemes
tt&c	Telemetry tracking & command
TU	Terminal unit
UAIS	Universal Automatic Identification System
UAV	Unmanned aerial vehicle
UF	Uncertainty function
UHF	Ultra-high frequency
UHFG	Ultra-high-frequency generator
ULS	Uplink station
UMAS	Unified Measurement Acquisition Stations
USNO	United States Marine Observatory
UTC	Universal Time Coordinated
VCO	Voltage-controlled oscillator
VDF	VHF direction finder
VDOP	Vertical dilution of precision
vessel POB	Vessel persons onboard
VHF	Very high frequency
VOR	Very-high-frequency omnidirectional range
VP	Vertical plane
VRC	Variable range circle
VTMS	Vessel traffic management system
WAAS	Wide Area Augmentation System
WDCMS	Wide-area differential correction and monitoring system
WNRS	Weather-navigational radar station
WP	Waypoints
WRS	Wide-area reference stations
WS	Wind shear

Part I
General Information on Radar Location and Radio Navigation Systems

Chapter 1
Basic Terms and Definitions

Radar location—It is the science on methods and means of information obtaining on objects based on receiving and analysis of radio waves, reflected or radiated by these objects. The obtained data comprises the **radar information**, to which a target coordinates, its velocity and qualification profile are referred.

Active, semi-active, active with passive respond and passive radar location are distinguished.

Radar location objects are called **radar targets**.

Aerodynamic (flying vehicles), space, land, underground, sea and underwater radar targets are referred to (artificial) **phantom radar targets**.

Ground and water surface, local (ground) features, clouds, meteorite, atmospheric irregularities, etc., are related to **natural radar targets**.

Technical equipment for radar information acquisition is called **radar stations** (RS), **radar locators, radars**.

Radio locators are differed by radio wave band used, by sounding signal shape, polarization type, amount of implemented channels, number and type of measured coordinates, radar deployment position.

An aggregate of radar stations and auxiliary technical aids integrated for fulfillment of some radar location task is called **radar system**.

Target data is received during observation.

The following stages can be can be notionally allocated at radar surveillance:

- detection of targets;
- measurement of moving targets coordinates and parameters;
- resolution of targets;
- discrimination and identification of targets.

Detection is provided on the base of analysis of received electromagnetic wave and reduced to decision making on presence or absence of target in observed airspace segment.

D. A. Akmaykin et al., *Theoretical Foundations of Radar Location and Radio Navigation*, Springer Aerospace Technology,
https://doi.org/10.1007/978-981-33-6514-8_1

Measurement concludes in coordinates and parameters estimation acquisition of radar targets movement.

Resolution—separates detection and position finding in airspace of several objects with little distinctive coordinates and parameters of movement.

Discrimination (classification)—object referring to some from the allocated classes. Partition into classes can be carried out randomly, e.g., by purpose, for example, passenger aircraft, bomber aircraft, satellite, motor ship, cruiser. Special attention is for target pertaining to one of "friend" or "foe" classes, i.e., so-called identification, usually considered as a separate task.

Radio navigation—science on radio-technical (engineering) methods and equipment of information acquisition on position of moving objects in airspace and movement of these objects from one point of airspace to another.

Radio navigation objects are aircraft, helicopters, ships, satellites, ground transport and other moving vehicles.

Main tasks of navigation are the following:

- determination of current position and parameters of object movement, in particular, a course (heading);
- ensuring of object movement in defined trajectory and lead out to specified point at assigned time.

Trajectory is a spatial curve, by which an object center of mass is shifted during movement.

Trajectory projection into ground surface is called **course line** (track or path).

Trajectory projection of flying vehicles into vertical plane—**flight profile**—is examined as well.

Ground surface point, above which a moving object is located, is called **position**.

Information acquisition on navigation is carried out using different technical aids. In radio navigation, these aids are radar locators and other radio-technical means: range-finding, goniometric-ranging (rho/theta), range-difference (hyperbolic) systems, onboard radar stations, Doppler ground velocity systems. Autonomous (independently from ground systems) measurement of navigational parameters using radio waves is conducted by **radio navigation devices**. Such devices are autonomous: It could be radio-altitude meters, Doppler ground velocity and drift angle metering systems.

Radio navigation system is called an aggregate of installed on moving object and out of it (on ground, in space) of interconnected radio-technical devices intended for measurement of navigational parameters. Special feature of radio navigation systems comparing with radar location systems concludes in information acquisition on objects itself using radio signals radiated from airspace points with the known coordinates—**radio navigation points**.

Both radar location and radio navigation represent an area of radio electronics. Radar location and radio navigation systems are referred to a class of information systems. They have a common task of information retrieval from picked electromagnetic oscillations, including an information content—coordinates and parameters of

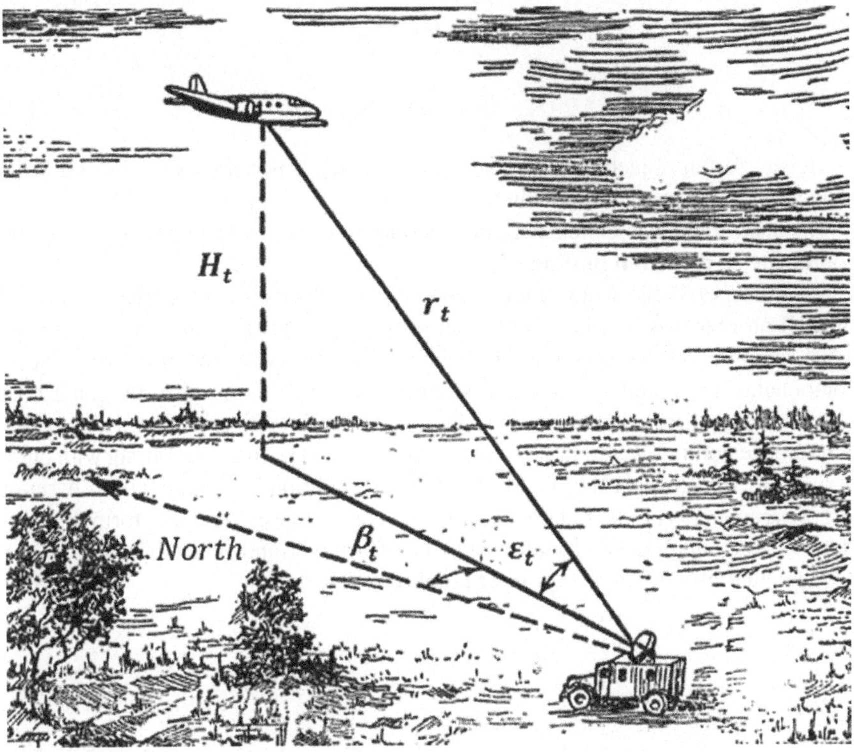

Fig. 1.1 Target parameters measurement in polar coordinate system

objects movements. Usually, coordinates are measured in polar or in cylindrical coordinate systems. Polar coordinates of an object (target) are slant distance (range) r_t, azimuth β_t and elevation angle of an object ε_t (Fig. 1.1).

Slant range—r_t is a distance from initial point 0 (RS) to an object T.

Azimuth β_t of an object—is called a clockwise read-out angle between direction to north and projection to horizontal plane of straight passing through initial point and an object.

Angle between this straight line and its projection is called **target elevation angle** ε_t.

In cylindrical coordinate systems, an object position is defined by the following coordinates: horizontal range r_h, target altitude (height) H_t and target azimuth β_t.

Horizontal range—r_h projection length of straight line segment, connecting initial point and an object, into horizontal plane. **Target altitude—H_t** is a distance from a target up to horizontal plane, passing through initial point, or from a target up to a ground surface (h_t).

Object coordinates are converted from a coordinate system into another according to formulas:

$$r_h = r_t\, cos\varepsilon_t, \tag{1.1}$$

$$H_t = r_t\, sin\varepsilon_t. \tag{1.2}$$

Hence, for an object position finding in airspace it is quite enough to know its slant range and angular coordinates.

Distance finding is called **ranging (radametry)** and determination of angular coordinates—**direction finding** (bearing).

Parameters of object movement characterize a change of its spatial position in time. If an object is a point, then its movement parameters characterize a change of its spatial coordinates and are described, as a rule, by an object velocity vector components V_t: radial V_r, oriented from a reference point; and orthogonal to it tangential V_{tg}, directed at tangent to a circle passing through an object with a center at reference point. Acceleration vector components can be measured along with other derivatives with respect to time of object spatial position change function. A set of target parameters is to be radar measured and can be represented in a form of target state vector. Supplemented with a number of a target and other data, this vector acquires the form of so-called **target label** (profile).

Chapter 2
Principal Physics of Radar Location and Radio-Navigation

A carrier for radar location and radio-navigation information is an electromagnetic (EM) field. Authors assume that the most corresponding definition of EM field is the following:

Electromagnetic field *is a one of the forms of matter existence capable to carry an energy and information.* Nothing more!

Within perceptions of comprehensive physics, without getting into micro-world and quantum effects, in order to describe electromagnetic field properties, i.e., quantitative characteristics of denoted form, it is necessary to use 52 independent of one another numbers in each space point and each time moment. Decades have been spent of Maxwell, Ampere, Volta, Hertz, Joule, Kirchhoff, Lenz, Lorentz, Ohm, Faraday, Oersted and many others genius efforts to observe that these are precisely 52 numbers. Exactly these scientists proposed a raw of perfect physic-mathematical models describing electromagnetic field.

In the most modern approaches, **four vectors** are used in these models as a fundamental notions,—electric field vector **E** (3 numbers) and magnetic field vector **H** (3 numbers), and electric displacement vector **D** (3 numbers) and magnetic induction vector **B** (3 numbers), describing electromagnetic field in Q given point with (x, y, z) coordinates.

These vectors are interrelated with Maxwell equations representing four (4) differential equations, two (2) from which are be inscribed in vector, two—in scalar forms; i.e., it is referred to eight (8) differential equations in which derivatives of each **E**, **H**, **D**, **B** vector component with respect to t time, i.e., its changing rate (in total $3 \times 4 = 12$ numbers), are present. Besides, equivalent terms of these equations are also derivatives of each from the components in each special coordinates x, y, z; i.e., another $9 \times 4 = 36$ numbers. Hence, this refers to $12 + 12 + 36 = 60$ variables interrelated with eight (8) Maxwell equations that gives ground to refer to $60 - 8 = 52$ independent of one another numbers, characterized an EM field in $Q(x, y, z)$ point at t time moment.

D. A. Akmaykin et al., *Theoretical Foundations of Radar Location and Radio Navigation*, Springer Aerospace Technology,
https://doi.org/10.1007/978-981-33-6514-8_2

Here, it is appropriate to rise a question that concludes into the following. Well, Maxwell described electromagnetic field basing on **E, H, D, B** vectors and its derivatives in all four (4) spatial–time coordinates. Technically, Maxwell equations represent eight (8) equations interrelated with 52 variables. Is it possible to pick out another combination from these variables and relate them with some another equations? We have an unambiguous answer: "Of course, it is possible." For complete description of electromagnetic field in some problems, it is more convenient to use another characteristics of a field such as, for instance, a vector **A** and φ scalar potentials, Hertz vector **Π**, which are easily converted into **E, H, D, B** classical vectors.

So why the preference was given to **E, H, D, B** vectors and what exists in nature? Historically it happened that electromagnetic field and, more precisely, its development, was observed in a form of some force actions on electric charges for what it was convenient to introduce electric field vector **E** (E vector) as a force acting on a unit charge. As for the second part of question: "What exists in nature?"... we can answer that in nature there are no **E** or **A**, or **Π**. There is electromagnetic field, and **E**, **A** and **G**—its model characteristics, tools of its properties description.

Further, we will base upon classical description of electromagnetic field via **E, H, D, B** vectors. It is important that Maxwell equations state the fact that any medium within electromagnetic theory is described using its all three characteristics—dielectric permittivity ε, conductivity σ and magnetic permittivity μ (the mentioned is not spread over electric and magnetic anisotropic medium).

For isotropic medium, we can express **D, B** vectors in terms of the rest two **E, H** vectors using equitation: $\mathbf{D} = \varepsilon\varepsilon_0\,\mathbf{E}$ и $\mathbf{B} = \mu\mu_0\,\mathbf{H}$, where $\varepsilon_0 = 8.85 \cdot 10^{-12}$ F/m and $\mu_0 = 4\pi \cdot 10^{-7}$ H/m—electric and magnetic constants correspondingly.

So, for an "observer" positioned in point Q with (x, y, z) coordinates, in t time moment, electromagnetic field is described by **E**(x, y, z, t), **H**(x, y, z, t), **D**(x, y, z, t), **B**(x, y, z, t) vectors and derivatives of each its vectors components in all x, y, z, t variables. Exactly these values interrelated between each other the Maxwell equations:

$$\begin{cases} \operatorname{rot} \boldsymbol{E} = -\frac{\partial B}{\partial t}, \\ \operatorname{rot} \boldsymbol{H} = \frac{\partial \boldsymbol{D}}{\partial t} + j, \\ \operatorname{div} \boldsymbol{D} = \rho, \operatorname{div} \boldsymbol{B} = 0, \\ \boldsymbol{D} = \varepsilon\varepsilon_0 \boldsymbol{E}, B = \mu\mu_0 H. \end{cases} \quad (2.1)$$

Assume Eq. (2.1) in a little expanded form, for which purpose expressions for $\operatorname{rot}\boldsymbol{E}$ and $\operatorname{rot}\boldsymbol{H}$, we can write in a form of $\boldsymbol{F}_E$ и $\boldsymbol{F}_H$ vectors with corresponding coordinates as:

$$\begin{cases} \operatorname{rot} \boldsymbol{E} \equiv \boldsymbol{F}_E\left(\dot{E}_y^z - \dot{E}_z^y, \dot{E}_z^x - \dot{E}_x^z, \dot{E}_x^y - \dot{E}_y^x\right) \\ \operatorname{rot} \boldsymbol{H} \equiv \boldsymbol{F}_H\left(\dot{H}_y^z - \dot{H}_z^y, \dot{H}_z^x - \dot{H}_x^z, \dot{H}_x^y - \dot{H}_y^x\right) \end{cases}. \quad (2.2)$$

In this case, the first six Maxwell equations will be as follows:

$$\begin{cases} \boldsymbol{F}_E\left(\dot{E}_y^z - \dot{E}_z^y, \dot{E}_z^x - \dot{E}_x^z, \dot{E}_x^y - \dot{E}_y^x\right) + \mu\mu_0 \dot{\boldsymbol{H}}\left(\dot{H}_x^t, \dot{H}_y^t, \dot{H}_z^t\right) = 0 \\ \boldsymbol{F}_H\left(\dot{H}_y^z - \dot{H}_z^y, \dot{H}_z^x - \dot{H}_x^z, \dot{H}_x^y - \dot{H}_y^x\right) - \varepsilon\varepsilon_0 \dot{\boldsymbol{E}}\left(\dot{E}_x^t, \dot{E}_y^t, \dot{E}_z^t\right) = J \end{cases}. \tag{2.3}$$

The rest two equations we represent in the following form of simple equation:

$$\begin{cases} \dot{E}_x^x + \dot{E}_y^y + \dot{E}_z^z = \rho/\varepsilon\varepsilon_0 \\ \dot{H}_x^x + \dot{H}_y^y + \dot{H}_z^z = 0 \end{cases}, \tag{2.4}$$

where ρ—is a volume charge density.

As we can see the Maxwell equations represent a system of first-degree linear differential equations with constant coefficients which are electro-physical characteristics of ε and μ medium, where electromagnetic filed is examined, relatively to derivatives of each component of **E**(x, y, z, t) and **H**(x, y, z, t) vectors in all x, y, z, t variables.

The situation where it is necessary to consider a charge density and when it differs from zero in radar location tasks is uncommon. Therefore we further regard $\rho = 0$. As we can see, the left parts of all equations are identical relatively to **E** and **H** vectors. The difference available in the right parts of equations. The volume charge density ρ is in the right part of the first Eq. (2.4), then this equation ascertains the presence of electrical charges. Zero in the right part of second Eq. (2.4) reads opposite, that there are no electrical charges. To conclude, one of the electrical field sources is electrical charges.

Equation (2.1) shows that if **H** vector does not change in time. i.e., $\dot{\boldsymbol{H}}_t \equiv 0$, then **E** vector has a same property, and consequently, electrical and magnetic fields are existed separately independently of one another.

So, what causes a magnetic field? Let it be in opposite way, **E** vector does not change in time. i.e., $\dot{E}_t \equiv 0$, then $\boldsymbol{F}_H\left(\dot{H}_y^z - \dot{H}_z^y,\ \dot{H}_z^x - \dot{H}_x^z,\ \dot{H}_x^y - \dot{H}_y^x\right) = \boldsymbol{I}^{\mathrm{cm}}$. As we can see, the electrical field source is direct current (DC).

What stands for electrical field source and how it is possible to synthetically generate it? The presence of alternate current (AC) leads to change in time of magnetic field that results the appearance of alternating electric field and so on, and this gives on opportunity to speak about electromagnetic field development. Electromagnetic field appears only when in course of time the change of electric charge density ρ happens, i.e., AC develops resulting in chain of **E** varying vector—**H** varying vector, etc.

Several comments are to be outlined on electric charges. Within most common models clarifying electromagnetic processes, an elementary charge definition is used which an electron features. However, such an interpretation does not arise from Maxwell equations. It states another kind of matter characteristic—a volume charge density. They assume that there are some points in space where a "charge" concentration can be very high and nothing more about it. Here, as a charge we regard again a form of matter existence with spatial none-uniformity. For all electrical engineering's and radio-engineering's, an electric model is a quite acceptable and hence a universally accepted. In quantum electrodynamics, such model of a charge unfortunately is inappropriate anyway; thus, it has quite other approaches.

Similar discussion can be done with respect to electrical current, and within common models, it can be examined as electrical charges movement. However, in some cases as for instance during current passage through capacitor such a current interpretation is not acceptable. To meet a requirement of continuity of current at segment between capacitor coating a "bias current" (electric induction current) term is introduced, equals to derivative of a magnetic field vector in time—$\dot{\boldsymbol{H}}_t$, which exactly turns to be numerically equal to conduction current in external circuit of capacitor.

Being within frames of classical model of a current as a motion of charged particles we should answer the question, what turns to be a source of electromagnetic field? An answer—"everything!" Since within the frames of admitted "electronic" model, all atoms contain electrodes which are in continuous movement, all material objects, the temperature of which differs from absolute zero, are the source of electromagnetic emission; i.e., each of continuously emits electromagnetic field. All surrounding furniture, walls and floor, tables and chairs, doors and windows have the same property. Certainly, the power of this emission is very small, but within our discussions this is not important to the story. The most powerful natural radiation source is a Sun. Thunderstorm lightings are powerful radiator. Finally, it is necessary to mention the space radiation, constantly effecting on our planet. The abovementioned demonstrates natural sources of electromagnetic radiation representing by nature continuously operating generators of electromagnetic field. From the point of view of an earth habitant, this is so-called background radiation which always exists at input of receiving device of any type of radar station generating a continuous noise to radar signal.

It is clear that with distance from electromagnetic field source, the **E** and **H** vectors length, i.e., its $|E|$ and $|H|$ modules, should decrease. This brings up the question on laws of such decrease. From electrodynamics, we know that power flux density, carrying by electromagnetic field at quite big distance from its source, is in proportion to $|E|^2$ and relation $|E|/|H| = 120\pi$.

Let us consider that $|E|^2$ depending on distance up to R source decreases according to law $|E|^2 = \frac{\alpha}{R^n}$, where α—is a some irrelevant coefficient and n—is a parameter to be determined.

In this case, energy, carrying by an electromagnetic field through any sphere, the surrounding medium (environment), should be a constant value, which demands an energy transfer condition in free space, i.e.,

$$|E|^2 \cdot S_{\text{sphere}} = \frac{\alpha}{R^n} \cdot 4\pi R^2 = 4\pi\alpha R^{2-n} = \text{const}$$

consequently $n = 2$. Hence, an energy transfer condition requires a dependency $|E|^2 \sim \frac{1}{R^2}$, and hence $|E| \sim \frac{1}{R}$.

Further we examine situation where an object of radar observation is located quite far away from an observer (far zone). In this case as it is known from electrodynamics, an electromagnetic field in homogeneous medium with a high degree of accuracy can be described using only one segment of **E** vector, for instance, E_x (further will be denoted as E), uniquely connected with perpendicular to it magnetic vector component H_y (further will be denoted as H) using a relation. As for the third components of **E** and **H** vectors, in this case $E_z = H_z = 0$. Then $E_x = \sqrt{\frac{\mu_0}{\varepsilon_0}} H_y = 120\pi H_y$.

According to abovementioned, Eqs. (2.2)–(2.4) modify to the following form:

$$\begin{cases} \boldsymbol{F}_E\left(-\dot{E}_x^z, \dot{E}_x^y - \dot{E}_y^x\right) + \mu\mu_0 \dot{\boldsymbol{H}}\left(\dot{H}_x^t, \dot{H}_y^t, \dot{H}_z^t\right) = 0, \\ \boldsymbol{F}_H\left(\dot{H}_y^z - \dot{H}_z^y, \dot{H}_z^x - \dot{H}_x^z, \dot{H}_x^y - \dot{H}_y^x\right) - \varepsilon\varepsilon_0 \dot{\boldsymbol{E}}\left(\dot{E}_x^t, \dot{E}_y^t, \dot{E}_z^t\right) = J, \end{cases}$$

$$\begin{cases} \dot{E}_x^z = \mu\mu_0 \dot{H}_y^t, \\ \dot{H}_y^z = \varepsilon\varepsilon_0 \dot{E}_x^t + J_x, \end{cases}$$

$$\frac{\partial E_x(x, y, z, t)}{\partial z} = \mu\mu_0 \frac{\partial H_y(x, y, z, t)}{\partial t},$$
$$\frac{\partial H_y(x, y, z, t)}{\partial z} = \varepsilon\varepsilon_0 \frac{\partial E_x(x, y, z, t)}{\partial t} + J_x. \tag{2.5}$$

Let us differentiate the first equation in z, and the second in t, then we obtain the following:

$$\begin{cases} \frac{\partial^2 E_x(x,y,z,t)}{\partial z^2} = \mu\mu_0 \frac{\partial^2 H_y(x,y,z,t)}{\partial t \partial z}, \\ \frac{\partial^2 H_y(x,y,z,t)}{\partial t \partial z} = \varepsilon\varepsilon_0 \frac{\partial^2 E_x(x,y,z,t)}{\partial t^2} + \frac{\partial J_x(x,y,z,t)}{\partial t}, \end{cases} \tag{2.6}$$

From the obtained relations, we have the following:

$$\frac{\partial^2 E_x(x, y, z, t)}{\partial z^2} = \varepsilon\mu\varepsilon_0\mu_0 \frac{\partial^2 E_x(x, y, z, t)}{\partial t^2} + \mu\mu_0 \frac{\partial J_x(x, y, z, t)}{\partial t}. \tag{2.7}$$

Notice that $c = \frac{1}{\sqrt{\varepsilon_0 \mu_0}}$.

Equation (2.7) describes a changing character of **E** vector in homogeneous medium characterizing by ε and μ parameters, in time t, at distance z from the source. For further analysis it is appropriate to do the following.

The Fourier theorem states that any physically realized function $g(t)$ can be represented in a form of the following integral transformation:

$$g(t) = \int_{-\infty}^{\infty} G(\omega) e^{-j\omega t} d\omega,$$

where $G(\omega) = |G(\omega)| e^{j\Phi(\omega)}$—function spectrum $g(t)$.

In other words, any function can be represented in a form of sinusoid sum of different frequencies ω with amplitude equals to $|G(\omega)|$, and initial phase, equals to $\Phi(\omega)$. This conclusion conditionally can be formulated as that "electromagnetic field is a set of electromagnetic waves (radio waves) of different frequencies."

All the above considered the (2.7) formula can be represented in the following form:

$$\frac{d^2 E_{x\omega}(x, y, z, t)}{dz^2} + \frac{\omega^2}{c^2}\varepsilon\mu E_{x\omega}(x, y, z, t) = -j\omega\mu\mu_0 J_{x\omega}(x, y, z, t), \qquad (2.8)$$

Sub-index ω means that it is question of E_x and J_x spectral components at ω frequency. Further, to avoid formulas blocking up, indexes and arguments in formulas won't be written down, then formula (2.8) will be as follows:

$$\frac{d^2 E}{dz^2} + k^2\varepsilon\mu E = -j\omega\mu\mu_0 J. \qquad (2.9)$$

As we can see a field at big distances from electromagnetic waves source in homogeneous medium with ε and μ electro-physical characteristics describes by common linear differential equation of a second order with constant coefficients with right-hand side.

Solution of equations of such class is formed from a general solution of homogeneous equation and partial solution of non-homogeneous equation. As it is known, the general solution describes free system behavior described by homogeneous differential equation without any external influence on it. As for the partial solution than it describes forced system behavior under external influence action described by the right-hand side of differential equation. Based upon above considered, we can arguable that electromagnetic wave source of ω frequency is a J alternative electric current of the same ω frequency of some source positioned in some V volume or some point $Q(x_0, y_0, z_0)$. Electromagnetic wave generated by this current and described by **E** vector, propagates in free space characterizing by ε and μ parameters, and its behavior is described by homogeneous equation.

What happens at achieving by electromagnetic wave of an area where abrupt change ε and μ takes place (e.g., object surface, boundary line of two medium, etc.) and at transition to it.

Electrodynamics formulates a quite natural assertion on value continuity of **E** vector component along boundary line by both of its sides. However, equations describing a behavior of **E** field will differ by presenting in it of medium parameters: ε_1 and μ_1 in first medium and ε_2 or μ parameters—in the second one. On a formal level, the presence of a new $\boldsymbol{J}_{\text{new}}$ field source can lead to this situation, naturally positioned exactly on the boundary line. Such radiation sources arising at achieving of boundary line by electromagnetic wave referred as "surface current," "induced current," etc. This current as it follows from Maxwell equations generates its own electromagnetic field propagating in all directions. In radar location, this field in observation point is admitted to term as "reflected wave," for other points of space it is interpreted as "scattered wave."

Radar and radio-navigational technologies of observation are based on use of such properties of electromagnetic waves and effects of its interaction with background-target environment as:

- constant velocity and propagation linearity (in homogeneous medium) of electromagnetic waves (by using for measuring of range and angular coordinates);
- Doppler effect (is used for measuring of velocity), Huygens–Fresnel principle (is used for generation and variation of physical or signal field posted around radar);
- signal coherency (is used for optimum processing of radar signals to increase energy potential and accuracy characteristics of observation), possibility to maintain harmonic structure during the process of radar transformations in narrow-band signals;
- focus on physical formation, radiation, receiving and processing of timely spatial radar signals using different radio electronic devices;
- use of possibilities delivering by adequate homomorphous coding of background-target environment parameters to characteristics of radar and radio navigation signal;
- use of procedures organization possibility of matching receiving signals with reference signals prepared for radiation that increases both quality of information contact and its ability of refuse to destruction during electronic countermeasures, and also qualitatively update a catalog of used signals, for instance, via abrupt upgrade of its bandwidth or even transition to video signals;
- possibility of effectively evaluation of only coordinated target characteristics as against of none-coordinate where real success is achieved only after evaluation of coordinate characteristics of target element (segments) due to procedures of high resolution;
- possibilities use of up-to-date computers for replaying of physical functions of signals formation and processing, and also for performing of virtual modes (e.g., synthesizing of antenna aperture, polarization scanning, interferometry observation, CCD—coherent change detection—mode of targets acquisition and evaluation of its none-coordinated characteristics according to changes in radar image), where based on fixed experimental results the unimplemented in experiment observation conditions are generated;
- possibility of direct instantaneous evaluation of a set of coordinate target parameters—range, radial velocity, directional cosines, derivative of directional cosines and polarization structure of scattered by target signals which form effective technologies of radar images acquisition cueing the information-intensive structures of background-target environment;
- possibility as a sounding signal (illumination) to use signals generated by outsourced radio electronic systems: radio transmission stations of digital and analog television, radio broadcasting, cellular communication systems, HF radio stations, global positioning systems signals, etc.

Operation on signals of ground-based digital television is of principal interest since signals of this system occupy a relatively wide band (up to 10 MHz) that permits to obtain a best targets resolution in range rather than during using of signals of other available radiating sources. The digital character of coding signal and as consequence its relatively uniform amplitude spectrum permits to achieve detection characteristics continuity. Due to the absence of owned transmitter, it is provided the emission security, green functioning at less economical and operational costs. However, the powerful primary illumination signal and generated by it background returns (clutters) are serious noises.

For radar location (RLS) and radio-navigation stations (RNS), it is assumed that characteristics of received radar signal are limited by the following: amplitude, phase, frequency, delay time, polarization, wave angle of arrival, time and spectral structure, change dynamics of mentioned characteristics in time and space. Modern RLS and RNS handle by six components of resource: space, time, frequency, polarization, energy (power), accumulated and delivered in a form of data and information knowledge on objects of background-target environment. The purpose of RLS and RNS is in decrease of uncertainty on objects of background-target environment induced in general due to the internal system features, external influence of foreign objects and parameters influence of ambient environment through which an interaction with background-target environment is performed within an available resource.

The non-coordinated (no-kinematic) parameters (characteristics) of target mean an information on type, class of observation object, its size, electro-physical properties of surface, structural properties, and on intention of those to whom this object "belongs." Sometimes it is called the attribute data of target. These parameters play an important role during target discrimination, or, wider, during monitoring of background-target environment, since inherently the monitoring task finitely includes an identification and interpretation of observed situation. Equally important to plan and perform various courses of action according based on non-coordinate objects characteristics. An opposite to non-coordinate parameters are of course the coordinated parameters meaning the typical characteristics of radar targets: presence fact, amount of elements in complex target, range, velocity, flight altitude, maneuvering parameters, acceleration, angular coordinates, spatial position, dynamics (flight pattern).

Chapter 3
Radar Targets and Its Reflecting Properties

3.1 Types of Radar Targets

The range of radar surveillance, measuring of coordinates and obtaining of other characteristics of target depend not only on technical characteristics of a radar, but on reflecting properties of a target itself.

The physical subject matter of reflection consists in that the radar electromagnetic wave excites high-frequency current in radiated object: conduction current—in metals; bias (displacement) current—in dielectric material. Herewith, the target itself becomes an independent source of electromagnetic energy radiation (or re-radiator [intermediate emitter] of electromagnetic wave radiating it) in space directions including into direction to the radar. An observer of this wave wherever positioned reads this wave as "reflected from a target," considering it as "scattered wave" in all other directions. Radio wave scattering (reflection) depends first of all on geometric dimensions and shaping of a target, its coating structure and material, and movement (deflection) behavior relatively to the radar as well.

Radar targets depending on geometric dimensions and surveillance types are divided into *point* and *distributed* targets.

To **point target** from a position of an observer refers a target, which depending on output signal in terminal (output) radar unit, does not permit to evaluate both dimensions and details of a target and its amount as well. In radar display, such target in the most cases gives a blip in a form of luminous dot that has defined its name—a "point" target.

It is clear that target classification relying on surveillance device characteristics has a far from absolute character. Jumping a little bit ahead (more closely this issue will be examined further), let us introduce a term of resolution capability of radar station.

The main approximation consists in that radar antenna radiates into a space a some "bunch" of electromagnetic energy of finite size representing a some of parallelepiped two sides length of which equals $a = D\Delta\alpha$ and $b = D\Delta\beta$, where D is a distance

D. A. Akmaykin et al., *Theoretical Foundations of Radar Location and Radio Navigation*, Springer Aerospace Technology,
https://doi.org/10.1007/978-981-33-6514-8_3

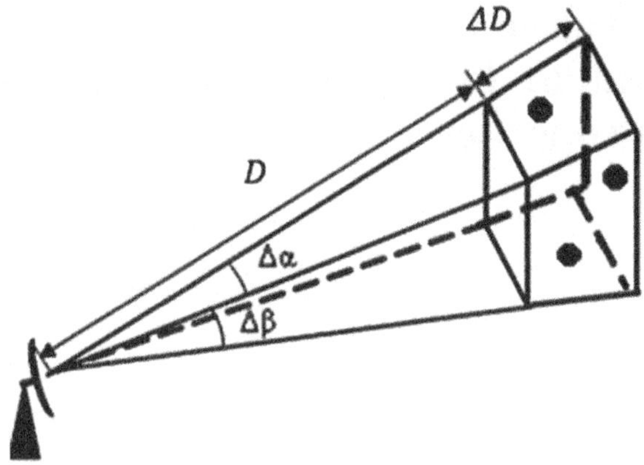

Fig. 3.1 Resolution element

to target, $\Delta\alpha$ and $\Delta\beta$—radar antenna beam (directional pattern) width in azimuth and elevation angle correspondingly (Fig. 3.1). The third part size does not depend on distance and is defined by duration of radiated pulse $\Delta D = \frac{c\tau}{2}$ at pulsed mode or some other finite value at non-pulsed radar location.

If special measures are not taken, then all that inside the parallelepiped with sides equal to $D\Delta\alpha$, $D\Delta\beta$ and ΔD, at radar output will be characterized as a single point target.

A term "point target" is closely related with radar resolution capability that could be however high in idealized absence conditions of uncertainty and random effects, but it in principle is limited by different random processes in radar, emission bandwidth (pulse duration or band value of frequency modulation at continuous radiation), radar antenna aperture width. In practice, frequency characteristics (response rate) of a receiver, terminal unit and other elements of reflected signal transmission path can also affect on resolution capability.

Point targets have r sizes smaller than space element size corresponding to radar resolution (Fig. 3.1):

$$r < \Delta D, r < D\Delta\alpha, r < D\Delta\beta. \tag{3.1}$$

An aggregate of stated conditions can serve as a definition of a point target for this radar at R range. If any one or more of the conditions are not met (3.1), this target is not a point target. A typical example of point target is an aerial vehicle (AV). Figure 3.2 shows indication of several AV characterized as point targets on radar display.

Based on radio waves' reflection pattern from point targets, they can be divided into elementary and complex targets. Reflecting objects of a simple geometrical shape—sheet, ball, exciter (dipole), angle reflector—can be referred to **elementary targets**. **Complex targets** are objects comprised of a big amount of elementary

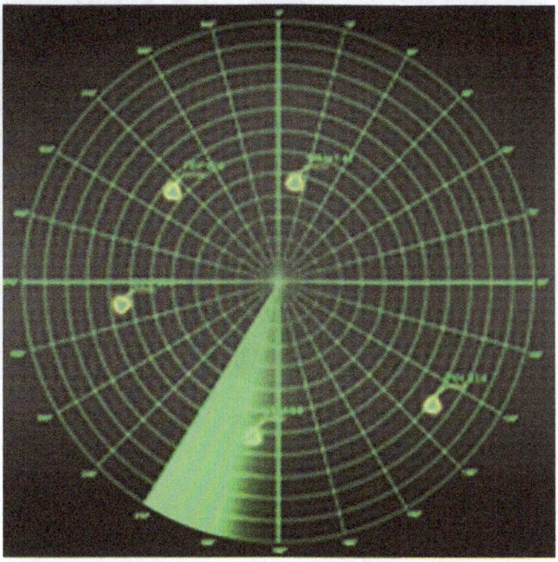

Fig. 3.2 Radar image of point targets

reflectors. Particularly, one of the primary radar targets—aerial vehicle—consists of different surfaces and details simultaneously participating in creation of reflected signal.

Distributed targets have sizes bigger than linear resolution capability of radar in one, two or three coordinates. Distributed target practically can be split into a set of targets (according to amount of resolution elements) that is impossible for a point target. Distributed targets can be divided into linear, area and volume targets.

Linear target is called a target, the length (l_a) of which exceeds the radar linear resolution probability even in one of coordinates, and width (l_b) is less than any linear resolution probabilities. Rivers, seashores, electric power lines, runaways and taxing lanes refer to linear radar targets.

Area target is called a target, two sizes of which exceed the radar resolution probabilities in two coordinates, and height is less than linear resolution probability in elevation angle. Parts of earth surface represent an area targets.

Figure 3.3 shows radar images obtained from linear target (river) and area target (mountainous terrain).

Volume target is called a target, all sizes of which exceed the radar linear resolution probabilities. Rain cloud or other hydrometeors, cloud of radar reflector devices (dipoles), ejected for enemy radar deception (radar passive jamming), are often volume-distributed targets. The radar images of volume meteorological targets obtained by Doppler weather radar system (DWRS) are represented in Fig. 3.4.

Therefore, target division into point and distributed (linear, area, volume) is based on relation between target dimensions and radar resolution probability.

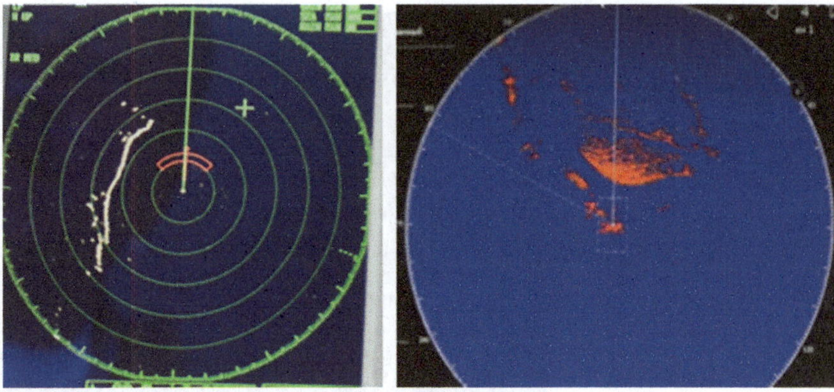

Fig. 3.3 An example of linear and area targets radar image

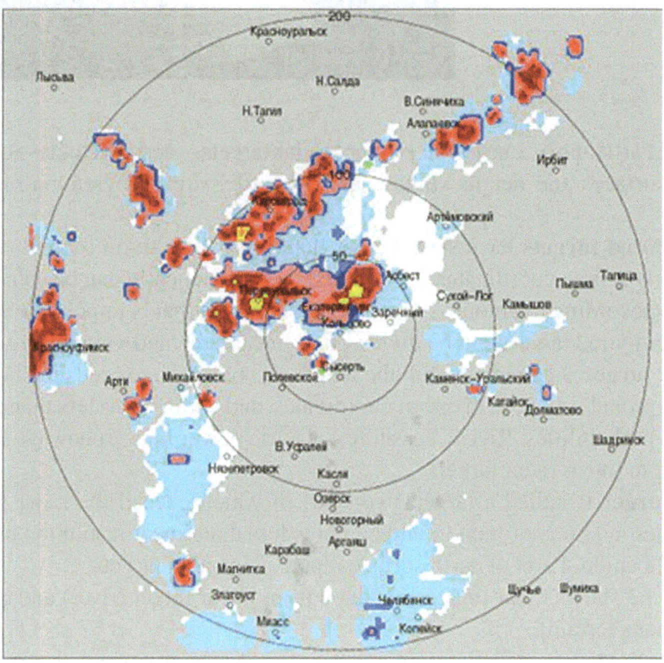

Fig. 3.4 Radar images of volume targets

At radar probing (sounding) of different objects, an important role is given to a relation between target sizes, which we will describe by a some generalized size r, and by operating wavelength λ, at which a probing is taken place.

1. An object size is comparable with wavelength ($r \sim \lambda$). This is a case of resonance reflection when incident wave excites great currents in radiated object serving as

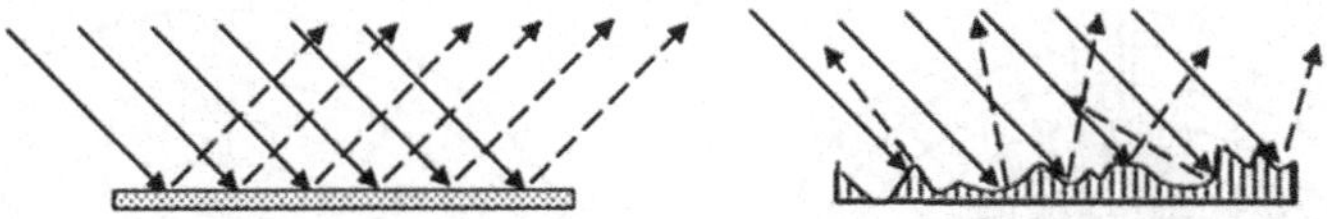

Fig. 3.5 Mirror and diffuse reflection of radio waves

a source of intensive reflected wave. A resonance reflection has a strip of dipole scatterer, the length of which equals to approx. half of the radar wavelength. A cloud, composed from a big amount of such reflectors, gives an intensive signal, cluttering the display screen and impairing the operation of the radar automatic terminal units.

2. An object size is much greater than wavelength ($r \gg \lambda$). This case is of prime importance for radar location, since the majority of radar targets, point and distributed, satisfy the relation ($r \gg \lambda$). At presence of large surfaces (in comparison with wavelength λ), the radio wave reflection by analogy with an optical reflection in a high degree depends on reflecting surface condition, i.e., relation between its irregularities p and wavelength λ. Rough surfaces give diffuse reflection and smooth surfaces—mirror reflection.

Mirror reflection of radio waves (Fig. 3.5.), providing the equity of incident and reflection angles, happens at simultaneous fulfillment of terms: $r \gg \lambda$ and $p \gg \lambda$. Energy of reflected radio wave, excepting a case of its vertical incidence on receiving antenna of the radar, is extremely small. Consequently, calm water surfaces (sea, lake, river) are indicated on displays of onboard radars in a form of dark segments, spots and strips.

Diffuse reflection, obeying the Lambert optical law (Fig. 3.5), happens at fulfillment of the following terms: $r \gg \lambda$ and $p > \lambda$. Directional pattern of reflected radiation has a form of sphere, tangent to reflecting surface.

This means that due to different orientations of rough surface separate elements, the incident electromagnetic waves are scattered in all directions, including to the direction of the radar. Accordingly, the radar display screen in locations corresponding to a field, tilled soil, forest and other analog objects with scattered (diffused) reflection happens to be cluttered by reflections to a greater or less degree which are often called ground (earth) reflections.

Besides these cases of mirror and diffuse reflection, there are intermediate cases when a reflection maintains scattered in a whole, but there is also a preferred direction of reflection corresponding to the mirror reflection optical law.

Most difficulties refer to directional pattern determination of reflected radiation in case of complex point target. If the most distributed targets are an aggregate of smaller similar to each other targets which in size correspond to resolute by the radar a space area, and each from these small (already point) targets represents an aggregate of a big number of reflectors with almost the same characteristics (trees, culms), then a signal of a complex point target generates as a result of wave interference reflected by different and diverse (in value, direction, surface condition) elements.

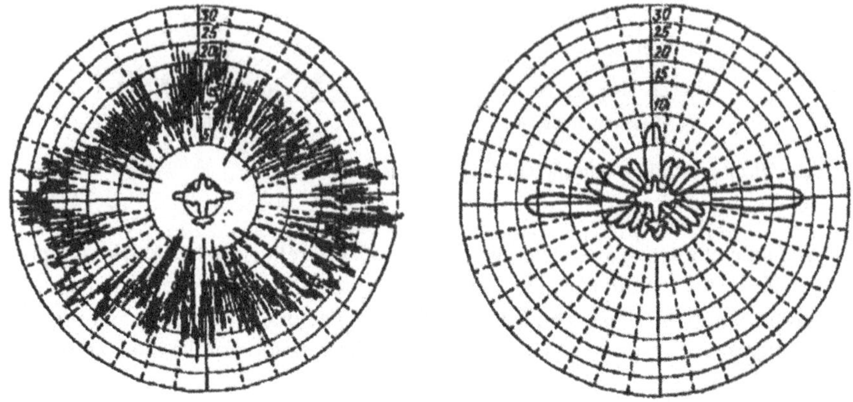

Fig. 3.6 Evolution of AV scattering pattern at increase of wavelength (left-3 cm, right-10 cm)

Directional pattern of reflected radiation of complex point target is a sort of intermediate case between body diagram of a simple geometric form which can be calculated via deterministic methods and diagram of a multiple-element distributed target which is relatively easy calculated via methods of probability theory. Directional pattern of reflected radiation of complex point targets (aircraft, ship, unmanned air vehicle, ballistic missile, etc.) is defined in most cases experimentally.

Particularly, due to complex configuration of aerial vehicle (AV) and its different construction elements with reflection properties, the directional pattern of reflected radiation is quite interference, multi-lobed. With decrease of the radar wavelength, a number of lobes in pattern are increased, and lobe width is decreased: Pattern becomes more interference. In super-high frequency band, the width of some separate lobes constitutes several angular degrees (see Fig. 3.6).

Additional calculation difficulty of signal intensity reflected from a complex point target consists in that directional pattern of reflections is not remained stationary relatively to the radar but moves together with a target. In direction to the radar, both lobes of a pattern and nulls between them are "functioning" that leads to abrupt changes of intensity and reflections phase—amplitude and phase fluctuations of target signal. Both translational motion of a target (motion of target mass center), especially at its maneuvering, and oscillatory movement of a target in course, roll and pinch (motion relatively to mass center) lead to a signal change of complex point target received by the radar. Signal changes will be as more rapid as reflection pattern is interference.

Besides a target itself, a set of radar parameters affects the changes of reflected signal from a target. Particularly, antenna rotation rate at scanning has an influence. The fact that modern transmitting devices often do not provide a constant sounding signal in amplitude and phase also has a case of influence. Radio waves, which radiate a target, are itself changed due to radar transmitter irregularity, and this change is transferred to reflected wave.

Thus, reflected from a real target signal due to influence of a set of factors is a random, fluctuated signal. On a fluctuated useful signal, an another random process is overlapped—noises and interference—which results in formation of random value "signal + noise," perceived by the radar receiving device. Therefore, for quantitative assessments of both a value of inherent reflected signal and results of its influence on the radar, a machinery of probability theory is initiated.

3.2 Radar Cross Section of Radar Targets

Let the radar radiates through antenna with G gain factor and electromagnetic wave of P_Σ power. Then, a value of power flow density near a target, located at R distance from antenna, will be as follows:

$$\Pi_{\text{rad}} = \frac{P_\Sigma G}{4\pi R^2}. \tag{3.2}$$

It is obvious that in direction to antenna, a wave will scatter the power of which P_{sct} is proportional to Π_{rad} value, defined by (3.2) equation. Herewith, a proportionality factor should have an area dimension. Let us denote it A, then $P_{\text{sct}} = A\Pi_{\text{rad}}$, and so, a power flow density of scattered wave Π_{sct} near antenna will be defined using the following equation:

$$\Pi_{\text{sct}} = \frac{P_{\text{sct}}}{4\pi R^2} = A\frac{\Pi_{\text{rad}}}{4\pi R^2} = A\frac{P_\Sigma G}{\left(4\pi R^2\right)^2}. \tag{3.3}$$

As a Π value is proportional to squared absolute value of electric vector, then from (3.2.) and (3.3) equations, we will have the following:

$$A = 4\pi R^2 \frac{\Pi_{\text{sct}}}{\Pi_{\text{rad}}} = 4\pi R^2 \frac{E_{\text{sct}}^2}{E_{\text{rad}}^2}. \tag{3.4}$$

An advantage of introduced A parameter concludes in that it does not depend on distance, though it included in (3.3) formula in an explicit form. This is explained by the fact that E_{sct} is proportional to E_{rad}/R value, so therefore A is defined only by a target. A factor (coefficient) A is called an **effective scattering cross section (area)/radar cross section (RCS)** or an **effective reflective area (ERA)** and is one of the main characteristics of radar targets.

At varying mutual position of antenna and target, the RCS will obviously be changing. Dependence of RCS from azimuth and elevation angle is called a **backscattering diagram (BSD)**. For the purpose to calculate RCS and BSD, it is necessary to solve a corresponding diffraction problem that is connected with calculation difficulties; hence, here, we found a wide application to different approximation methods. Let us examine them.

RCS calculation in most cases narrows to solution of corresponding diffraction problems, determination of necessary polarization components of scattered field and further calculation of RCS according to abovementioned formulas. Let us make a brief review of some presently used calculation methods of RCS.

It is commonly known that to find a scattered one in designated space point, it is necessary to solve Maxwell equations with respect to corresponding boundary conditions. This way leads to so-called rough methods of diffraction theory. It is not recommended of course to think that solution obtained via rough method gives a more precise answer for real targets than obtained by the approximation method. The point is that in all cases during RCS calculation, the radar target is changed with necessity by the approximate idealized model based on such models for which it is possible to obtain a rigorous solution. Unfortunately, it has a limited number: sphere, spheroid, disk, thin wire and a number of some others. An error arises due to this change on frequent occasions which happens to be so big that knocks the bottom out of all advantages of rigorous solutions. In this connection, the approximate methods can give a more precise solution as at modeling stage, it permits to consider more factors determining a scattering of a field. Nevertheless, at rigorous methods, it is always a possibility to rigorously estimate an error of obtained solutions at all stages of problem solution, while use of approximate methods there is not such a possibility as it is mathematical difficulties make due using one or another physical hypothesis.

Approximate methods have become widespread in radiolocation for targets RCS calculation: geometrical optics method and aperture-field method. Let us examine it in detail.

The basis for geometrical optics method lies in the following physical assumptions:

- electromagnetic field wavelength is considerably less than characteristic dimension of bodies with which a field is interacted.
- boundary lines (interface) of interacted with electromagnetic field bodies and medium are completely smooth, and interface curvature is insignificant; then, within small areas, the boundaries refract and reflect electromagnetic waves in accordance with Fresnel's equations for planar interface.
- energy propagation of electromagnetic field happens along the rays.
- in case of electrically non-uniformity medium, the wavelength of electromagnetic field in medium is considerably less than a distance at which medium parameters are notably changed.

As for the most radar targets mentioned above, the conditions are quite accepted, and therefore, solutions obtained within geometrical optics are quite satisfactory. In fact, this method is applicable to large highly conductive bodies and gives an opportunity to calculate the RCS without use of any other laws of electromagnetic besides Snell law of reflection and refraction.

Let us illustrate the abovementioned example of RCS calculation of flat and convex bodies, the sizes of which considerably exceed the wavelength. Let the flat electromagnetic wave incidents on such body (Fig. 3.7).

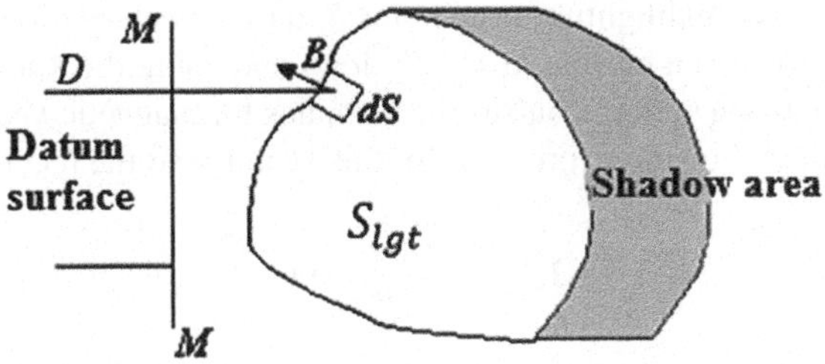

Fig. 3.7 Illustration of geometrical approximation

At geometrical approximation, the body surface is divided into two zones: shadow area and lightened area. Herewith, it is necessary to consider that under the influence of incident wave, there will not be any induced surface currents in area of shadow. Despite such admission is physically unproven as there are always induced currents in shadow area, the contribution of generated field into combined scattering is very small, and therefore, the mentioned approximation is quite proper.

Let us examine neighborhood of point B on Fig. 3.7, which, due to assumption that body curving radius is considerably more than wavelength λ, can be changed by the flat big (in comparison with λ) highly conducted ground (Fig. 3.8). As it is known from electrodynamics, on an examined interface, the relative magnetic vector $\dot{H}_n$ should be equal to zero. This is with necessity results in lengths relation of magnetic vectors of incident $\mathbf{H}_{\text{inc}}$ to reflected $\mathbf{H}_{\text{rfl}}$ waves, under the effect of which (Fig. 3.8) the tangential component of resulting magnetic vector $\mathbf{H} = \mathbf{H}_\tau$ arises, equals to duplicated tangential component of magnetic vector of incident wave $\mathbf{H}_{\tau\text{inc}}$. The tangential component of vector $\mathbf{H}$ leads to electrical current origination, the density $\mathbf{I}$ of which will be determined by the known relation: $\boldsymbol{I} = [\boldsymbol{n}, \boldsymbol{H}] = 2[\boldsymbol{n}, \boldsymbol{H}_\tau]$, where $\mathbf{n}$ is a normal vector to the surface.

Let us introduce some datum (reference) surface MM (Fig. 3.7) and designate through D_0 a distance from the radar up to this plane, and via D a distance from point B; a value of incident wave magnetic vector in MM plane will be written as $\mathbf{H}_{0\text{inc}}$. As a target is located on a quite a distant range from the radar, then incident wave can be considered as a flat; therefore, $\mathbf{H}$ vector value in B point will be: $\mathbf{H}_{\text{inc}} = \mathbf{H}_{0\text{inc}} e^{-jkr}$, where $k = 2\pi/\lambda$ is the wave number (wavelength constant), $\Delta D = D - D_0$.

By this means, the current density at lighted part of surface can be represented as follows: $\mathbf{I} = 2[\mathbf{n}, \mathbf{H}_{0\text{inc}}]\mathrm{e}^{-jkR}$.

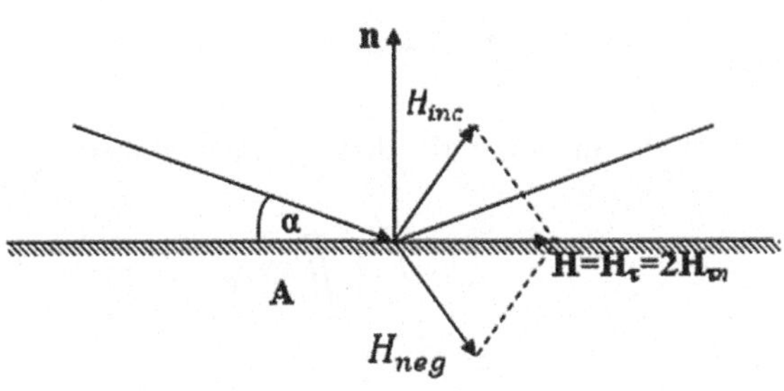

Fig. 3.8 Radio wave reflection at interface

By highlighting near point *B*, an elementary dipole of *dl* length and *db* width, in which a current $dl = ldb$ flows and using the known from electrodynamics and antenna system courses the formulas for magnetic vector in far zone, we will obtain the following expression for this vector near the receiving antenna:

$$d\mathbf{H}_{\text{sct}} = j\frac{\mathbf{H}_{0\text{inc}}}{\lambda R}\exp(-jkD_0)\exp(-2jk\Delta D)\cos\alpha dS. \tag{3.5}$$

The sense of α angle is clear from Fig. 1.8, from where we can see that the product $\cos\alpha dS$ is a projection of lightened area to the plane perpendicular to propagation direction of incident wave.

Let us introduce a designation: $\cos\alpha dS = dS'$ and integrate a Eq. (3.5) in lightened area: $\boldsymbol{H}_{\text{sct}} = j\frac{\boldsymbol{H}_{0\text{inc}}}{\lambda^2 R^2}\int_{S_{\text{lgh}}} e^{-jk\Delta D}\cos\alpha dS'$.

As in flat wave, the electric and magnetic vectors are proportional, and then, we obtain the following expression for RCS of flat and convex bodies, the sizes of which are considerably more than λ:

$$A = \frac{4\pi}{\lambda^2}\left|\int_{S_{\text{lgh}}}\exp(-jk\Delta D)\cos\alpha dS'\right|^2. \tag{3.6}$$

Under interval index, there is a fast oscillating function, and hence, at small change of incident angle α, the RCS will experience the more oscillations the more will be relation of body linear sizes to wavelength.

Formula (3.6), despite on external simplicity, turns to be in some cases inconvenient as it leads to calculation necessity of complex integral equations. However, with used assumption that body sizes and its curve radii are much more than wavelength, we can obtain a more compact and convenient relation.

Let us introduce two typical (characteristic) body dimensions ρ_1 and ρ_2 (this could be curve radii of surface) and imagine exponent index as follows:

$$2k\Delta D(x, y) = 2k\sqrt{\rho_1\rho_2}\frac{\Delta D(x, y)}{\sqrt{\rho_1\rho_2}} = \rho\varphi(x, y), \tag{3.7}$$

where:

$$\rho = 2k\sqrt{\rho_1\rho_2} = \frac{4\pi}{\lambda}\sqrt{\rho_1\rho_2} \gg 1;\ \varphi(x, y) = \frac{\Delta D(x, y)}{\sqrt{\rho_1\rho_2}}$$

With this in view, write down an integral including to (3.6) formula:

$$J = \iint_{S_{\text{lgh}}}\exp(-j\rho\varphi(x, y))\mathrm{d}x\mathrm{d}y. \tag{3.8}$$

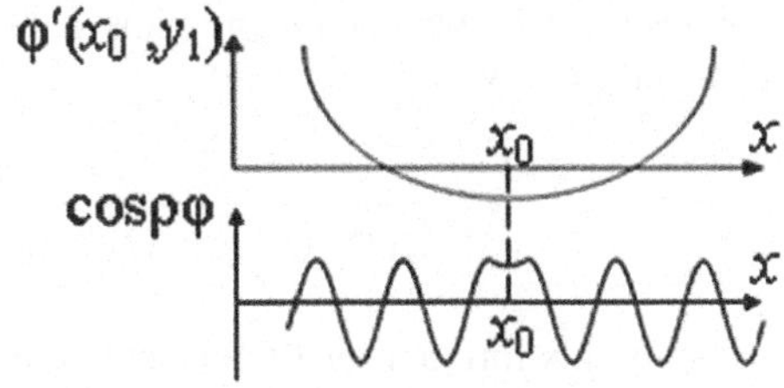

Fig. 3.9 Explanation of stationary phase point

Fast oscillating functions $\cos[\varphi(x, y)]$ are under integral. All expression under integral sign has an oscillatory character changing in phase according to $\varphi(x, y)$ law (Fig. 3.9).

First, fix some value $y = y_1$. Then, we can see that two neighboring oscillating half-waves (positive and negative) have almost equal areas and almost completely suppress each other for which reason an integral value roughly drops with ρ increasing. However, this compensation of neighboring half-waves becomes not effective in neighborhood of the point $x = x_0$, where $\varphi'(x_0, y_1) = 0$. In such points, called **points of stationary phase**, "oscillation frequency" $\rho y''(x)$ tends to zero, and oscillating process is terminated. The same situation takes place during y change. So, an integral value depends primarily on behavior of subintegral function in the vicinity of (x_0, y_0) points, where $\varphi'_x = \varphi'_y = 0$, besides while ρ increasing the essential parts of integration domain are decreased.

Let us expand a function $\varphi(x, y)$ by Taylor series in neighborhood of the point of stationary phase ($\varphi'_x = \varphi'_y = 0$):

$$\varphi(x, y) = \varphi'(x_0, y_0) + \frac{1}{2}\varphi''_{x^2}(x_0, y_0)(x - x_0)^2 + \frac{1}{2}\varphi''_{y^2}(x_0, y_0)(y - y_0)^2 + \frac{1}{2}\varphi''_{xy}(x_0, y_0)(x - x_0)(y - y_0) + \ldots \quad (3.9)$$

and substitute its representation in (3.8) formula:

$$J(\rho) \cong e^{-j\rho\varphi(x_0, y_0)} \int \exp\left\{-\frac{j\rho}{2} - \left[a^2(x - x_0)^2 + 2b(x - x_0)(y - y_0) + c^2(y - y_0)\right]\right\}, \quad (3.10)$$

where designations are introduced: $a^2 = \varphi''_{x^2}(x_0, y_0)$; $b = \varphi''_{xy}(x_0, y_0)$; $c^2 = \varphi''_{y^2}(x_0, y_0)$.

By performing a reaggregation in exponent index and by considering the known value of Poisson integral: $\int_{-\infty}^{\infty} e^{ju^2} du = \sqrt{\pi} e^{-j\pi/4}$, we obtain:

$$J(\rho) \cong 2j\frac{\pi}{\rho}\frac{e^{-j\rho\varphi(x_0, y_0)}}{\sqrt{a^2c^2 - b^2}}. \quad (3.11)$$

Substitute the obtained expression in (3.6) equation with regard to expressions for ρ and $\varphi(x, y)$:

$$A = \frac{\pi}{\Delta D''_{x^2}(x_0, y_0)\Delta D''_{y^2}(x_0, y_0) - \left[\Delta D''_{xy}(x_0, y_0)\right]^2}. \quad (3.12)$$

Apply this formula for RCS calculation of convex surface of double curvature. Examine a paraboloid, the surface equation of which in Cartesian coordinates can be expressed as follows:

$$\Delta D(x, y) = \frac{x^2}{2\rho_1} + \frac{y^2}{2\rho_2}, \quad (3.13)$$

where: ρ_1 and ρ_2 are principal (largest and smallest) radius of curvature in apex of paraboloid $x = y = 0$.

We shall find a stationary phase point:

$$\begin{cases} \Delta D'_x = \frac{x_0}{\rho_1} = 0 \\ \Delta D'_y = \frac{y_0}{\rho_2} = 0 \end{cases}$$

Hence, $x_0 = y_0 = 0$. Calculate the second differential coefficients in these points:

$$\Delta D''_{x^2}(x_0, y_0) = \frac{1}{\rho_1}; \; \Delta D''_{y^2}(x_0, y_0) = \frac{1}{\rho_2}; \; \Delta D''_{xy}(x_0, y_0) = 0. \quad (3.14)$$

By substitution of obtained values in (3.12) formula, for RCS, we will get:

$$A = \pi \rho_1 \rho_2. \quad (3.15)$$

As a surface of the second order in neighborhood of surveillance point on convex surface describes it with a quite good manner, then (3.15) formula has a quite general character. The same conclusion can be reached based on the following simple arguments. Select on a surface body, a small element is perpendicular to incidence direction and limited with arches of principal radii of curvature dS_1 and dS_3. A mirror (reflection) point (stationary phase point or as it admitted in radar location, glare point) is located in the center of element. The incident wave power accounted for small element $dP = \Pi dS_1 dS_2$ (Π—flow density of incident wave power near a target), after reflection according to geometrical optics law, will be distributed in solid angle $d\Omega = 2\frac{dS_1 dS_2}{\rho_1 \rho_2}$.

The power flow density of scattered field in neighborhood of antenna will be as follows:

$$\Pi_{\text{sct}} = \frac{dP}{D_0^2 d\Omega} = \frac{\Pi_{\rho_1 \rho_2}}{4D_0^2}. \quad (3.16)$$

From which, using (3.13) formula, we will receive (3.15) expression. Thus, RCS of smooth convex body of doubled curvature does not rely upon wavelength. Calculation results by (3.15) formula represent a fact with an error nor more than 20% at condition of $\rho_1 \rho_2 > 2\lambda$.

As an example, find RCS of a ball with ρ radius. Surface equation can be represented as $\Delta D(x, y) = \rho + \sqrt{\rho^2 - (x^2 + y^2)}$. Here, the point of stationary phase also will be a point $x_0 = y_0 = 0$, and the second differential coefficients of ΔD will be equal $D''_{x^2}(0, 0) = D''_{y^2}(0, 0) = \frac{1}{\rho}$, $\Delta D''_{xy}(0, 0) = 0$.

Substitution of these equations in (3.12) formula will determine the ball RCS:

$$A = \pi \rho^2. \tag{3.17}$$

which equals to just an area of its cross section.

Clarify a physical sense of stationary phase points.

As a condition $\Delta D'_x = \Delta D'_y = 0$ is executed for them, then this means that in these points, a tangent plane to body surface will be parallel to datum plane MM, i.e., perpendicular to direction of wave propagation. Consequently, these segments of researched body give the main contribution into a field, scattered into direction to receiving antenna.

It is possible to easily calculate within an examined RCS method of flat target located perpendicularly to direction of wave propagation. In this case, $\Delta D = \text{const}$ and the whole target surface used to be a "glare." For this reason, it is more reasonable to directly use (3.4) formula from where another formula arises:

$$A = \frac{4\pi S^2}{\lambda^2}. \tag{3.18}$$

A (3.18) formula testifies on possibility of some another approach to target RCS calculation. At examining of a target location, i.e., perpendicular to Poynting vector-Flux density of electromagnetic energy, on its surface besides narrow area of a size about λ near a border, the in-phase currents of equal amplitude are excited. This means that a target can be examined as antenna with in-phase uniform amplitude field distribution (the abovementioned narrow area near a border does not play a serious role as it is admitted that targets are extremely larger than λ), directional gain G of which can be calculated according to the known formula:

$$G = \frac{4\pi S}{\lambda^2}. \tag{3.19}$$

Field power, scattered by perfectly conducting target, will be equal to power of wave incident on it (with regard to the fact that current irregularity at target edges can be neglected), which obviously will be written as $P_\Delta = \Pi_{\text{rad}} S$; accordingly, the power flow density of antenna scattered wave with regard to target directional properties will be as follows:

$$\Pi_f = \frac{\Pi_{\text{rad}} SG}{4\pi D^2}. \tag{3.20}$$

Generalize obtained results for the case when Π_f vector forms some θ and γ_0 angles with Z and X coordinate system axis connected with a target. In this case, the current phase induced in point A will be left from corresponding current phase in point O at $\Delta\psi = kAG = kr \sin\theta_0$ value, where $r = OA$.

If coefficients of point A designate through (x, y), then:

$$\Delta\psi = kr \sin\theta_0 \cos(\gamma - \gamma_0) = kr(x \cos\gamma_0 + y \sin\gamma_0) \sin\theta_0. \tag{3.21}$$

As for current amplitude, then within examined assumptions, it should be considered as uniform along the whole surface of a target. A field in neighborhood of receiving antenna will represent a sum of elementary fields created by elementary currents flowing on target surface; herewith, a phase of elementary field will equal $2\Delta\psi$.

In this case, (3.20) formula will still be correct, but G should means that as it is admitted in antenna techniques, the product of maximum value of directional gain $G_{\max}$ by normalized directional pattern (in power) of antenna, which has a uniform distribution of amplitude currents and its phases distribution according to (3.21) law, herewith a flat aperture of such antenna coincides with a target surface.

$G_{\max}$ value will be determined by (3.19) formula, and normalized directional pattern $F(\theta_0, \gamma_0)$ is defined as:

$$F(\theta_0, \gamma_0) = \frac{1}{S^2} \cos^2\theta_0 \left| \iint\limits_S \exp\{-2jk(x \cos\gamma_0 + y \sin\gamma_0) \sin\theta_0\} \mathrm{d}x\mathrm{d}y \right|^2$$

So, considering the said, we obtain:

$$F(\theta_0, \gamma_0) = \frac{4\pi}{\lambda^2} \cos\theta_0 \left| \iint\limits_S \exp\{-2jk(x \cos\gamma_0 + y \sin\gamma_0) \sin\theta_0\} \mathrm{d}x\mathrm{d}y \right|^2, \tag{3.22}$$

i.e., we obtained a formula in form convenient for flat targets RCS calculation.

The examined relations give wide possibilities for RCS calculation of quite a big variety of targets.

3.3 Radar Targets Scattering Matrix

As it was mentioned above, one of the main tasks of radar location concludes in conducting of classification of detected target, i.e., in determining of its reflecting characteristics, dimensions, configuration and spatial orientation.

Radar location can be examined as a tool of remote sensing of radar targets, which can be any research objects including sea and aerial vessels, unmanned aerial vehicles (UAVs), space vehicles, and ground surface and hydrometeorologicals. An information carrier on radar targets is a scattered electromagnetic wave registered by a receiving radar device integrated in a radar station (radar). At any processing technique of radar signals, the desired information can be retrieved and then interpreted only via comparison of transmitted and received electromagnetic waves.

During solving of air traffic control (ATC) tasks, the radar and targets in the most cases are located on a quite far distance from each other that permits to consider incident wave in neighborhood of a target and scattered near an antenna—as a flat one—which can be described using only one electric vector **E**. Limiting by a case of single-position radar location, when receiving and transmitting antennas are combined in a space that is specific to ATC radar, finally, the problem of data acquisition on radar targets leads to comparison of electric vector of radiated $\boldsymbol{E}_{\mathrm{rad}}$ and received $\boldsymbol{E}_{\mathrm{rcv}}$ electromagnetic waves of an antenna.

At such comparison, the essential role is for range from a radar to a target, D, which is easily considered; based on that, radiation field falls as $1/D$; accordingly for the purpose to avoid overloading of resulting relations, a field scattered by a target in direction to a target (reflected wave) can interpret as $\boldsymbol{E}_{\mathrm{rcv}}$.

In radar location, as it will be explained in detail further, in the most cases, the narrow-band signals $S(t)$ are used, for which, according to Hilbert, the following representation is admitted:

$$S(t) = A(t)\cos(\omega t + \varphi(t) + \varphi_0), \tag{3.23}$$

where $A(t)$ and $\varphi(t)$ are some slowly varying functions per high frequency period $T = 2\pi/\omega$, φ_0 is initial phase, which at this review does not play an important role; consequently, we will read it as $\varphi_0 = 0$.

In teaching materials on radio-engineering, particular on radar location, it often uses its formulation in a complex form instead of (3.23) representation relying on the fact that: $A\cos(\omega t + \beta) = \mathrm{Re}\left(Ae^{j\omega t}e^{j\beta}\right)$.

Since all operations have a linear character, then a functional operator describing the real part of a complex number, Re, is dropped in formulation, and instead of (3.23), expression (3.24) is as follows:

$$S(t) = A(t)e^{j\omega t}e^{j\varphi(t)}, \tag{3.24}$$

With such an approach, we can firmly assert both on amplitude and phase of examined signals.

Since, in general case, an antenna radiates an elliptically polarized wave (in particular, linear, circular wave), and reflected is also turns up to the same, then it is reasonable instead of **E** vector use its two orthogonal components—a horizontal E^{hor} and vertical E^{ver}. It is clear that each of the E_{rad} radiated wave components "engenders" both components of $E_{\text{rsv}}^{\text{hor}}$ and $E_{\text{rsv}}^{\text{ver}}$ scattered wave; herewith, each of them will differ from engendered components of radiation field both in amplitude and in phase. This particular difference characterizes an observed target.

In this relation, the scattering process of electromagnetic waves is linear; a linear dependence exists between independent components of scattered and radiated waves.

This permits us, due to abovementioned, to relate components of electrical vector of radiated wave in neighborhood of a target with corresponding components of scattered wave near antenna using the following equations:

$$\begin{cases} E_{\text{rsv}}^{\text{hor}} = S_{hh}\exp(j\Psi_{hh})E_{\text{rad}}^{\text{hor}} + S_{hv}\exp(j\Psi_{hv})E_{\text{rad}}^{\text{ver}}, \\ E_{\text{rsv}}^{\text{ver}} = S_{vh}\exp(j\Psi_{vh})E_{\text{rad}}^{\text{hor}} + S_{vv}\exp(j\Psi_{vv})E_{\text{rad}}^{\text{ver}}, \end{cases} \tag{3.25}$$

where $E_{\text{rad}}^{\text{hor}}$ and $E_{\text{rad}}^{\text{ver}}$ horizontal and vertical components of radiated wave and $E_{\text{rsv}}^{\text{hor}}$ and $E_{\text{rsv}}^{\text{ver}}$ are horizontal and vertical components of received wave, $S_{m,n}$ some coefficients (m, $n = h$, v), showing an amplitude change of incident wave, and $\Psi_{m,n}$ is phase change which exists during reflection.

As we can see, introduction of complex coefficients $\dot{S}_{mn} = S_{mn}e^{j\psi_{mn}}$, simplifying the relation (3.25) formulation, looks like naturally:

$$\begin{cases} E_{\text{rsv}}^{\text{hor}} = \dot{S}_{hh}E_{\text{rad}}^{\text{hor}} + \dot{S}_{hv}E_{\text{rad}}^{\text{ver}}, \\ E_{\text{rsv}}^{\text{ver}} = \dot{S}_{vh}E_{\text{rad}}^{\text{hor}} + \dot{S}_{vv}E_{\text{rad}}^{\text{ver}}, \end{cases} \tag{3.26}$$

Relation (3.26) can be represented in vector-matrix form:

$$\mathbf{E}_{\text{rsv}} = \dot{\mathbf{S}}\mathbf{E}_{\text{rad}}, \tag{3.27}$$

where $E_{\text{rsv}} = \begin{bmatrix} \frac{E_{1\text{rsv}}}{E_{2\text{rsv}}} \end{bmatrix}$, $E_{\text{rad}} = \begin{bmatrix} \frac{E_{1\text{rad}}}{E_{2\text{rad}}} \end{bmatrix}$ are column matrix of received and radiated waves (index 1 corresponds to x-component, index 2 corresponds to y-component);

$$\dot{S} = \begin{bmatrix} \dot{S}_{11} & \dot{S}_{12} \\ \dot{S}_{21} & \dot{S}_{22} \end{bmatrix} = \begin{bmatrix} S_{11}e^{j\psi_{11}} & S_{12}e^{j\psi_{12}} \\ S_{21}e^{j\psi_{21}} & S_{22}e^{j\psi_{22}} \end{bmatrix}, \tag{3.28}$$

scattering matrix.

Scattering matrix $\dot{S}$ is a main characteristic of a radar target. Further, for the purpose not to aggregate formulas, we will write just S instead of $\dot{S}$.

As we can see from relation (3.28), in each time moment, a radar target is characterized by four complex or eight real numbers. However, as it follows from aerodynamics laws, off-diagonal elements of scattering matrix happen to be equal between each other, i.e., $\dot{S}_{12} = \dot{S}_{21}$ or $S_{12} = S_{21}$ u $\psi_{12} = \psi_{21}$. Consequently, a mount of parameters, characterizing a target, decreases up to three complex or six real numbers.

Let us review a physical sense of these parameters. For this purpose, return to (3.25) system. Let the horizontally polarized wave be radiated, i.e., $E_{\text{rad}}^{\text{ver}} = 0$, then we have the following:

$$\begin{cases} E_{\text{rsv}}^{\text{hor}} = S_{hh} \exp(j\Psi_{hh}) E_{\text{rad}}^{\text{hor}}, \\ E_{\text{rsv}}^{\text{ver}} = S_{vh} \exp(j\Psi_{vh}) E_{\text{rad}}^{\text{hor}}, \end{cases} \tag{3.29}$$

from which we obtain for S_{hh} and S_{vh}:

$$\begin{cases} S_{hh} = \left|E_{\text{rsv}}^{\text{hor}}\right| \cdot \left|E_{\text{rad}}^{\text{hor}}\right|^{-1}, \\ S_{vh} = \left|E_{\text{rsv}}^{\text{ver}}\right| \cdot \left|E_{\text{rad}}^{\text{hor}}\right|^{-1}, \end{cases} \tag{3.30}$$

and for Ψ_{hh} and Ψ_{vh} phases:

$$\begin{cases} \Psi_{hh} = \arg \frac{E_{\text{rsv}}^{\text{hor}}}{E_{\text{rad}}^{\text{hor}}}, \\ \Psi_{vh} = \arg \frac{E_{\text{rsv}}^{\text{ver}}}{E_{\text{rad}}^{\text{hor}}}. \end{cases} \tag{3.31}$$

Thereby, relation argument of horizontal component of received wave near antenna to analog value of radiated wave near a target equals to Ψ_{hh} phase and module of this relation to S_{hh} value. Relation argument of vertical component of received wave near antenna to horizontal component of radiated wave near a target ($E_{\text{rad}}^{\text{ver}} = 0$) equals to Ψ_{vh} and module of this relation S_{vh} value. The same meaning for S_{vv} and S_{hv}, where a radiation case of vertically polarized wave ($E_{\text{rad}}^{\text{hor}} = 0$) should be examined.

Scattering matrix elements S_{11} and S_{22} are called **primary** and S_{12}—**switched** or **crossing** elements.

Measurements of absolute values Ψ_{11}, Ψ_{12}, Ψ_{22} present severe difficulties; therefore, as a rule, measurements are limited to its relative values Ψ_{12}–Ψ_{11}, Ψ_{22}–Ψ_{11}, i.e., equivalent to the case when Ψ_{11} phase is admitted to be equal to zero. Severe difficulties arise at measuring of S_{11}, S_{12}, S_{22} absolute values, and therefore, relative measurements are also conducted and relations S_{12}/S_{11} и S_{22}/S_{11} are checked, i.e., equivalent to a case when S_{11} is assumed equal to one. At such an approach, a target describes not by six, but by four parameters.

In real conditions of radar operation due to continuous change of relative position of antenna and target, all scattering matrix elements in general case are changed randomly, and hence, the most complete characteristic of a radar target will be corresponding to multi-dimension distribution laws.

We will return to further analysis of scattering matrix properties latter, and now, let us examine the most commonly encountered nowadays case when radar operates only on one fixed polarization as both in transmission mode and in receiving mode.

Herewith, generally, there is a possibility to conduct a signal measuring only proportional to S_{11}. It is clear that in such situation, a target is described only by one number S_{11}. It was mentioned before that S_{11} links fields measured in different points (one is near a target and another is near an antenna); consequently, it is obvious that S_{11} depends on range (distance) that causes certain inconvenience.

3.4 Scattering Cross Section of Simple Point Targets

Let us find RCS of simple point targets which both cannot be as models of some real targets and be as its component parts.

Ball RCS Scattering and diffraction problems of flat electromagnetic wave on a sphere are examined to the fullest extent possible comparing to all other bodies of simple and complex forms. The special meaning of this problem implies for radar location, since a sphere is one of the bodies of a simple form for which a string solution is available. For this reason, metal spheres are widely used as RCS standard sample. Besides, a sphere has a unique property: This is a single-body scattering of energy in all directions uniformly. In other words, a sphere is an all-directional reflector as in case when receiving antenna coincides with transmitting one (so-called single-positioned radiolocation) and when receiving and transmitting antennas are located in different positions (two-positioned radiolocation). Omnidirectional property of a sphere at single-positioned radiolocation is obvious and does not require explanations. As for omnidirectivity of a sphere at two-positioned radiolocation, it puzzles sometimes and needs in proof, which we perform in a form given in V.O. Kobak "Radar reflectors" monography.

Let a flat wave incidents on perfectly conducting sphere, the radius of which $\rho \gg \lambda$, along negative direction of OZ axis, herewith the power flow density equals to Π_{rad} (Fig. 3.10).

Define a power flow density scattered by a sphere at an angle to direction of incident. For this purpose, describe around a sphere a second auxiliary concentric sphere of $R_0 \gg \rho$ radius. Highlight a strip corresponding to specularly reflected ray scattering at β and $\beta + d\beta$ angles. Based on geometry, a strip will have ρ sin $(\beta/2)$ radius and $(\rho/2)d\beta$ width. The full power for a strip equals to $\Pi_{\text{rad}} dS_1$, where $dS_1 = \frac{1}{2}\pi\rho^2 \sin\beta d\beta$—a strip projected area to incident wave front. After reflection, an energy distributes on a surface of a second sphere in the same way along an annular strip, an area of which at $R_0 \gg \rho$ condition equals to $dS_r = 2\pi R_0^2 \sin\beta d\beta$. From here, we determine a power flow density of scattered wave $\Pi_{\text{sct}} = \Pi_{\text{rad}} \frac{dS_1}{dS_2} = \Pi_{\text{rad}} \frac{\rho^2}{4R_0^2}$.

Therefore, in fact, a power flow density of scattered wave does not depend on β angle and is a constant value in all directions. Exclusion for this is $\beta = \pi$ direction, where application of optical zoom is impossible.

Fig. 3.10 Proof of sphere omnidirectivity

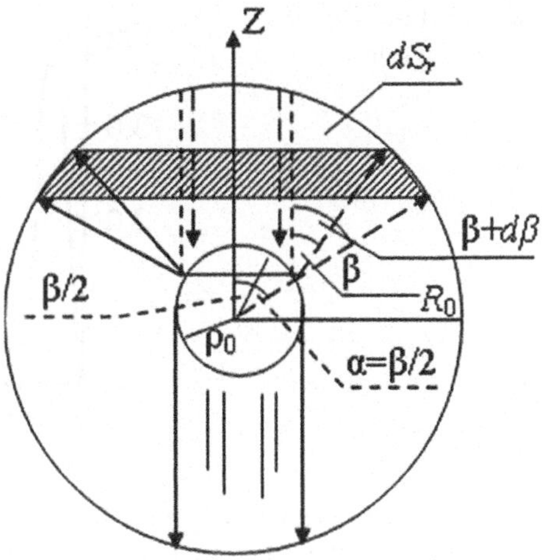

As for RCS numerical value then in case of perfectly conducted ball, the radius ρ of which much greater then λ wavelength. Ball RCS can be calculated with quite a high accuracy according to (3.17) formula. If the ball has a finite conductivity and its material has a complex dielectric constant (capacitance) ε, then it is quite obvious that the RCS in comparison with the previous case decreases by a factor of reflection. A rigorous theory confirms this assertion, i.e.,

$$A = \left|\frac{1 - \sqrt{\varepsilon}}{1 + \sqrt{\varepsilon}}\right| \pi\rho^2.$$

Significant importance in radar location places a case when a sphere serves as a target, the radius of which is much smaller than the wavelength that perfectly simulates rain droplets and some phantom targets. However, the obtained formulas do not permit to use them for this case. Let us give formulas without conclusions for RCS of perfectly conducted ball with $\rho \ll \lambda$ radius: $A = 144\pi^2\rho^6\lambda^{-4}$ and dielectric ball: $A = \frac{\pi^5\sigma^4\rho^6}{\lambda^4}\left|\frac{\varepsilon-1}{\varepsilon+2}\right|^2$.

For rain droplets, when $|\varepsilon| = 80 \gg 1$, we get:

$$A = 60\frac{\pi^5\rho^6}{\lambda^4}. \tag{3.32}$$

In area, where ρ has an order λ or several λ, formulas for RCS calculation are quite lengthy; thus, we exploit a graphic representation.

In Fig. 3.11, a dependency of $A/\pi\rho^2$ relation of perfectly conducted ball on ρ/λ relation is performed.

What we have now is clear-cut oscillating character of this dependency. The maximum RCS happens only when a ball becomes a sort of half-wave dipole and the current half-wave lies along its semi-circle of $\pi\rho$ length, i.e., $\rho/\lambda^3\pi/2$; herewith,

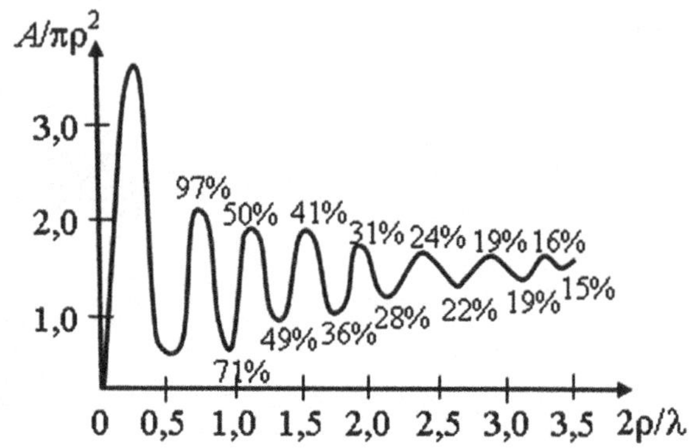

Fig. 3.11 Energy sphere scattering function

the RCS value itself exceeds more than 3.5 times of a ball cross-sectional area. This is explained by that in super-high frequency (SHF) band at increased rain intensity, when raindrops expand, fast growth of reflections from rain takes place that is more evident in millimeter wave band.

In summary, RCS of a big metal ball of ρ radius in comparison with λ wavelength of perfectly conducted ball equals to its big circle area: $A = \pi\rho^2$, i.e., does not depend both on wavelength and direction of radiation. In fact, exactly, a big ball physically illustrates a concept content on scattering cross section: visible from the radar side a ball area $\pi\rho^2$ seeming to intercept in space such power of radio waves which after uniform reradiation in all directions, produces near the radar a real power of radio wave reflected from a ball. This is precisely why a ball is reasonable to use as isotropic (non-directional) reflector of radio waves. In radar application practice, one can meet ball-shaped objects: sounding balloons, structural elements of different targets, etc.

Reflection from a ball is a private case of electromagnetic waves scattering by curved surface of RCS of any convex surface: $A = \pi\rho_1\rho_2$ assuming that $\rho_1 \gg \lambda$ and $\rho_2 \gg \lambda$, where ρ_1 and ρ_2 are primary (largest and smallest) radii of curvature in glittering point, i.e., in that point of reflection surface, where normal to a surface coincide with radiation direction (direction to the radar).

Practically, it is important that RCS of convex surfaces approximately satisfies the mentioned formulas and in cases where scatterer represents an open surface (e.g., a part of ball surface).

Rotation ellipsoid RCS Equation of ellipsoid surface (Fig. 3.12) is written as follows:

$$\frac{x^2}{a^2} + \frac{y^2}{b^2} + \frac{z^2}{c^2} = 1, \tag{3.33}$$

where a, b, c are semi-axes of ellipsoid.

Find out ellipsoid RCS while its observing at such an angle wherein Poynting (power flow) vector is perpendicular to ellipsoid surface in a point with (x, y, z)

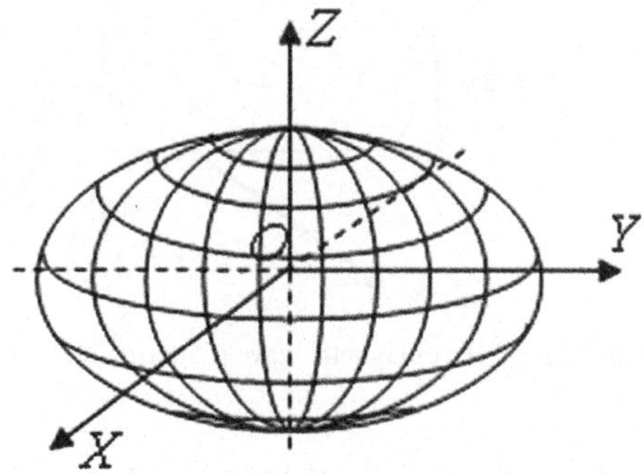

Fig. 3.12 Ellipsoid of rotation

coordinates. For this purpose, we need to find out a curvature primary radii in this point and exploit (3.12) formula. Primary radii of curvature are the values reciprocal to the second-order derivatives in X and Y of Z function taken in (x, y) point.

Lengthy brute force computation leads to the following expression for RCS:

$$A = \frac{\pi}{a^2b^2c^2}\frac{\left(a^2b^2 - b^2x^2 - a^2y^2\right)}{\left(a^2 - x^2\right)\left(b^2 - y^2\right)}. \tag{3.34}$$

If ellipsoid is outstretched along Z-axis and semi-axes, *X-axis* and Y-axis are equal to each other, i.e., $c > a = b$; then to find its RCS, while observing along Z-axis, it is necessary to suppose $x = y = 0$. As a result, we obtain: $A = \frac{\pi a^4}{c^2}$.

If ellipsoid is outstretched along X-axis, semi-axes c and b are equal to each other, i.e., $a > c = b$. In this case, while observing along Z-axis, RCS will be: $A = \pi a^2$, i.e., just to cross section the area.

If all three semi-axes are not equal between each other, then $A = \frac{\pi a^2 b^2}{c^2}$.

Cylinder RCS Rigorous problem solution on scattering of electromagnetic waves on cylinder is known only for a case of its infinite length at wave incident perpendicularly to generatrix. Cylinder characteristics of finite wave are determined only via approximate methods.

At incidence angles, different from normal, scattering by side cylinder surface in backward direction at physical optics approximation does not depend on polarization of incident electromagnetic wave. Circular-shaped cylinder RCS (Fig. 3.13) at θ radiation angle is expressed by the following formula:

$$A = \frac{2\pi}{\lambda} r l^2 \sin\theta \left[\frac{\sin\left(\frac{2\pi}{\lambda} l \cos\theta\right)}{\frac{2\pi}{\lambda} l \cos\theta}\right]^2, \tag{3.35}$$

where r is cylinder radius and l is its length.

Within a ray-optical approximation, the values for RCS result as somewhat different. Let us employ a formula (3.10). Since dependence from y is absent, then

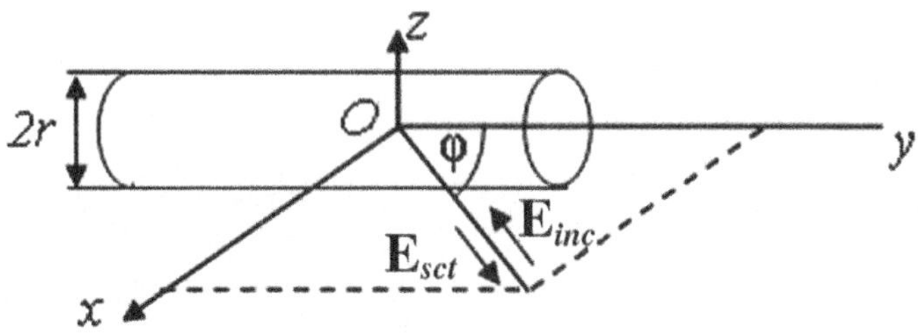

Fig. 3.13 Electromagnetic wave scattering on cylinder

an integral is calculated more easily, and we obtain:

$$A = \frac{2\pi}{\lambda}\frac{rl^2}{\Delta D''_{x^2}(x_0)}.$$

Simple geometrical consideration shows that $\Delta D''_{x^2}(x_0) = \operatorname{cosec}\theta$. Therefore, we have:

$$A = \frac{2\pi}{\lambda}rl^2 \sin\theta. \tag{3.36}$$

Geometric approximation gives good results only at θ angles close to 90°. Observing from the end within an examined approximation, its RCS should coincide with circular disk RCS, the formulas for which will be obtained latter.

For cylinders, the radius of which is small comparing with wavelength and cylinder length, i.e., $r \ll \lambda$ and $r \ll L$, the formulas turn to be correct which are related to fine wires and listed without proof:

$$A = \frac{2\pi}{\lambda}rL^2V(kr)V(kL), \tag{3.37}$$

where auxiliary function $V(kr)$ is expressed as follows:

$$V(kr) = \begin{cases} 1, & at\ r < 0.2\lambda \\ \frac{2\lambda r^{-1}}{\pi^2+\left[2\ln(0.178)\lambda a^{-1}\right]^2}, & at\ r < 0.2\lambda \end{cases}$$

Function graph $V(kr)$ is given in Fig. 3.14.

RCS values calculated via this graph have an error of not more than 10% at r change within limits of $\lambda/5000 \le r \le \lambda/100$.

We can see from Fig. 3.14 that the largest relative value of RCS which equals $16.4\frac{2\pi}{\lambda}rl^2V(kr)$ has a half-wave dipole.

Deviations of posterior maxima and minimum expressed in percentage from a limiting value, equal to $\frac{2\pi}{\lambda}rl^2V(kr)$, are depicted on a curve. RCS maxima correspond to $l = (2n+1)\lambda/2$ condition, where $n = 0, 1, 2, \ldots$. At $l = n$, RCS

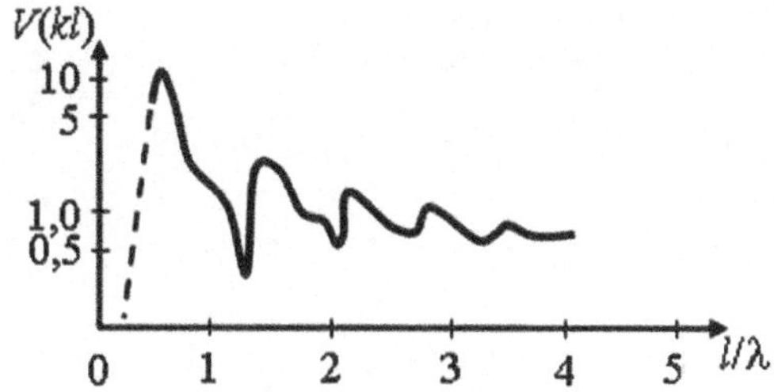

Fig. 3.14 Energy scattering function of fine wire

equals to its limiting value. RCS minimum approximately corresponds to condition $l = (n + 0.35)\lambda$. For cylinders, the radius of which lies within $(0.01\ldots0.5)\lambda$ limits and length within $(0.05\ldots5)\lambda$ limits λ; scattering characteristics at field polarization along cylinder axis are the most complicated since here two resonant sizes exist simultaneously. Information for this case is almost absent in the published literature.

For cylinders, the radius of which lays within $(0.01\ldots0.5)\lambda$ limits, and length exceeds 0.5λ; a relation $A/\left(\frac{2\pi}{\lambda}rl^2\right)$ as a function $2r/\lambda$ is expressed in Fig. 3.15.

Infinite cone RCS In general case, a rigorous problem solution on scattering of electromagnetic waves on a cone of finite size has not been obtained yet. Let us find, within ray-optical approximation, the RCS values for a cone of H finite height with angle near an apex 2α at its observing along an axis.

We use formula (3.21) and Fig. 3.16 for calculations, from where we can see that $\Delta D = \sqrt{x^2 + y^2}\text{ctg}\alpha$.

By transferring to polar coordinates, we obtain:

$$A = \frac{4\pi}{\lambda^2}\left|\int_0^{2\pi} \mathrm{d}y \int_0^{h\text{tg}\alpha} \mathrm{e}^{-2jkr\text{ctg}\alpha}\mathrm{d}r\right|^2. \tag{3.38}$$

Direct calculation result in:

$$A = \frac{\lambda^2\text{tg}^4\alpha}{16\pi}\left(1 + 4k^2H^2e^{-4\beta H} + 4kHe^{-2\beta H}\cos kH\right), \tag{3.39}$$

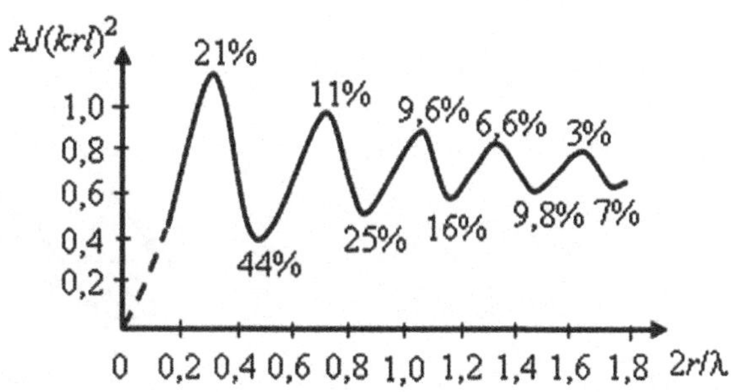

Fig. 3.15 Energy scattering function of disk

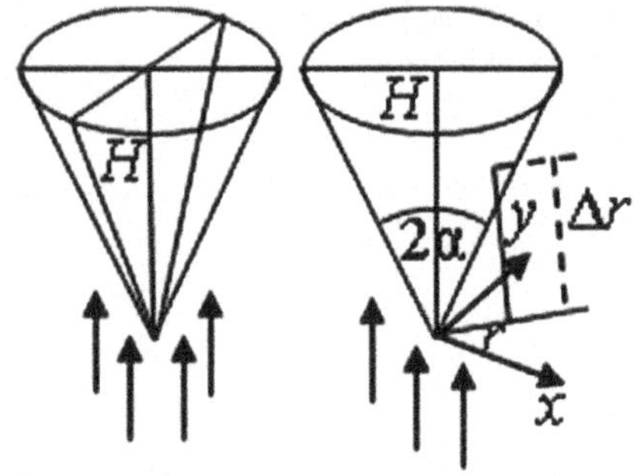

Fig. 3.16 Electromagnetic wave scattering on a cone

where β is attenuation coefficient.

For a finite cone, we can easily obtain:

$$A = \frac{\lambda^2 \mathrm{tg}^4 \alpha}{16\pi}. \tag{3.40}$$

Flat rectangular plate RCS To find a RCS of flat rectangular plate of $a \times b$ size (Fig. 3.17), we can at once use a formula (3.6) which determines a RCS at radiating from a direction defined by (θ_0, γ_0) angles

$$A = \frac{4\pi}{\lambda^2}\cos^2\theta_0 \left| \int_0^a \int_0^b e^{-2jk(x\cos\gamma_0 + y\sin\gamma_0)\sin\theta_0} \mathrm{d}x\mathrm{d}y \right|.$$

Direct calculations result in the following:

$$A = \frac{4\pi}{\lambda^2} a^2 b^2 \cos^2\theta_0 \left| \frac{\sin(ka\cos\gamma_0\sin\theta_0)}{ka\cos\gamma_0\sin\theta_0} \right|^2 \left| \frac{\sin(kb\sin\gamma_0\sin\theta_0)}{kb\sin\gamma_0\sin\theta_0} \right|^2. \tag{3.41}$$

As we can see, a maximum RCS value exists at vertical observation, when $\theta_0 = 0$, and its value equals:

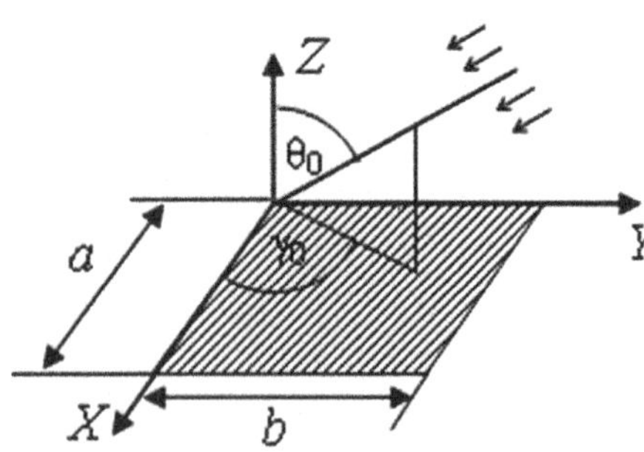

Fig. 3.17 Electromagnetic wave incidence on flat rectangular plate

$$A\text{max} = \frac{4\pi}{\lambda^2} a^2 b^2 = \frac{4\pi S^2}{\lambda^2}. \tag{3.42}$$

Pay attention that the last formula is correct for flat plates of any shape at normal beam incidence. It is of interest to note that RCS of a flat plate opposed to RCS of a ball depends both on wavelength λ and direction to the radar (γ_0 и θ_0 angles); besides, it is proportional not to a geometrical area but to its square. The last is connected with directivity of plate reflections contrary to indirect ball reflection. At sliding incidence angles, we can obtain a good result for RCS using the following formula:

$$A = \frac{ab^2}{\lambda} \left[\cos^2 \alpha + \frac{2\pi^3}{(ka)^2} \sin^2 \alpha \right], \tag{3.43}$$

where $\alpha = ka - 0.3\pi$.

Area of sliding incidence angles distinct in that at changing of a longitudinal size, the RCS oscillates with $\lambda/3$ period.

The given considerations show that metal sheets are unreasonable to use as phantom targets imitating real objects as far as a little deviation of radiation direction from a normal to a surface decreases a sheet RCS up to a zero. A phantom radar target should have a considerable RCS value which is a little dependant on incidence angle. Then, it would be well observed by radar. Angled reflectors with equal facet forms, Luneburg lens etc., are used as radar phantom targets.

Flat circular plate RCS To find a RCS of a flat circular plate of ρ radius, in direction determined by θ_0 angle, we can also use (3.22) formula where we should assume that $\gamma_0 = 0$ and transfer to polar coordinates (r, γ):

$$A = \frac{4\pi}{\lambda^2} \cos^2 \theta_0 \left| \int_0^{2\pi} \int_0^{\rho} e^{-2jkr \sin \theta_0 \cos \gamma} r \, dr d\gamma \right|^2. \tag{3.44}$$

Employing a Bessel function representation, we obtain:

$$A = \frac{4\pi}{\lambda^2} \left(\pi \rho^2\right)^2 \cos^2 \theta_0 \left[\frac{2J(2k\rho \sin \theta_0)}{2k\rho \sin \theta_0} \right], \tag{3.45}$$

where $J(\cdot)$ is Bessel function of a first order.

Formula (3.45) gives satisfactory results only in $\rho > 2\lambda$ and $\theta_0 < 45°$ condition; at $\theta_0 > 45°$, it gives not only numerically but qualitatively wrong results. In this case, we can use the following representation:

$$A = \frac{4\pi}{\lambda^2} \left(\pi \rho^2\right)^2 \left\{ \left[\frac{2J(2k\rho \sin \theta_0)}{2k\rho \sin \theta_0} \right]^2 + \left[\frac{2J_2(2k\rho \sin \theta_0)}{2k\rho \sin \theta_0} \right]^2 \right\}, \tag{3.46}$$

which gives a good compliance with an experiment at $\rho > 2\lambda$ and $\theta_0 < 90°$ condition.

At sliding angles $(\theta_0, 90°)$, the following formula gives good results:

$$A = \frac{4\rho}{k}\cos^2(2k\rho - 3\pi/4), \tag{3.47}$$

that leads to that in area of sliding angles, the RCS dependency from $k\rho$ has an oscillating character.

In area of small radii of disk RCS when ρ is considerably less than wavelength, RCS can be calculated as follows:

- for horizontal polarization:

$$A = \frac{4\pi}{\lambda^2}\left(\pi\rho^2\right)^2\left(\frac{8k\rho}{3\pi}\cos^4\theta_0\right)^2, \tag{3.48}$$

- for vertical polarization:

$$A = \frac{4\pi}{\lambda^2}\left(\pi\rho^2\right)^2\left[\frac{4k\rho}{3\pi}\left(2+\sin^2\theta_0\right)\right]^2. \tag{3.49}$$

At normal incidence, the following formula is correct:

$$A = \frac{1024}{9}\frac{\pi^3\rho^6}{\lambda^4}. \tag{3.50}$$

Figure 3.18 shows a dependence of $A/\left[4\pi\lambda^{-2}\left(\pi\rho^2\right)^2\right]$ relation from $2\rho/\lambda$.

Maximum relative value of RCS, equals to $3.8\frac{4\pi}{\lambda^2}\left(\pi\rho^2\right)^2$, is achieved at $2\rho \cong 0.48\lambda$ that is very close to $\rho = \lambda/4$ condition.

Triangular plate RCS Let us examine a finite thin identically conducted plate having a shape of right isosceles triangle with base of $2a$ and b height at intrinsic admission $a, b \gg \lambda$ and find its RCS in direction defined by (γ_0, θ_0) angles.

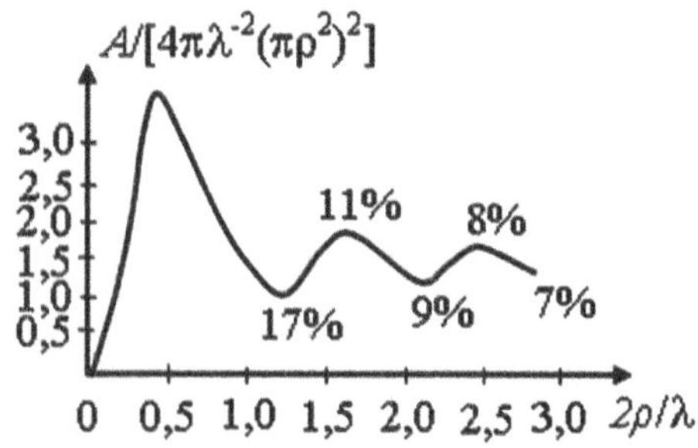

Fig. 3.18 Energy scattering function of fine wire

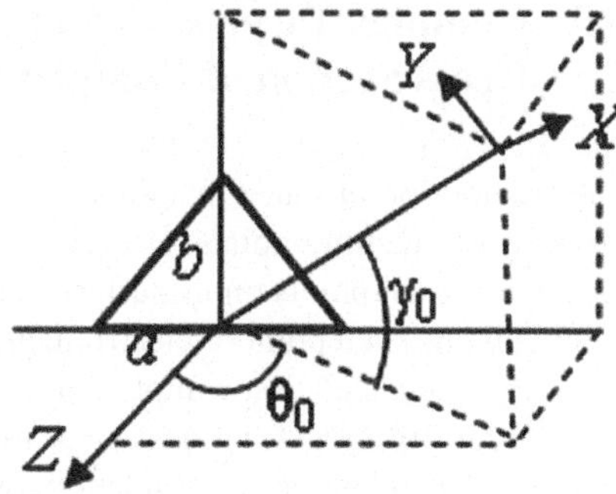

Fig. 3.19 Electromagnetic wave scattering on a flat triangular plate

For $\gamma_0 = \theta_0 = 0$ case, we obtain $A = \frac{4\pi}{\lambda^2} a^2 b^2$ regardless of polarization type of incident wave (Fig. 3.19).

To find a RCS of a triangular plate at γ_0 and θ_0 arbitrary angles, it is reasonable to use (3.22) relation. However, due to extreme awkwardness of obtained relations, we confine to RCS examination at its determination in two mutually orthogonal planes, perpendicular to plate surface $\gamma_0 = 0$ and $\gamma_0 = \pi/3$.

For the first case (horizontal plane, $\gamma_0 = 0$), we have:

$$A = \frac{4\pi}{\lambda^2} \cos^2\theta_0 \left| \int_0^b \mathrm{d}y \int_{-(a-\frac{ay}{b})}^{a-\frac{ay}{b}} \mathrm{e}^{-2jkx\sin\theta_0} \mathrm{d}x \right|^2$$
$$= \frac{4\pi}{\lambda^2} (ab)^2 \cos^2\theta_0 \left[\frac{\sin(ka\sin\theta_0)}{ka\sin\theta_0} \right]^4. \tag{3.51}$$

For the second case (vertical plane, $\gamma_0 = \pi/2$), we have:

$$A = \frac{4\pi}{\lambda^2} (ab)^2 \cos^2\theta_0 \left\{ \left[\frac{\sin(kb\sin\theta_0)}{kb\sin\theta_0} \right] + \left[\frac{1 - \sin 2(kb\sin\theta_0)/2(kb\sin\theta_0)}{kb\sin\theta_0} \right] \right\}. \tag{3.52}$$

The obtained formulas quite good correlate with experimental data.

In horizontal plane, the backscattering diagram (BSD) represents a quite rapid (at change of observing angle) oscillating function; herewith, a width of main BSD lobe in 0.5 level comprises $18\lambda/a$ [deg] and level of the first side lobe—26.6 dB. In vertical plane, an opposite case, the BSD represents a monotone function with no lobes. Width of the main lobe in 0.5 level at $b > 6\lambda$ and $a > 2\lambda$ condition equals to $31.5\lambda/b$ [deg]. If $kb\,\sin\theta_0 > 2$, then $A = \frac{4\pi}{\lambda^2}\left(\frac{\mathrm{ctg}\theta_0}{kb}\right)^2$; however, an application area of this formula is limited by angles $\theta_0 < 35°$.

Besides plates in shape of isosceles triangle, we can frequently meet in practice plates in shape of right-angled triangle. Any triangles with *a* base and *b* height in vertical plane also would have no-lobes BSD.

3.5 Complex Targets. Calculation Technique of Scattering Cross Section of Complex Bodies

The previous point stated that the main contribution into scattered by a target field is given by some glitter points. However, even for the simplest real radar targets, there are a lot of such points on its surface. This leads to abrupt change as opposed to the case with one such point of electromagnetic wave scattering character and naturally to significant calculation complication of reflecting characteristics.

For a **group target,** let us regard a target which should be modeled by not less than two dependent or independent between each other point scatterers.

Firstly, examine the simplest one from a group target, the system from two point scatterers C_1 and C_2, by which we already can see a complex character of RCS dependence from an angle, at which such a target is observed (Fig. 3.20).

From Fig. 3.20, we can see that resulting scattered field E, which will be near a receiving antenna of radar, can be found by formula:

$$E = \frac{E_u}{\sqrt{4\pi}D}\left(\sqrt{A_1}e^{-j\phi_1} + \sqrt{A_2}e^{-j\phi_2}e^{-2jkd\sin\theta}\right)e^{-jkR}e^{j\omega t}, \tag{3.53}$$

where E_u is electric vector of incident wave of a group target; D is distance between group target and radar; φ_1 and φ_2 are phase shifts, arising at reflection from C_1 and C_2 targets; A_1 and A_2 are RCS of C_1 and C_3 targets.

For RCS, a group target A will have the following representation:

$$A = A_1 + A_2 + 2\sqrt{A_1 A_2}\cos(\phi_1 - \phi_2 - 2kd\sin\theta). \tag{3.54}$$

As we can see, depending on observing angle θ, the RCS of a target is changed from its maximum value $A(\sqrt{A_1} + \sqrt{A_2})^2_{\text{max}}$ up to the minimum one $A(\sqrt{A_1} - \sqrt{A_2})^2_{\text{min}}$. Thereby, BSD of two-point target is having a multi-lobed character; herewith, a lobe width near an angle $\theta = 0$ has meaning at $d \gg \lambda$, equals to

Fig. 3.20 Two-point target

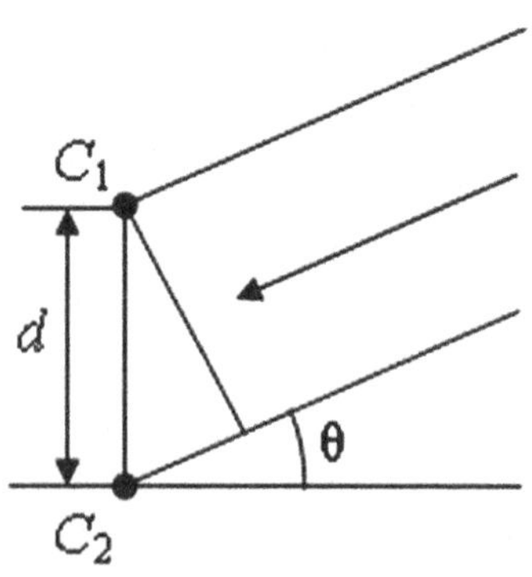

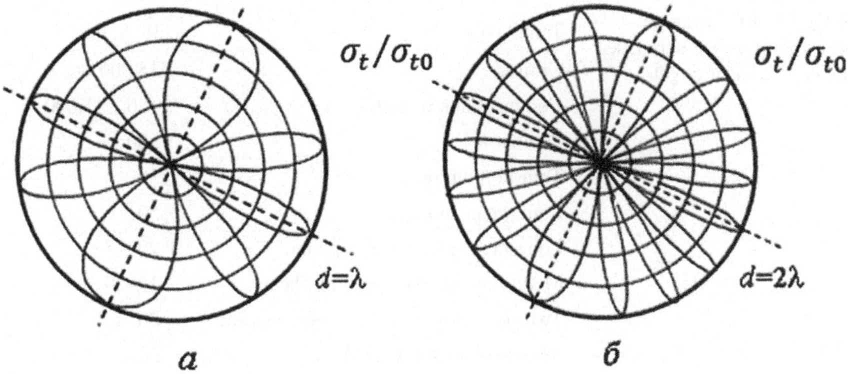

Fig. 3.21 Backscattering diagram of two-point target

λ/2*d*. The BSD of two-point target is depicted in Fig. 3.21, consisting from two similar targets ($A_1 = A_2, \varphi_1 = \varphi_2 = 0$), for which the following formula is correct:

$$A_1 = 4A_1^2 \cos^2(kd \sin\theta).$$

If a group target represents a set of point targets located on a same distance between each other, then its RCS can be found according to the following formula:

$$A = \sum_{j=1}^{N} A_1 + 2\sum_{j\neq k}^{N} \sqrt{A_j A_k} \cos\psi_{jk}, \tag{3.55}$$

where $\psi_{jk} = 2kd \sin\theta_{jk} + \varphi_i - \varphi_k$.

For real conditions, such models of group targets are too idealized. This is explained in that during observing process, as a rule, a continuous change of θ angle, $\varphi_1 - \varphi_2$ differences, distances between points C_1 and C_2, RCS of separate targets happens. Herewith, all these changes are having a random disordered character. In this connection, a meaning of a certain value of RCS in some fixed time point is out of particular interest. The statistical laws define problem which is brought to the forefront, characterizing a conformity of RCS change or its statistic parameters—mathematical expectation, dispersion, etc.

Both for a two-point target and for mentioned above multi-point target, the assumptions on uniformity of distribution law θ and independence of random variables θ and A_j between each other are seemed to be intrinsic. Therefore, average value of RCS A_m and d_A^2 dispersion can be easily found:

$$A_m = \sum_{j=1}^{N} (A_j)_m; \quad d_A^2 = \sum_{j=1}^{N} d_{A_j}^2, \tag{3.56}$$

Table 3.1 RCS average values for different aerial, sea and ground radar targets

Target type	RCS, m^2
Cruiser	15,000
Commercial sea craft	150…15,000
Submarine	100
Transport aircraft	50
Long-range aircraft	50
Long-range passenger airplane	20–25
Medium-range passenger airplane	15–20
Two-propeller small passenger airplane	20–25
Single-rotor light aircraft	5–10
Helicopter	1–3
Stealth aircraft	0.1–1
Cruise missile	0.01–0.03

where d_A^2 is dispersion of RCS of *j*-target.

As we can see, even in simplest problem, we have to inevitably deal with statistical approach to calculation of RCS.

Table 3.1 shows RCS average values for different target types.

The calculation technique of complex body RCS is given bellow.

The sequence of approximate calculation of complex body RCS is composed of the following stages:

- body division into elements;
- its surface segments approximation by simple bodies;
- calculation of its bodies RCS and results integrating for receiving of RCS value of a body as a whole.

Body division into elements is carried out on the base of its structural features and physical representation on reflective capability of different surface segments. For example, the elements of the aircraft are engines, wings, antenna under radiotransparent radome, fuselage, cockpit, tail assembly. In its turn, on each of the mentioned segments, there could be more selected small parts that are segments of local reflection. It is obvious that an amount of selected elements, joined into a table, depends on direction of observing. So, complex body elements are defined in some (little) sector of observation angles: For another sector, there will be different table.

At approximation of selected element by a simple body, it is guided not only by requirements of geometric similarity as much as by aiming to provide value coincidence of element RCS and its approximated simple body in the examined sector of observation angles. A part of elements is not modeled by geometrically similar bodies, an equivalent to it in RCS disks, angular constructions and other bodies in observation angles sector which are selected for them.

Another part of moments during modeling simplifies and is approximated by simple bodies.

Element RCS values, approximated by simple bodies, are easily calculated according to abovementioned formulas. The main calculation difficulties arise at RCS determination of complex elements (air inlet, engine nozzle). They are defined, as a rule, experimentally or via calculation on computer.

The next calculation stage concludes in RCS calculation of a target according to the known RCS of its parts. Let it be N number of such elements, RCS each of them in direction $\gamma_i \left(i = \overline{1, p}\right)$ will be $A_n(\gamma_i)$, $n = \overline{1, N}$.

It is obvious that required RCS can be found using the following equation:

$$A(\gamma_i) = \left| \sum_{n=1}^{N} \sqrt{A_n(\gamma_i)} \exp\left[-j\psi_n(\gamma_i)\right] \right|^2, \tag{3.57}$$

where $\psi_n(\gamma_i)$ is field phase, reflected from element in direction γ_i.

Calculations according to (3.57) formula are quite easily fulfilled if a body RCS is changed a little in the selected sector of observation angles. In this case, an amount of directions can be selected as equal to one, two, and required determination accuracy of reflected field phase is comparatively low.

If complex body dimensions comprise hundreds of wavelength, and its BSD is strongly interference in this sector of observation angles, then calculations according to (3.57) formula become almost non-executable.

If we assume in (3.57) formula that reflected field phases from complex body elements are interdependent random values and density of probability distribution $\psi_n(\gamma_1)$ for all n is uniform in interval $[0, 2\pi]$, then RCS average value of such body will be:

$$\overline{A} = \frac{1}{N} \sum_{n=1}^{N} A_n(\gamma_i). \tag{3.58}$$

At fulfillment of mentioned conditions, we can find standard deviation value σ of complex body RCS value from its average value:

$$\sigma \cong \sqrt{\overline{A}^2 - \left(\overline{A}\right)^2} = \frac{1}{N} \sqrt{\left[\sum_{n=1}^{N} A_n^2(\gamma_i)\right]^2 - \left[\sum_{n=1}^{N} A_n(\gamma_i)\right]^2}.$$

Since a hardly determined phase value $\psi_n(\gamma_1)$ of fields, reflected from separate elements, is not included in calculating formula, calculations on examined technique happen to be not complicated for arbitrary bodies, and hence, it has become widely used in engineering practice.

The knowledge of RCS value spread of complex body, determined via examined technique, is of vital importance. Representation on its value can be obtained, assuming that RCS of all elements is similar, i.e., $A_n(\gamma_i) = A_0$. Then, the required (confidential) interval can be quite easily calculated and seemed to be equal to:

$\overline{A} \pm \sigma = A_0\left[N \pm \sqrt{N(N-1)}\right]$. With $N = 2$, this interval comprises (0.59–3.4) A_0, and for $N = 3$ – (0.55–5.45)A_0 correspondingly.

For elements which differ by its RCS values and comprised of two or three equal segments of local reflection, interval values are given in Tables 3.2 and 3.3. The table data give an opportunity to fulfill an estimation of complex bodies RCS.

Based on the stated technique, the RCS calculation of long-range passenger airplane of Il-96-300 type is presented. In accordance with calculation sequence at the first stage, we divide an airplane into separate elements. For observation angles in a small sector, we can select the following elements:

(a) engine;
(b) wing panels;
(c) fuselage;
(d) tail assembly.

On each of the listed elements, we should mention segments of local reflection, i.e., to conduct an additional division of a surface into more detailed small parts.

The division results and further approximation of both the elements itself and its parts by simple bodies are presented in Table 3.4.

Table 3.2 RCS calculations of two-element body

RCS of the first element A_1	RCS of the second element A_2	Calculation results		RCS minimum values
		Average RCS $\overline{A}_n(\gamma_i)$	Interval of RCS values $\overline{A}_n(\gamma_i) \pm \sigma$	
1	1	2	0.59…3.4	0…4.0
1	1.5	2.5	0.77…4.2	0.05…4.9
1	2	3	1.0…5.0	0.17…5.8
1	4	5	2.2…7.8	1.0…9
1	9	10	5.8…14.2	4.0…16

Table 3.3 RCS calculations of three-element body

RCS of the first element A_1	RCS of the second element A_2	RCS of the third element A_3	Calculation results		RCS minimum values
			Average RCS $\overline{A}_n(\gamma_i)$	Interval of RCS values $\overline{A}_n(\gamma_i) \pm \sigma$	
1	1	1	3	0.55…5.46	0…9
1	1	9	11	4.84…17.2	1…25
1	1	49	51	36.9…65.1	15…81
1	2	2	5	1.0…9.0	0…15
1	2	9	12	4.38…19.6	0.36…29
1	2	49	52	34.7…69.3	21…89
1	4	4	9	2.07…15.9	0…25
1	4	9	14	4.10…23.9	0…36
1	4	49	54	31.5…76.5	16…100

Table 3.4 RCS of airplane separate parts

Element	RCS of element separate parts нта	Designation of element separate parts and its substituted bodies
Engines	A_1 A_2 A_3 A_4	First engine Second engine Third engine Fourth engine
Wing panels	A_5 $A_6 = A_5$ A_7 $A_8 = A_7$ A_9 $A_{10} = A_9$	Leading edge part of left wing is substituted with circular cylinder of 6.25 length and 0.15 m radius The same leading edge part of right wing Mid-part of left wing edge is substituted with circular cylinder of 6.75 length and 0.125 m radius The same edge part of right wing Leading edge end-part of left wing is substituted with circular cylinder of 6.75 length and 0.075 m radius The same edge part of right wing
Fuselage	A_{11}	Front part of airplane is substituted with prolate spheroid with semi-axis, equal to 1.92 m and 8.25 m
Tail assembly	A_{12} $A_{13} = A_{12}$ A_{14}	Leading edge of left horizontal plane is substituted with circular cylinder of 6.5 m length and 0.1 m radius. Cylinder is oriented at 133° angle to fuselage axis The same edge of right horizontal plane Leading edge of vertical plane is substituted with circular cylinder of 7.5 m length and 0.1 m radius

This table data show that we managed to approximate the most elements by simple bodies, the approximated analytical RCS expressions of which are known. In view of this, we should consider $A_5 \dots A_{14}$ as known values for each value of observation angle from nose heading angles (forward of the beam) sector. Values $A_1 \dots A_4$ are not expressed in analytical form; however, there are special calculation programs on computers for their determination.

The final calculation stage concludes in joining of RCS values of separate elements for determination of resulting RCS of an airplane. If we conduct joining according to random phase method, then airplane RCS average value is defined as follows:

$$\overline{A}_n(\gamma_i) = \frac{1}{14}\sum_{i=1}^{N} A_n(\gamma_i), \quad i = \overline{1, P}. \tag{3.59}$$

At the same time, an interval of assumed RCS values of airplane model equals to:

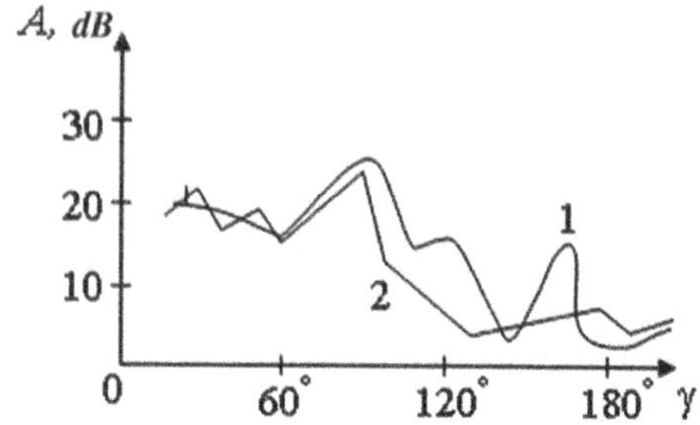

Fig. 3.22 Experimental dependence A from γ for airplane

$$\overline{A}_n(\gamma_i) = \pm\left\{\frac{1}{14^2}\sum_{n=1}^{14} A_n^2(\gamma_i) - \left[\frac{1}{14}\sum_{n=1}^{N} A_n(\gamma_i)\right]\right\}^{0.5}, \quad i = \overline{1, P}. \tag{3.60}$$

In (3.59) and (3.60) formulas, the observation angles from examined sector of "nose" observation angles are denoted by γ_i sign.

Figure 3.22 shows calculated (1) and experimental (2) curves, expressing dependence of average RCS of long-range passenger airplane from observation angle. The curve comparison shows that using random phase method, we can correctly approximate the RCS sequence of complex body.

3.6 Surface-Distributed Targets

The most widespread type of surface-distributed targets is ground surface. The radar reflections depend on it as both on radar parameters (wavelength, radiation power, radiated area sizes, radiation direction, wave polarization) and on surface itself (complex dielectric constant, ground roughness grade, non-uniformity of surface layer).

Effective surface permittivity extremely depends on soil moisture content. However, for the most land covers, the energy reflection coefficient, calculated for smooth surfaces, is always practically included in interval from 0.15 up to 0.25. Except for grass, forest and agricultural crops, the mentioned value is changed from 0.03 up to 0.45.

Surface irregularities are quite difficult to mathematically describe, but are easily characterized in quality. In the most scattering by ground surface models the statistical models of its description are used since an interest in radar location pays not to some certain surface, but to some of its class. In the most models of ground surface, only two to three parameters (standard height deviations, average slopes, correlation interval) are used.

Let us examine a ground surface with a simplest Lambert model which sometimes is called an absolute-roughened surface. Introduce it, as it was performed for volume-distributed targets, a specific RCS a_0, as an average RCS of a unit of area. Smooth flat plane can serve as electro-dynamical introduction of examined surface, each surface element (elementary area) of which changes the phase of incident wave by independent random way. The scattering field in this case represents a sum of partial waves with random phases. As it is known, in this case, the power summing of these waves is taken place. This means that incident wave at θ angle induces on elementary area ΔS a current, proportional to $\cos\theta$, and since BSD of this area also is $a\cos\theta$, then area RCS of S square will be $A = a_0\cos^2\theta\Delta S$. For surface-distributed targets, another parameter is introduced—back-reflection coefficient (BRC)—which is defined as $\gamma(\theta) = a_0\sec\theta$, and the most tables have been made exactly for it. The dependence $\gamma(\theta)$ from θ for different types of surface is depicted in Fig. 3.23.

Experimental measuring shows that many of statistical smooth surfaces stand the function $\cos^n\theta$, where n number changes from 1 up to 20 and even 50–60 as a good approximation to normalized BSD.

Sometimes, the Rayleigh criteria can be useful by which all surfaces are divided into smooth and rough. Examine a wave incident on rough surface at θ angle (Fig. 3.24).

Elementary examination shows that 1 and 2 beam path difference on plane MN will be $\Delta\psi = 2kh\sin\varphi$. Rayleigh proposed to consider a surface as smooth at $\Delta\psi \le \pi/4$. This means that if observation is carried out at φ angles, satisfying the condition: $\sin\varphi \cong \varphi \le \lambda/16h$ or, if average height of irregularities h: $h \le$

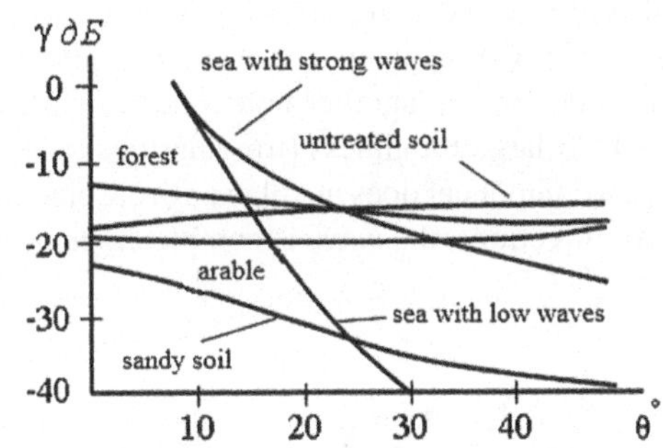

Fig. 3.23 Dependence of back-reflection coefficient from wave incident angle

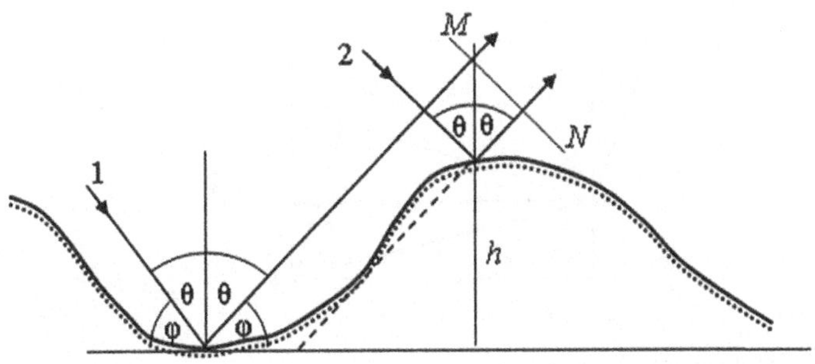

Fig. 3.24 Rayleigh criteria

$\lambda/16 \sin\varphi$, the surface can be examined as a smooth. At extremely small gliding angles φ, any surface will be as this one. For example, at $\lambda = 3$ cm and $\varphi = 30°$, the maximum admittable height of irregularities will be 0.75 cm, and at $\varphi = 1°$, we have 21.5 cm.

Examine in detail the phenomena of scattering (effect) on a rough surface. For this purpose, we turn to Fig. 3.25*a*.

Highlight two beams 1 and 2, incident at angle θ to *X*-axis, which travel into *A* and *B* points. After scattering on areas near these points, the energy scatters into all directions that is depicted in a form of rays going out from *A* and *B*. Highlight from it two parallel 1´ and 2´, forming an angle θ with normal line. Find difference of beams path 1–1´ and 2–2´ on plane *MN*, perpendicular to 1´ and 2´ beams (find a difference Fig. 3.25b), which, obviously, will be equal to segments difference (AD-CB). From Fig. 3.25c, we can easily see that $AD = AE \sin\theta_1 = x \sin\theta_1 - y\cos\theta_1$. The same is for $BC = x\sin\theta + y\cos\theta$. Consequently, the desired path difference and corresponding difference of phases will be as follows:

$$\Delta\psi = kx(\sin\theta_1 - \sin\theta) - ky(\cos\theta_1 + \cos\theta) = \Delta\psi_1 + \Delta\psi_2. \tag{3.61}$$

As we can see, $\Delta\psi$ consists of two addends, the first of which when approaching θ_1 to θ, i.e., to mirror direction, becomes zero, and the second addend has no such property. As selected beams 1 and 2 are arbitrary, this means for all points (x, y) the adherence to Eq. (3.61). Consequently, at $\theta = \theta_1$, there are two components in scattered form: one coherent and another non-coherent which vanishes only for smooth surface ($y = 0$). At large roughness (irregularities), extremely increasing a wavelength, the coherent component does not play any role practically even in mirror direction. Out of mirror directions, the wave is completely non-coherent.

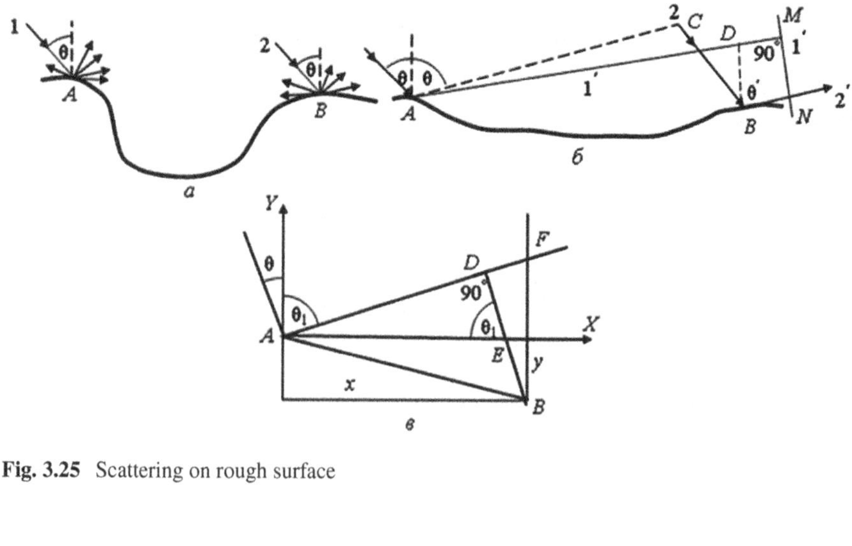

Fig. 3.25 Scattering on rough surface

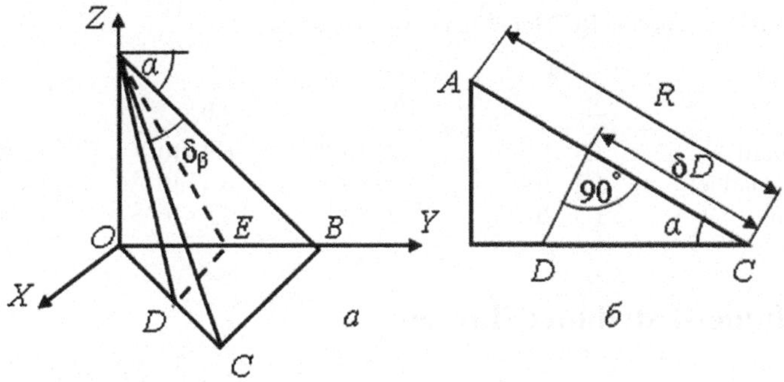

Fig. 3.26 RCS of ground surface

Let us transfer to RCS finding of ground surface observed at some θ angle to horizon by radar antenna with fan-shaped directional pattern of $\Delta\beta$ width (Fig. 3.26).

Based on figure, it is easy to find that resolvable square S_P can be calculated using the following equation:

$$S_p = BC \cdot CD = \frac{H\delta\beta}{\sin\alpha}\frac{\delta D}{\cos\alpha} = \frac{2H\delta\beta\delta D}{\sin 2\alpha}, \tag{3.62}$$

where δD is resolution capability in range, $\delta D = c\tau/2$.

Hence, RCS of ground surface can be found by formula:

$$A = \gamma(\alpha)S_p \sin\alpha. \tag{3.63}$$

For surfaces, close to diffused one, $\gamma(\alpha) = \gamma_0$ and with regard to (3.62) equation, we obtain:

$$A = \gamma_0\frac{H\delta\beta\delta D}{\cos\alpha} = a\frac{H\delta\beta\delta D}{\cos^2\alpha}. \tag{3.64}$$

With a not so high range resolution capability, a square will limited by width of antenna directional pattern $\Delta\psi$ in a second plane as well, then the equation

$$A = \frac{\gamma(\varepsilon)\pi H^2\delta\beta\delta\psi}{4\sin\alpha}$$

is correct. As is obvious, RCS of extended (area extensive) target is proportional to square range.

For surface-distributed targets, the value of specific RCS depends on wavelength and type of surface. This is illustrated by data stated in Table 3.5.

Table 3.5 Values of specific RCS for different types of surface

Surface	λ, cm			
	23	10	5.6	3.2
Smooth, forestless	$1.3 \cdot 10^{-4}$	$3.2 \cdot 10^{-4}$	$6.3 \cdot 10^{-4}$	$1 \cdot 10^{-3}$
Forest, irregularities	$1.3 \cdot 10^{-3}$	$3.2 \cdot 10^{-3}$	$6.3 \cdot 10^{-3}$	$1 \cdot 10^{-2}$
Mountainous	$1.3 \cdot 10^{-2}$	$3.2 \cdot 10^{-2}$	$3.3 \cdot 10^{-2}$	$1 \cdot 10^{-1}$

3.7 Volume-Distributed Targets

We have examined up to the moment targets, dimensions of which are less than the radar resolution element, and we called them point targets. There is another target class which completely fills in a resolution volume of radar and called volume-distributed targets. Among these are, first of all, hydrometeors (clouds, rain, hail, snow), atmospheric dust and combustion products. Here, we can refer artificial clouds, composed of dipoles (chaffs), angular or other types of reflectors.

We can put volume-distributed targets into a correspondence with a model, representing a system of a big amount of individual reflectors. For this purpose, the introduced earlier RCS determination happens to be inconvenient, as a value of reflected from it energy is clearly depends on resolution volume, i.e., on radar parameters. In this connection, the definition of volume unit RCS is introduced, which is called a specific RCS or reflectivity. Let designate it as a_0. As it is known, separate reflectors of volume-distributed targets are statistically independent between each other, and thus, a power of reflected signal from unit volume equals to a sum of signal power reflected from each of the reflectors located in this volume. From all has been said, it follows that average value a_0 will represent a sum of average RCS $\overline{A}_i$ of each reflector, i.e.,

$$a_0 = \sum_{i=1}^{N} \overline{A}_i = NA.$$

In the first approximation, we can select RCS of dielectric sphere as A_i. If we limited by examination of rain, then the experiment confirms a good coincidence with data, obtained in (3.24) formula. By this means, we obtain:

$$a_0 = \frac{60\pi^5}{\lambda^4} \sum_{i=1}^{N} \overline{\rho_i^6}. \tag{3.65}$$

Admitting that droplets particles radii ρ_i are distributed by a normal law with an average value ρ_0 and dispersion σ, we find an average value ρ^6:

$$\overline{\rho^6} = \frac{1}{\sqrt{2\pi}\sigma}\int_0^\infty \rho^6 e^{-\frac{(\rho-\rho_0)^2}{2\sigma^2}} d\rho. \tag{3.66}$$

Since a dispersion σ is not too large as a rule, then $\overline{\rho^6} = \rho_0^6$. Thus, (3.65) formula will be written as:

$$a_0 = \frac{60\pi^5}{\lambda^4} N\rho_0^6. \tag{3.67}$$

Expression $Z = \sum_{i=1}^{N} \overline{\rho_i^6} = N\rho_0^6$ in meteorology bears a name of **reflectivity multiplier**, which is connected with rain intensity rate I, i.e., with an amount precipitation mm/h, with $Z = 200I^{1,6}$ formula, and for real rain ranges from 150 to 500 mm^6/m^3. For hail $Z = 3 \cdot 10^6$ mm^6/m^3, and for snow $Z = 1000I^{1,6}$ (I—now a snowfall intensity rate in mm/h in terms of water, i.e., after thawing). Typical average radius of rain droplets $\rho_0 \cong 1$ mm, and water droplets in a cloud—$\rho_0 \cong 0.01$ mm. Amount of rain droplets in m^3 has a multiple of $N \cong 10^3$, and in cloud—$N \cong 10^9$. A useful can be a Z multiplier representation via liquid water content (LWC) M(g/m^3). For average rain, it is given as $Z = 0.03M^{1,8}$ mm^6/m^3. The dependence of specific RCS from I and M is given in Fig. 3.27.

Average RCS value of volume-distributed targets A can be calculated according to formula:

$$a = a_0 \delta V, \tag{3.68}$$

where δV is a volume, filled with elementary reflectors, which can be both less and more of radar resolution volume. In case equality of these volumes with the presence of radar cone-shaped beam, its value can be easily calculated:

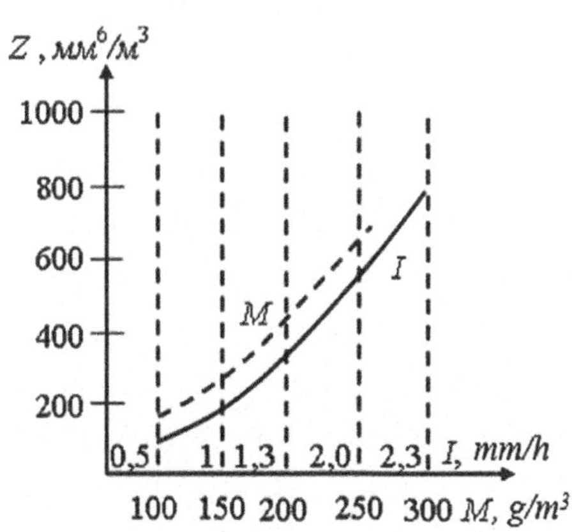

Fig. 3.27 Dependence of specific RCS from intensity and water content

$$\delta V = \frac{c\tau}{2}\pi\left(D\frac{\Delta\theta}{2}\right)^2, \tag{3.69}$$

where $c = 3 \cdot 10^8$ m/s is light speed, τ sounding-pulse duration, D range to element, $\Delta\theta$ width of directional diagram.

The most distinctive feature of (3.68) and (3.69) formulas is that RCS of volume-distributed targets depends not only on a_0, but on radar parameters and range to examined object. Magnitude order of RCS for rain in S-band at 50 km range comprises a value of 0.5–1.5 m^2, and in millimeter wave band—40–60 m^3. In case, if a volume-distributed target contains a volume δV_t, less of resolvable δV_p, we should substitute in (3.68) formula δV_t or product of fill factor q to δV_p, where $q = \delta V_t/\delta V_p$.

To volume-distributed targets also refer targets, which bear a name of optically observed. These could be atmospheric inhomogeneity, turbulence and wind shear leading to abrupt change of refraction coefficient and flock of insects, birds and electrical charges in atmosphere. The specific RCS of such targets, as a rule, enormously less than a rain RCS, though in individual cases can be equal or even large. It is possible to reduce RCS of hydrometeors by improving the radar resolution capability. Then, the interested for us desired (useful) signal of a point target on display screen (or in automatic finite device) will be associated with less intensive masking signal of hydrometeor. Observability of a point target will improve.

Chapter 4
Measuring Principles and Techniques of Objects Movement Parameters and Coordinates

4.1 Range Measurement Techniques

Range and angular coordinates measurements of objects are based on radio waves rectilinear propagation in homogeneous medium, and uniformity of radio waves propagation velocity equals to light speed $c = 3 \times 10^8$ m/s.

Pulse range measurement method. The pulse method is used nowadays in the most radar stations by measuring a distance to a target D according to elementary formula $D = \frac{ct_D}{2}$, i.e., according to dwell time t_D of reflected or reradiated (retransmitted) pulse relatively to transmitted (sounding) pulse.

The functional scheme of pulse range measurement method and voltage oscillogram (surge) on its different points are performed in Fig. 4.1.

The radar synchronizer generates pulses providing simultaneous initiation of a transmitter composed of a modulator and high-frequency generator (oscillator) and dwell time meter. Oscillations form is determined by the radar modulator (2). These are pulses of τ duration, following each other with T period. High-frequency generator "fills in" pulses of modulator with a carrier frequency (3). Pulsed high-frequency oscillations are formed by a transmitting antenna into indirect (sky) beam. When a target gets into this beam, it scatters an incident on it radio waves in all directions including in direction to the radar. Picked up by receiving antenna, a weak reflected signal (4) is amplified and detected (5) by radar receiver. In the most pulse radars, the detection is carried out regardless of phase of high-frequency fill-in. In other words, high-frequency fill-in is used only as a carrier frequency, and contained in it information on a target is often not used. We should also mention that up-to-date pulse radars are usually represent a single-antenna radars, meaning that transmitting and receiving antennas are joined, and antenna switching from transmitting mode into receiving one and vice versa is provided by fast-response device called polyplexer or antenna switch. The pulse, after being reflected from a target (target pulse), goes from receiver output to meter forcing at this moment to conduct dwell time reckoning t_D of reflected pulse relatively to sounding pulse. Dwell time meter t_D for reading

D. A. Akmaykin et al., *Theoretical Foundations of Radar Location and Radio Navigation*, Springer Aerospace Technology,
https://doi.org/10.1007/978-981-33-6514-8_4

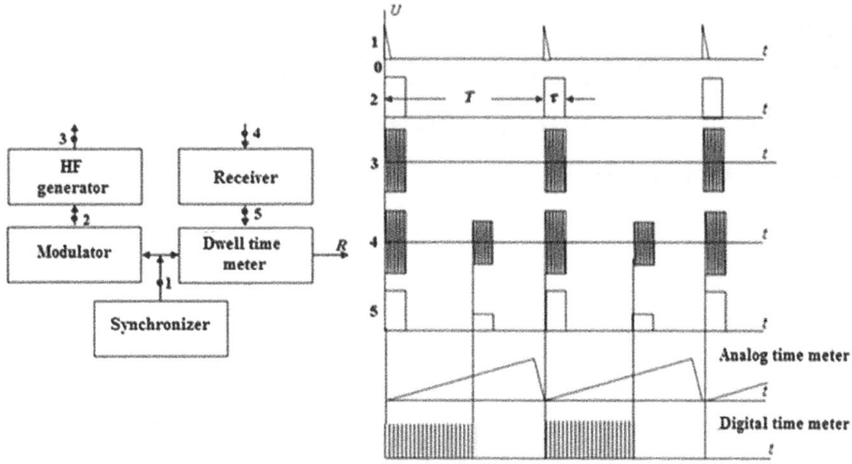

Fig. 4.1 Scheme of pulse range measurement method

can be constructed on analog principle (continuous time countdown from a radiation moment) or on digital principle (discrete time countdown) using fast-response electronic devices. Two lower positions in Fig. 4.1 illustrate both principles of time countdown. Each of meters starts at the moment of sounding pulse transmitting and performs time countdown at the moment of receiving of reflected signal.

In analog measuring instrument the reading represents a recording of instantaneous value of saw-tooth voltage at receiving moment of target pulse ($U = k_1 t_D$), and in digital instrument—the reading of a number of uniformly coming pulses, accumulated by arrival moment of target pulse ($n = k_2 t_D$). In both cases, timely uniform processes (saw-tooth voltage or pulses counting) are associated with uniform time flow, as it takes place in any watches. Definitely, that dwell time meters can be directly graduated in range digits, i.e., permits to read (pick up) a target distance D. Regular, with T period, repetition of range measuring process of the same target (Fig. 4.1) gives a more stable measuring result and reduces a measuring error.

It should be noted that the ground radar directly measures not a horizontal range (distance) to a target D_g, but slant distance D, since exactly slant distance serves as a trace of radio waves passing to a target and back. These distances are connected with a relation $D_g \cos\beta = \sqrt{D^2 - H^2}$, where H—target height; β—target elevation angle.

Besides, for ground and sea radars, the range depends on antenna installation height H_A. This limits a detection range of a target at assigned altitude H_t, which in assumption of standard refraction is defined by the following formula:

$$D(km) \approx 4.12(\sqrt{H_t(m)} + \sqrt{H_A(m)}.$$

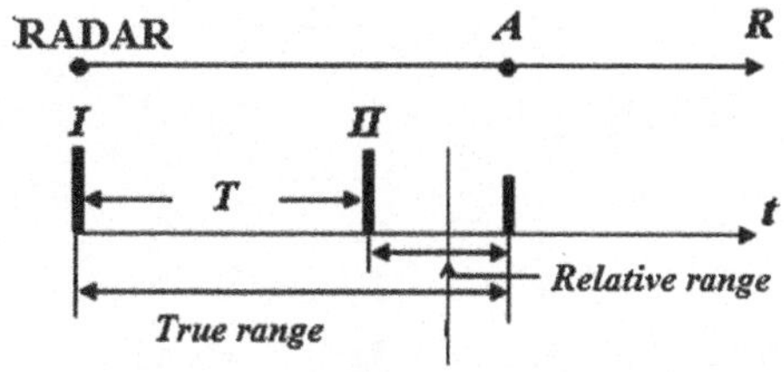

Fig. 4.2 Single-valued range measurement

In idealized noise-free conditions and other random interference, a target detection range, accuracy and range resolution capability can be obtained (conceptually) arbitrary high. For detection range, this is correct because a real sensitivity of receiving device in the mentioned conditions is arbitrary high, and long-distance reception is possible at any signal energy.

Single-valued condition of range measurement at pulse method of space sounding is remained to discuss. Ambiguity situation connects with a value of selected pulse repetition period (interval), if a period is short, then the reflected from a target *A* pulse, corresponding to sounding pulse *I*, shows up in radar receiver not in this but in the next period *T*, when range count happens relatively to sounding pulse II (Fig. 4.2).

The recorded at this dwell time will differ exactly on *T* period from a true one. The range measurement will be fault. To eliminate a phenomenon of ambiguity, it is necessary to increase a pulse repetition period of radar in such a way as to make it more than radio wave propagation time up to the most distant target and back. So, a single-valued condition of range measurement in pulse radar consists in the following:

$$T > \frac{2D_{\max}}{c} \tag{4.1}$$

Pulse method of range measuring in radar has become the most widespread. One of advantages of pulse method is a division simplicity of direct and reflected oscillations (waves): During radiation, a receiver is closed due to antenna switch, and during receiving mode, the transmitter does not radiate. Indeed, use of antenna switch, as it was mentioned before, is not permitting to provide measuring of very small distances up to objects, i.e., does not permit to eliminate dead area in range.

The minimum nominal detection range via pulse method can be estimated by the following formula:

$$D_{\min} = \frac{c(\tau + t_{\text{rcv}})}{2}.$$

where t_{rcv}—recovering time of receiver sensitivity.

Moreover, so-called **dead area** effects the real value of minimum target detection range. As an example, a dead area of ship-borne radar is illustrated on Fig. 4.3.

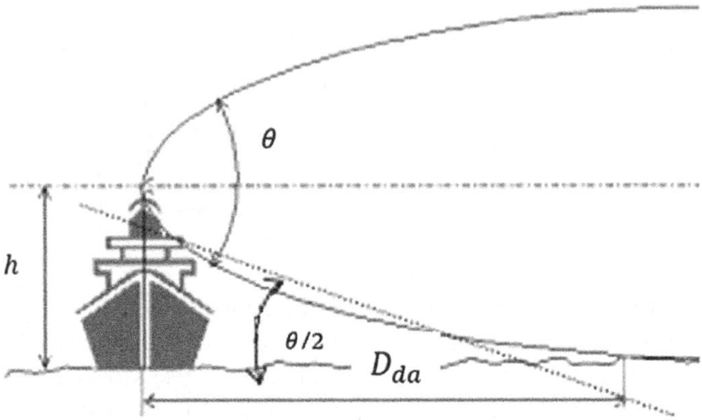

Fig. 4.3 Radar dead area calculation

The radar dead area D_{da} depends on antenna installation height h of radar and width of its directional pattern in vertical plane θ:

$$D_{\mathrm{da}} = \mathrm{hctg}\left(\frac{\theta}{2}\right).$$

On ship-borne radars screen, a circle is usually plotted around ship mark, the radius of which is determined by the radar dead area.

Use of high powers in pulse should relate to disadvantage of pulse method that is connected with a danger of breakdown in antenna-waveguide path and with use of huge pulse modulators.

Clear pulse measuring method is a non-coherent method in the sense that the pulse envelope represents a carrier of information on a target, and high-frequency fill-in is used only as a carrier frequency. Naturally, that non-use of information of high frequency as a carrier, somehow influenced by radiated target, leads to some losses in detection range and range measuring accuracy. For example, at radiation of stationary target, it is principally possible to measure its range with an accuracy of wavelength order of high-frequency oscillations; however, that required here, is carrying out of coherence (in-phase coincidence) of transmitted and received oscillations is a hard technical task when using the pulse method.

On grounds that pulse envelope detection is carried out in receiver of non-coherent pulse radar, rather than phase detection; it is also impossible to carry out direct measuring of instantaneous radial rate of a target using of Doppler effect. An average velocity of a target is detected indirectly, as a distance change to it in a time unit.

Frequency range measurement method. Naturally, that traveling (delay) time of radio wave, and, consequently, a target range can be measured, by "denoting" not an amplitude as it is carried out at pulse method, but a frequency (if it is, of course, not remained constant), fixing a difference in carrier frequency between received

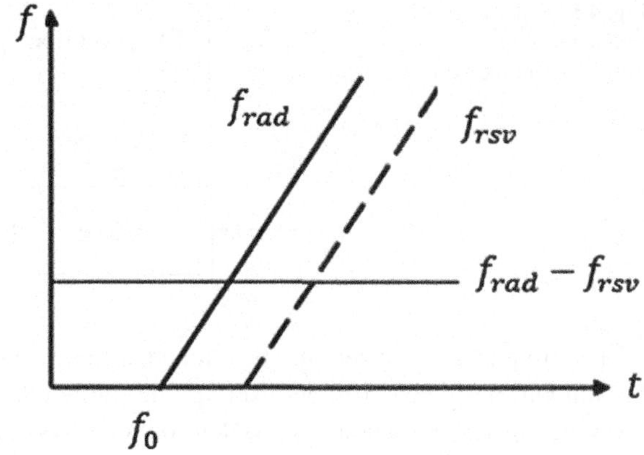

Fig. 4.4 Principle of frequency range measurement method

and transmitted oscillations. The above said results to frequency range measurement method requiring a frequency modulation (FM) of oscillations.

At frequency range measurement method (Fig. 4.4), a frequency modulation with linear law of frequency change (shift) is most commonly used:

$$f_{\text{rad}} = f_0 + kt \tag{4.2}$$

Herewith, a frequency of received oscillations (input frequency) equals to frequency of sounding oscillations sent earlier at a delay time t_D:

$$f_{\text{rcv}} = f_0 + k(t - t_D). \tag{4.3}$$

Equation (4.3) is correct at measurements on a steady target or a target moving in direction perpendicular to radial direction. In other words, to use formula (4.3) in order to determine a distance to a target, it is necessary that its radial velocity equals to zero. If not, an additional Doppler frequency F_d will appear in a frequency of received oscillations. There are special methods permitting to detect which part of frequency change in received oscillation is stipulated by a delay time (range) and which—by Doppler effect (radial velocity).

At the output of mixer where oscillations continuously coming from frequency transmitter f_{rad} and from frequency receiving antenna f_{rcv}, a signal on difference frequency (beat frequency) is formed:

$$F_D = f_{\text{rad}} - f_{\text{rcv}} = kt_D = k\frac{2D}{c}, \tag{4.4}$$

giving a possibility, using a frequency analyzer, to determine a desired distance to a target $D = k_1 F_D$. Functional diagram of frequency range measurement method is performed in Fig. 4.5.

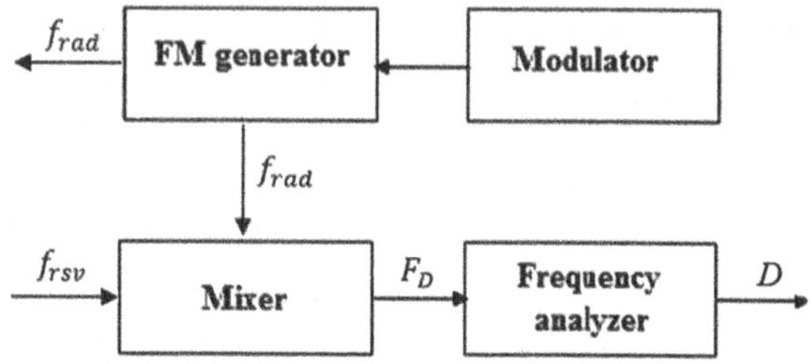

Fig. 4.5 Scheme of frequency range measurement method

The frequency analyzer, that positioned at the scheme output, represents a set of big amount narrowband filters overlapping the whole frequency band of beats. In one of the filters, a frequency F_D is determined according to a maximum signal, and, consequently, a distance to a target.

Maximum operational range (detection range) of radar with frequency modulation at application of analyzer of described (parallel) type and other equal conditions not worse than operational range of pulse radar with the same average power (energy) of radiation.

Disadvantage of continuous-wave radar consists in difficulty of effective division of radiated and received oscillations since unlike to pulse radars in this case a transmitter and receiver operate simultaneously. Transmitter oscillations, influencing on receiver input, generate a constant background reducing radar detection range. Decoupling (isolation) of transmitter and receiver in continuous-wave stations is one of the most serious technical problems.

Minimum operating range of continuous-wave stations is less than pulse radiation radars; in other words, a frequency method permits to measure very short ranges. This is determined by the absence of powerful pulse of transmitter and connected with it a dead area in range.

Single valuedness of range read-out in frequency-modulated radar is provided in the same way as it is carried out in pulsed radars: FM period is selected in such a way that it exceeds a signal delay time reflected from the most remote target.

Besides accurate measuring of short ranges, the pulse method has the following advantages:

- relatively small radiating power that eliminates a breakdown danger in antenna-waveguide path and reduces transmitter weight and dimensions;
- possibility of direct and single-valued measuring of target instantaneous velocity based on Doppler effect that permits to conduct discrimination (selection) of objects in velocity.

Discrimination in velocity capability, which opposed to pulsed radar, has a continuous-wave radar, gives a possibility to avoid returns from local features as having a zero velocity relatively to radar (as opposed to detected targets) and discriminate separate targets located in one resolution element with different velocities. In aircraft airborne FM radars, it is also a possibility to suppress background returns (clutters) based on differences in target velocities and ground relatively to airborne radar platform.

In naval FM radars, the same principle permits to suppress clutter reflections from sea disturbance for detection of a shore and target ships.

Phase range measurement method. Phase method of range detection is based on measurement of phase difference of radiated and received oscillations.

Radiating oscillations of constant frequency are having a phase:

$$\varphi_{\text{rad}} = \varphi_0 + 2\pi f_0 t. \tag{4.5}$$

Phase of receiving in t moment reflected from stationary target oscillations should be equal to a phase of earlier radiated oscillations, i.e., phase of oscillations radiated in a time moment $(t - t_D)$:

$$\varphi^*_{\text{rcv}} = \varphi_0 + 2\pi f_0 (t - t_D). \tag{4.6}$$

Since a process itself of radio waves reflection is connected with additional change of reflected wave phase ϕ_{rfl}, a relation (4.6) is due to be corrected:

$$\varphi_{\text{rcv}} = \varphi_0 + \varphi_{\text{rfl}} + 2\pi f_0 (t - t_D). \tag{4.7}$$

Difference of phases of radiated and received oscillations will comprise:

$$\Delta\varphi = \varphi_{\text{rad}} - \varphi_{\text{rcv}} = -\varphi_{\text{rfl}} + 2\pi f_0 t_D = -\varphi_{\text{rfl}} + 2\pi f_0 \frac{2D}{c}. \tag{4.8}$$

Range to a target:

$$D = \varphi_{\text{rfl}} \frac{c}{4\pi f_0} (\varphi_{\text{rfl}} + \Delta\varphi). \tag{4.9}$$

At condition of phase change consistency arising due to reflection φ_{rfl}, we have linear relationship between target range and phases equality of radiating and receiving oscillations:

$$D = \text{const} + \frac{c}{4\pi f_0}\Delta\varphi = \text{const} + \frac{\lambda_0}{4\pi}\Delta\varphi. \tag{4.10}$$

As we can see from formula (4.10), phase meter can be graduated directly in range units.

Considerable disadvantage of phase method is non-single-valued range measurement. In fact, phase meter will give out the same range readout both as at measuring of phase $\Delta\varphi$, and phase $\Delta\varphi + 2\pi n$, (n—is any whole (integral) number), despite that these are oscillations phases received from different ranges.

It is known that condition of single-valued range measurement via phase method will consist in $\Delta\varphi_{\max} < 2\pi$. By substituting it into equality, a shift of phases for the most remote target based on relation (4.9), we have: $-\varphi_{rfl} + 4\pi \frac{D_{\max}}{\lambda_0}$, from which we obtain a condition of the largest single-valued range measurement:

$$D_{\max} = \frac{\lambda_0}{2}\left(1 + \frac{\varphi_{\text{rfl}}}{2\pi}\right). \tag{4.11}$$

Formula (4.11) demonstrates that by phase method we can measure ranges comprising only fractions of operating wavelength. To obtain a single-valuedness, it is required by some means to narrow a band of simultaneously measured ranges in such way that maximum possible phases shift will not exceed 360° in this band. Another disadvantage of the method connects with difficulties of influence exclusion of unknown reflection phase from a target $\Delta\varphi_{\text{OTP}}$.

From the other hand, the phase method is distinguished by a high accuracy and a possibility to directly measure a radial velocity of targets based on Doppler effect.

It is interesting to mention a principal difference in a technique of dwell time reading (countdown) between pulse method from one hand, and a frequency and a phase one—from another hand. At pulse method, for dwell time reading, the reflected oscillations are compared in time with those which were radiated t_D before. At frequency and phase methods, the reflected oscillations are compared with those which are generated at this moment of time, i.e., in a moment of reading.

4.2 Angular Coordinates Measurement Methods

All types of measurement methods of angular coordinates (azimuth and elevation angle) of a target are based on creation using an antenna, a directional reception of radio waves, radiated, reradiated or scattered by an object. In radar location and radar navigation, the three base methods of angular coordinates measurement are widely used: maximum method, when an angle is registered by the maximum signal amplitude at passing of radiofrequency beam (directional pattern) through an object; minimum method, when a direction to an object is registered by the smallest amplitude in a receiver; target amplitudes comparison method, obtained from two mixed in a space radio rays.

Amplitude angular coordinates measurement methods. Fundamental distinctive feature of amplitude methods of angular coordinates measurement in radar location and radar navigation concludes in a cause (source) of electromagnetic emission of an objects. In radar navigation, an object itself is a source of electromagnetic radiation, in radar location, in the most cases, an electromagnetic radiation of an object is a consequence of sounding oscillations sending.

Functional measurement diagram of a target-bearing α relatively to selected direction via maximum method is given in Fig. 4.6. In radar location and radar navigation, an angular coordinates of an object are counted from some reference direction, e.g., a direction to the north or direction of an object longitudinal axis.

As we can see on a diagram, a target angular coordinate at **maximum method** is an angle between a reference direction and direction of antenna peak pattern to a target (beam angle). The purpose is to fix the last direction. This is achieved by rotating of directional pattern in space via mechanical rotation of antenna or

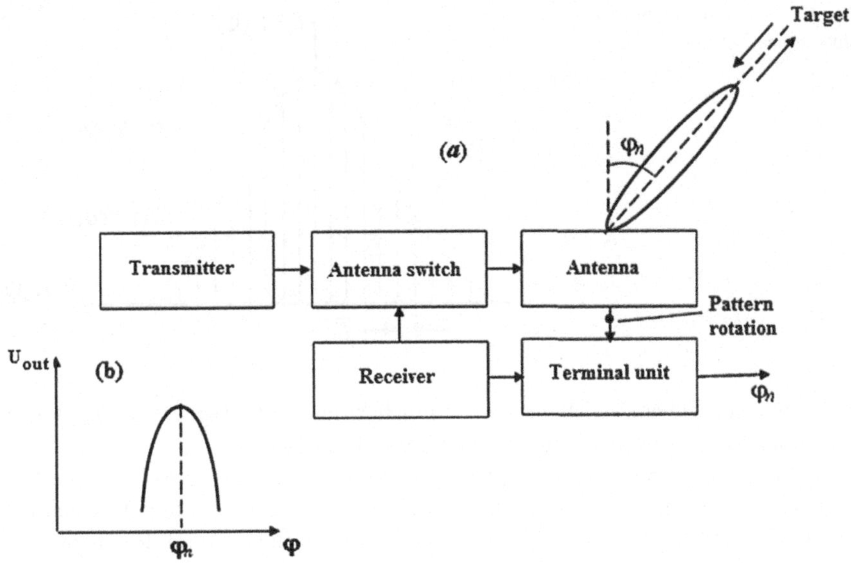

Fig. 4.6 Target-bearing measurement diagram

through electrical scanning and recording of a pattern angular position in a moment when output voltage of a receiver gains a maximum. Such recording is carried out in a finite device where both target signals from a receiver output are coming to, and corresponding to its angular positions of a pattern (radiofrequency beam). At mechanical rotation of antenna, its angular position is transferred to a terminal unit via servo selsyn-type or other systems (e.g., systems based on step motors are used in naval radars). The most radars use the same directional pattern for transmission and receiving (common antenna with TR switch), though for angular coordinate measuring it enough to have a unidirectional pattern. Bidirectional pattern provides a long-range operation of radars, including at measuring of angular coordinates.

The direction-finding (boresight) characteristic corresponding to the maximum method is depicted in Fig. 4.6. Bore sight characteristic is called a dependence of voltage at the output of angular instrument from rotation angle of antenna directional pattern axis. Type of boresight characteristic is stipulated by, mainly, a type of antenna directional pattern.

Particularly, in Radar location a narrow beam is frequently used, and consequently, a narrow boresight characteristic of the following form:

$$F(\varphi) = U_{\text{out}}(\varphi) = \frac{\sin k\varphi}{k\varphi}, \tag{4.12}$$

where k coefficient is defined by relation between antenna aperture d_a and operating wavelength λ, and a pattern maximum (peak) corresponds $\varphi = 0$. With increase

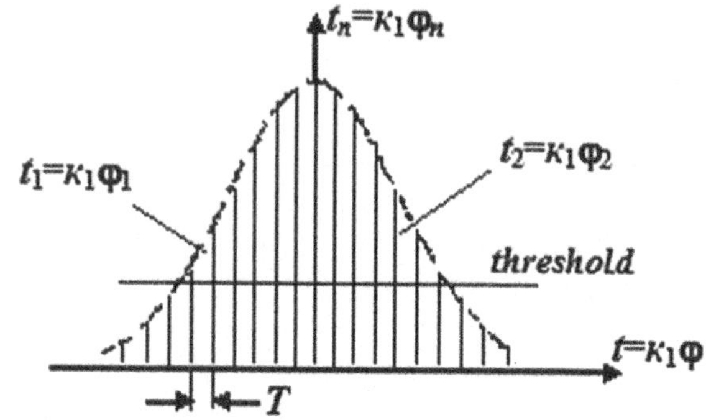

Fig. 4.7 Maximum method illustration

of k, a boresight characteristic (directional pattern) becomes higher, creating better conditions for measuring of angular coordinate. Usually $k = \pi \frac{d_a}{\lambda}$.

Within passing time of radiofrequency beam through a target (greatly slower than radio waves movement to a target and back) at the receiver output of pulse radar not one, but a train of reflected pulses are generated (as a rule, several dozens of), the amplitude of which regenerates a directional pattern of antenna (Fig. 4.7). Pulses train reflected from a target per one radiofrequency beam passing is called burst or pulses packet. On two-dimensional display with plan position indicator (PPI) sweep (radial-circular scanning) when electronic radius rotates synchronously with antenna directional pattern rotation, reflected pulse burst form a lighten arch, in the middle of which (central pulse of burst) an operator reads off direction to a target. In radar with automatic coordinates detection, the middle is determined via corresponding processing of received burst. For example, after the coordinates of the first and the last burst pulses were defined, exceeding the threshold (Fig. 4.7) via discrete count method, an angular coordinate of a target is counted out.

In real conditions of measurements, i.e., in conditions of interfering random influence, one should somehow consider all pulses exceeding the threshold in favor of measurements accuracy. In other words, all reflected by a target energy should be considered in measurements. The maximum method as for application in radar location is frequently called an **analysis method of envelope**.

Maximum method differs by simplicity of implementation, single-valuedness of reading (count) and often is used in surveillance modes of large space sectors (up to all-round surveillance). As it was mentioned before, a vital importance is for direction-finding (boresight) characteristic curvature (steepness) in clutter conditions. At long-range measuring, an essential value has a direction-finding characteristic curvature near the point where high-energy pulses are received (peak point of direction-finding characteristic). However, at maximum method, a curvature of direction-finding characteristic near a high-energy pulse of the central point is small. So, the maximum method is almost used only in the case when requirements to measuring accuracy of angular coordinates are low.

Calculate a directional pattern width of radar antenna. We are interested in width at 0.7 level from a maximum voltage value or, that is the same, at 0.5 level from a square

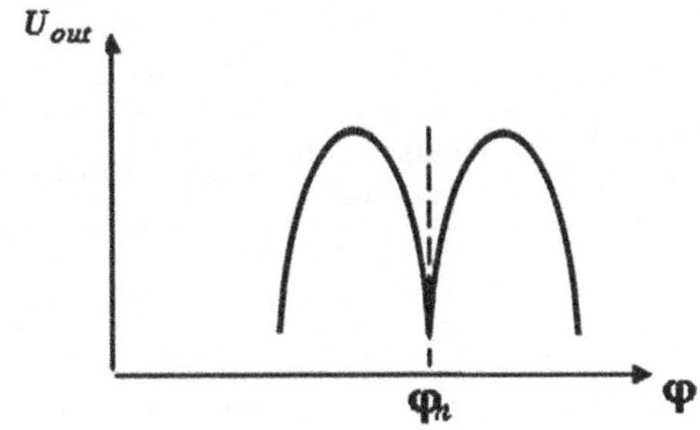

Fig. 4.8 Direction-finding characteristic (minimum method)

of this value (0.5 level in power). Then, we have: $\left[\sin\left(\pi \frac{d_a}{\lambda}\frac{\theta_{0.7}}{2}\right)/\pi \frac{d_a}{\lambda}\frac{\theta_{0.7}}{2}\right] = 0.7$. Equitation of $\frac{\sin x}{x} = 0.7$ type is satisfied at $x \cong 1.4$ rad. From here, we have: $\theta_{0.7} = 2.8\lambda/\pi d_a$. Hence, a width of radar antenna beam in radian can be quite precisely obtained, by dividing an operating wavelength into antenna aperture:

$$\theta_{0.7} \approx \frac{\lambda}{d_a}. \tag{4.13}$$

Minimum method (mathematically imprecise term), the direction-finding characteristic of which is depicted in Fig. 4.8, requires for its implementation in radar a presence of two beams in common directional pattern of antenna.

This can be achieved, for example, by installation of two emitters, correspondingly shifted relatively to focal point of antenna parabolic mirror (reflector). Angle α readout is carried out according to antenna position in a moment when amplitude of receiving signal achieves a lowest value. The method features accuracy, in comparison with maximum method, due to a large curvature of direction-finding characteristic near the lowest value of output voltage. A serious disadvantage of the method is a reducing of a radar operational range in directions close to bearing direction. So, the method can be implemented only at presence of strong signals.

Signals, received by radio rays, are compared to each other (e.g., by subtract circuit), and in a moment of signals equity the read-out of target direction is conducted—equisignal direction (*a-a*). Shift of directional patterns is usually achieved through the shift of radiators (exciters) relatively to focus of antenna parabolic mirror.

If an angular patterns shift relatively to equisignal direction comprises ε (Fig. 4.9), then residual (difference) signal at comparison circuit output can be expressed as follows.

$$U_{\text{out}}(\alpha) = F(\alpha + \varepsilon) - F(\alpha - \varepsilon). \tag{4.14}$$

At $\alpha = 0$, we have an equisignal direction when $U_{\text{out}}(\alpha)$ also equals to zero. A direction-finding characteristic curvature in equisignal direction:

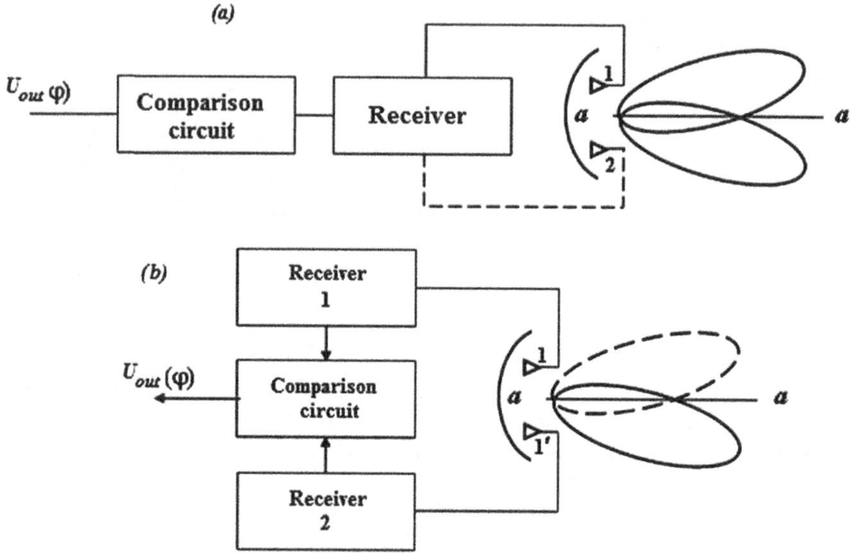

Fig. 4.9. Schemes of amplitudes comparison method

$$S = \left.\frac{U_{\text{out}}(\alpha)}{d\alpha}\right|_{\alpha=0} = \left.\frac{F(\alpha+\varepsilon)}{d\alpha}\right|_{\alpha=0} - \left.\frac{F(\alpha-\varepsilon)}{d\alpha}\right|_{\alpha=0}. \tag{4.15}$$

By considering both beams absolutely equal, we obtain that two addends in the right side (4.15), representing a curvature of each beam in equisignal direction, are differ only by sign. Then, a direction-finding characteristic curvature in equisignal direction equals to a doubled curvature of one beam in this point, i.e.,

$$S = 2\left.\frac{F(\alpha+\varepsilon)}{d\alpha}\right|_{\alpha=0}. \tag{4.16}$$

Circuits of comparison methods that illustrated in Fig. 4.9 at other equal conditions have the same direction-finding characteristic (Fig. 4.10), permitting to determine a target shift direction relatively to equisignal direction on sign of output voltage of comparison circuit. This means that it is a possibility for automatic return of antenna axis to equisignal direction that is fulfilled in acquisition-and-tracking radars (or in radars special modes). Comparison method practically permits to obtain a high measuring accuracy of angular coordinates. If common errors of the maximum method are measured in degrees, then at comparison method, it is fulfilled in angular minutes.

Identity of direction-finding characteristics of consequent and parallel comparison methods means its equivalence in idealized conditions of external random influence absence. In real, the radar operation follows by fluctuations of reflected from a target signal and external interferences of natural and artificial character. The result of such

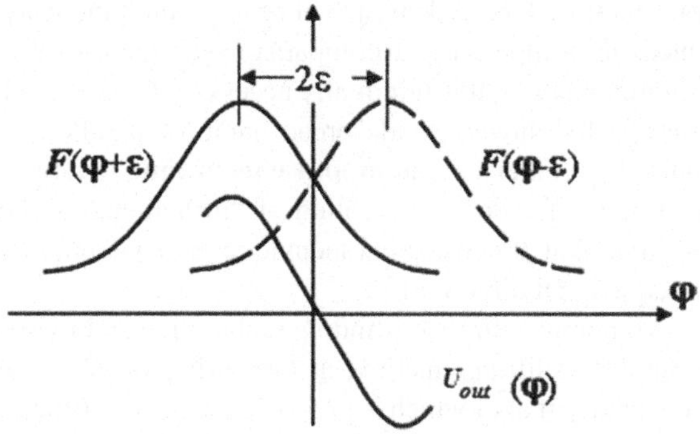

Fig. 4.10 Illustration of amplitudes comparison method

processes is a different target signal intensity (more precisely, signal-to-noise ratio) in different time periods. In conditions of random interference, a parallel comparison circuit has fundamental advantages before consequent circuit. Indeed, at forming of equisignal direction by consequent (single channel) method, some time passes T_s from a moment of signal receiving by one beam up to a moment of signal receiving by another one (Fig. 4.11a). Switching time T_s is required for antenna exciter (feed) shift from one position into another in a plane where an angle measuring is performed. To create a spatial equisignal direction (angles measuring in two perpendicular planes), in this case an exciter is rotated around antenna focal axis.

As for pulse radar in one beam position, a one pulse train is received, and in another beam—another one. To create an equisignal direction, it is minimally necessary to have two pulses of reflected energy which will be apart at a time of switching T_s and are to be compared in intensity using inertial circuit. In conditions of fluctuations, these two pulses could not be equal.

At forming of equisignal direction by parallel (dual-channel) method (Fig. 4.11b), two stationary radiators exist, symmetrically shifted in a bearing plane relatively to a focal axis of antenna mirror. Two simultaneously acting directional patterns are originated, on each of which the same "sample" of high-frequency reflected energy is received, separately amplified by two receiving channels. As for pulse radar, the same

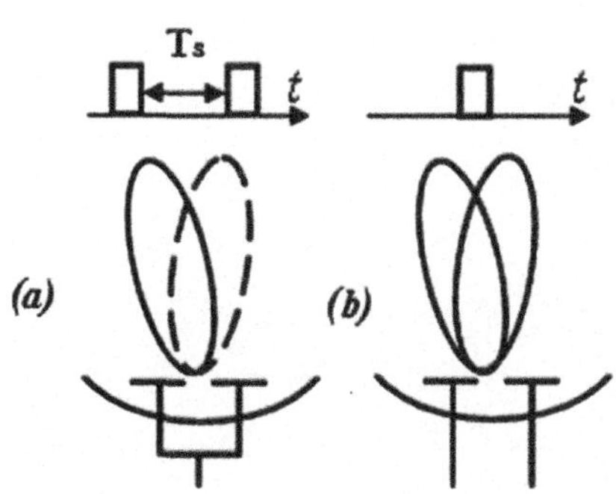

Fig. 4.11 Consequent and parallel radio direction-finding circuits

pulse train is received on each of beams simultaneously and external random interferences are compensated at comparison (subtraction). To create a equisignal direction at parallel method, it is minimally necessary to have not two, but one pulse of reflected energy. It follows a widespread name of parallel comparison technique—single-pulsed, or, otherwise, **monopulse technique** of target direction-finding (lobing). Its distinctive feature is a creation of "instantaneous" equisignal direction (it will be recalled that at consequent technique the switching time T_s is required for creation of equisignal direction).

Monopulse direction-finding technique has important advantages before pattern scanning (shifting) method at measuring of its angular coordinates of a target, a reflected signal of which is grossly changes in a time (fluctuated), and at operation in radio interference conditions as well. To create a equisignal direction simultaneously in two mutually perpendicular planes (in azimuth and elevation angle), it is necessary to have four (4) antennas (or four emitters in one antenna system), creating the same intercrossing directional patterns positioned at low angle to each other.

Phase angular coordinates measurement methods. Phase methods of targets angular coordinates measurement are based on measuring of signals phase difference, received by antennas, spaced at distance l (Fig. 4.12) and installed the way that its axis should be parallel to each other. Consequently, directional patterns axis will be parallel as well. A length l of antennas diversity (spacing) is called a base. Path (propagation) difference of radio waves will be $\Delta D = l \sin \alpha$, i.e., will depend on bearing angle α. Phases difference, corresponding to a distance ΔD and picked out in comparison circuit, is easily determined since a phase difference, equaling to 2π, corresponds to length λ (wavelength):

$$\Delta\varphi = \frac{2\pi}{\lambda}\Delta D = \frac{2\pi}{\lambda} l \sin \alpha. \tag{4.17}$$

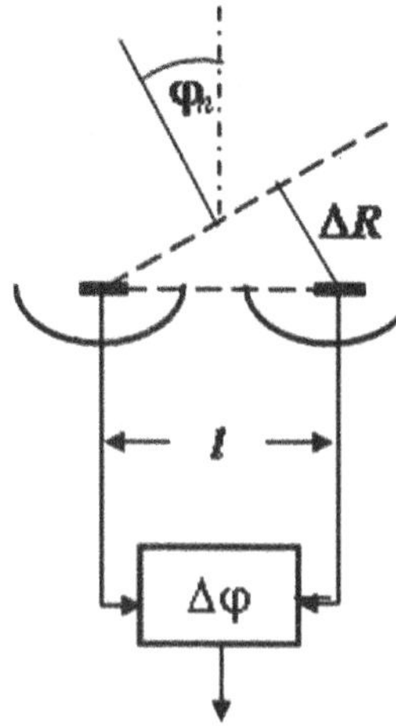

Fig. 4.12 Phase angular coordinates measurement method

Now then, according to picked out in comparison circuit phases shift in two spaced antennas, an angular target coordinate can be determined:

$$\alpha = \arcsin\left(\frac{\lambda}{2\pi l}\Delta\varphi\right). \tag{4.18}$$

Phase difference $\Delta\varphi$, usually picked out in a form of voltage at the phase detector output of comparison circuit, acquires a zero value when $\alpha = 0$.

A curvature of direction-finding characteristic will be defined via differentiation of $\Delta\varphi$ in α that leads to the next equitation:

$$S = \frac{\mathrm{d}(\Delta\varphi)}{\mathrm{d}\alpha} = \frac{2\pi l}{\lambda}\cos\alpha, \tag{4.19}$$

which at small angles α becomes:

$$S = \frac{2\pi l}{\lambda}. \tag{4.20}$$

As we can see, a curvature of direction-finding characteristic of phase direction finder, and, consequently, accuracy of direction-finding in real conditions depend on base relation to wavelength l/λ.

As at range measurement, a clear phase method of angular coordinates measuring features a readout ambiguity. The indistinguishable readouts values ($\Delta\varphi$ и $\Delta\varphi+2\pi n$) can correspond to different angular coordinates. Condition of measurement single valuedness of angular coordinate at phase direction-finding consists in $\Delta\varphi < 2\pi$. This condition of ambiguity requires a relation reducing between base value and operating wavelength.

$$\frac{l}{\lambda} < \frac{1}{\sin\alpha_{\max}}. \tag{4.21}$$

For instance, for intention of angular target position within ±30°, the relation l/λ should be less than 2.

Formulas comparison shows that requirements to accuracy and ambiguity of phase direction-finding are inconsistent. Thus, in radar engineering in some cases, the combination of amplitude and phase methods is used for angular coordinates measuring.

Temporal angular coordinates measurement method. Temporal measurement method of object angular coordinates found a use in goniometric radio navigation systems (RNS). Azimuth determination of an object relatively to ground-based beacon in such systems is based on measuring of time interval between receiving moment of reference signal (i.e., readout moment) and receiving moment of azimuthal signal. The sense of this method of measuring angular parameter consists in the following.

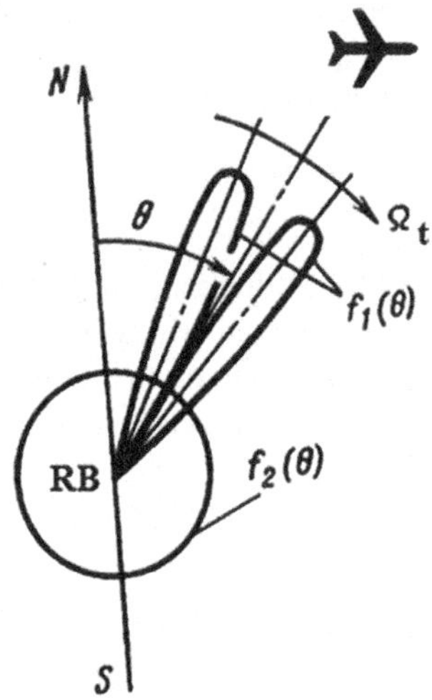

Fig. 4.13 Temporal azimuth measurement method

Very high-frequency (VHF) omnidirectional radio beacon has two transmitters operating in the same carrier frequency (aviation goniometric radio navigation systems use 873.6–935.2 MHz frequency band). One of the transmitters (azimuthal signals transmitter) generates continuous oscillations radiated by azimuthal pencil-beam (high-gain) aerial which has in horizontal plane a two-lobe directional pattern $f_1(\alpha)$ (Fig. 4.13). Azimuth antenna rotates with constant speed Ω. Another transmitter operates in pulse radiation mode and is called a reference signals transmitter. It has a non-directional (omnidirectional) antenna with directional pattern $f_2(\alpha)$. Reference signal, radiated by non-directional aerial, includes two sequences of radio pulses which, based on its amount of code packages (impulses) for one rotation of directional antenna, are called "35" and "36" reference signals. Each code package concludes two radio frequency pulses. Duration of each RF pulse of these sequences equals to 5.5 μs.

Reference pulses have repetition frequencies, correspondingly $F_{35} = 58.1$ Hz и $F_{36} = 59.76$ Hz.

At passing moment of north direction of geographic meridian by directional pattern minimum of azimuthal antenna, a coincidence of one of the "35" train pulses with one from "36" train pulses takes place for radio beacon position point.

Such coincidence is called a north coincidence. A coincidence moment of reference pulses of "36" and "35" train is fixed in object onboard equipment and designated as origin of azimuth reading t_N (Fig. 4.14). This moment of time does not depend on object angular position relatively to a beacon. In figure $T_A = \frac{1}{\Omega}$, $T_{35} = \frac{1}{F_{35}}$, $T_{36} = \frac{1}{F_{36}}$.

Azimuthal signal at receiving device input of an object has a shape of tandem bell-shaped pulse with sharp (peak) minimum, by which an arrival moment of azimuthal signal t_α is determined. Moment t_α is related with a measuring azimuth α by the following relation:

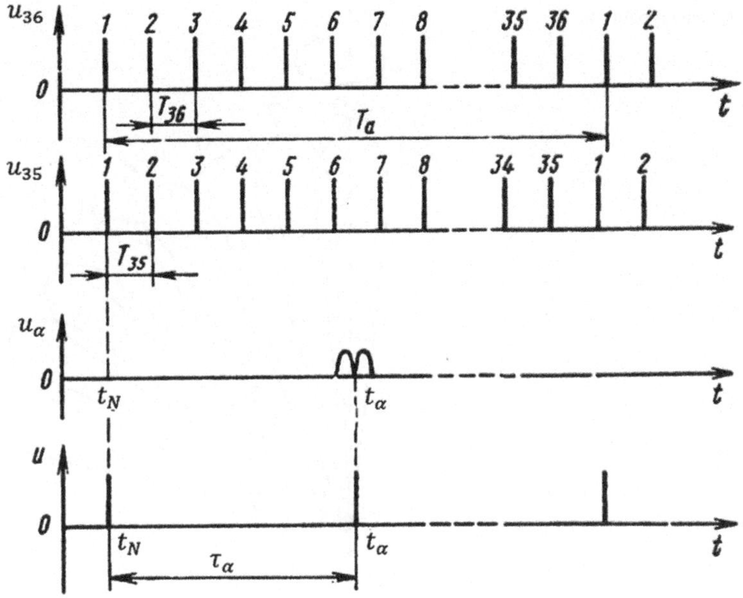

Fig. 4.14 Timing diagrams, clarifying a temporal method of azimuth measurement

$$t_\alpha = t_N + \tau_\alpha = t_N + \frac{\alpha}{\Omega}, \tag{4.22}$$

where τ_α—time delay of azimuthal signal relatively to t_N.

Time delay τ_α of azimuthal signal relatively to a north coincidence t_N is determined via different methods using an analog and digital meters. The most widespread nowadays is for digital meters.

As it is follows from (4.22), the following common requirements are specified for goniometric RNS:

- high stability of angular rotation rate Ω of azimuthal aerial;
- high accuracy of measuring device (meter) of arrival moment of azimuthal signal;
- high stability of azimuthal signal constant delay in onboard equipment path;
- high stability of north coincidence moment.

Angular coordinates measurement using partial patterns method. In radar location, the angular coordinates are frequently measured in different planes (in azimuth and elevation angle) in a similar manner. For instance, an azimuth and elevation angle in radar systems (or in target auto tracking modes) are measured via equal methods. However, in some cases, especially in ground long-range radars, the measurement is performed by different methods in order to provide rapid surveillance of assigned space sectors, i.e., in favor of scan period reducing. Herewith, azimuthal coordinates of a target are usually defined via antenna shift in azimuth (using of

Fig. 4.15 Partial patterns

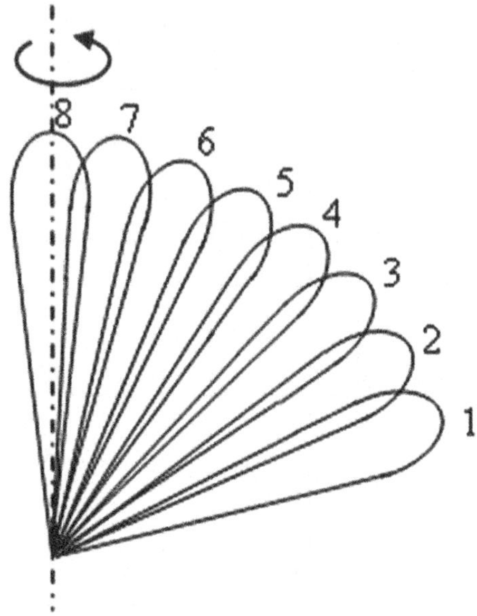

one earlier examined method of angular coordinates measuring), and a coordinate in vertical plane is defined in special way—by method of partial directional patterns (Fig. 4.15).

Radar antenna generates in space a complex ray consisting of a big amount of narrow beams fanning out in vertical plane. At rotation of fanned beam around vertical axis, the range and azimuth are measured, exploiting beam narrowness in horizontal plane. Signals received by separate beams come to corresponding to these beams receivers. Target elevation angle is roughly determined by beam number (receiver), by which a reflected signal was received. Updating (refinement) of elevation angle within a beam is conducted via one of comparison methods examined earlier.

The method disadvantage of partial patterns consists in a complexity of equipment, stipulated by the following: presence of special units, implementing a comparison method and accurate measuring of elevation angle; multi-channeling of receiving equipment; positioning of a huge amount of radiating elements in aperture of common antenna mirror (reflector).

4.3 Object Velocity Measurement Methods

Doppler velocity measurement method. There are two assigned objectives for measuring of velocity:

1. Measuring of locating object velocity (in area of radar location).
2. Measuring of actual object velocity (in area of radar navigation).

In both cases, a majority part of tasks of velocity measuring are solved through use of Doppler effect. Doppler effect—a frequency change of observed signal in comparison to radiated one at reciprocal displacement of a source of radiation and receiver—has been taken as a basis for measuring of relative speed of target movement, and also is a basis for methods of target discrimination and resolution in range of its radial movement velocity.

Examine in first turn a case when a radiated signal is a monochromatic, and a target moves away (approaches) relatively to a radar with constant speed. In this case, a range to a target is a function of time which can be performed as follows:

$$D(t) = D_0 + V_r t, \tag{4.23}$$

where $V_r = \frac{\mathrm{d}D(t)}{\mathrm{d}t}$—radial velocity of target motion.

Since a distance $D(t)$ is linear in time, then a frequency of received signal f_{rcv} can be represented as:

$$\begin{aligned} f_{\text{rcv}} &= \frac{1}{2\pi}\frac{\mathrm{d}\varphi_{\text{rcv}}(t)}{\mathrm{d}t} = \frac{1}{2\pi}\frac{\mathrm{d}}{\mathrm{d}t}\left[\omega_0\left(t - \frac{2D(t)}{c}\right) + \varphi_0\right] \\ &= f_0\left[1 - \frac{2}{c}\frac{\mathrm{d}D(t)}{\mathrm{d}t}\right] = f_0\left(1 - \frac{2V_r}{c}\right) = f_0 - \frac{2V_r}{\lambda}. \end{aligned} \tag{4.24}$$

As we can see from (4.24), a frequency of received signal f_{rcv} differs from frequency f_0 of radiated signal at value $F_d = -\frac{2V_r}{\lambda}$, called a Doppler frequency shift (Doppler frequency). It is necessary to mention that at target distant $D'(t) > 0$; hence a Doppler frequency shift F_d is negative, and, consequently, a frequency of received signal is smaller than radiation frequency. At target approaching to radar—vice versa: A frequency of received signal is bigger than a frequency of radiated signal, i.e., $D'(t) < 0$.

In this way, by comparing a frequency of received signal f_{rcv} with frequency of radiated one f_0, we can determined a Doppler shift F_d

$$f_{\text{rcv}} = f_0 \pm F_d,$$

and, consequently, a radial velocity of target movement as:

$$V_r = \frac{2}{\lambda}F_d. \tag{4.25}$$

The radar systems are usually use signals which have different modulation (amplitude or angular). At reflection of such signals by moving targets a distortion of spectrum shape and time signal structure result, as each component in spectrum is changed in accordance with (4.24), i.e., as:

$$f_i = f_{0_i}\left(1 - \frac{2V_r}{c}\right), \tag{4.26}$$

where f_{0_i}—frequency of i-component of spectrum of radar radiating signal; f_i—frequency i-component of receiving reflected signal. Hence, a spectrum width of the whole receiving signal also changes, compared to a radiated one. Where at oncoming movement of a target and radar platform the spectrum expands, and at distant movement—narrows. However, in practice, when $V_r \ll$ c, these changes usually are small. For this reason, the main influence effect on spectrum of velocity relatively to radar and target displacement is a shift of the whole spectrum in F_d value for carrier frequency of a signal. The spectrum width is left almost unchangeable.

In radar navigational systems, a signal with f_0 frequency is radiated by the system, located on moving object, which performs receiving functions as well. A signal f_{rcv} comes to input of measuring system. Let us examine different situations, when Doppler receive–transmit system, mounted on object, moving with V velocity.

1. No wind, beam of the system is directed to ground surface at γ_1 angle, positioned in vertical plane passing through an object axis. Then:

$$F_{d1} = \frac{2V\cos\gamma_1}{c} f_0. \tag{4.27}$$

2. No wind, beam of the system is directed to ground surface at γ_1 angle, positioned in vertical plane which is deviated at γ_2 angle relatively to vertical plane passing through an object axis. In this case:

$$F_{d2} = \frac{2V\cos\gamma_1\cos\gamma_2}{c} f_0, \tag{4.28}$$

3. Horizontal wind U is presented, object velocity with reference to wind V_1, object drift angle α_d. Beam of the system is directed to ground surface at γ_1 angle, positioned in vertical plane which is deviated at γ_2 angle relatively to vertical plane passing through an object axis. In this case:

$$F_{d2} = \frac{2V_1\cos\gamma_1\cos(\gamma_2 + \alpha_d)}{c} f_0. \tag{4.29}$$

In third situation, there is a possibility to measure an object moving velocity and drift angle, stipulated by the wind. However, by one measuring, this is practically impossible to do, since it is impossible uniquely determine an object velocity and drift angle. This ambiguity is increased when an object moving not only in horizontal plane. To eliminate an ambiguity in onboard velocity and drift angle measuring Doppler systems, measurements by three or four beams are conducted.

Correlation velocity measurement method. Let us examine an object, moving in parallel to ground surface with V velocity. Three antennas: radiating antenna A_r

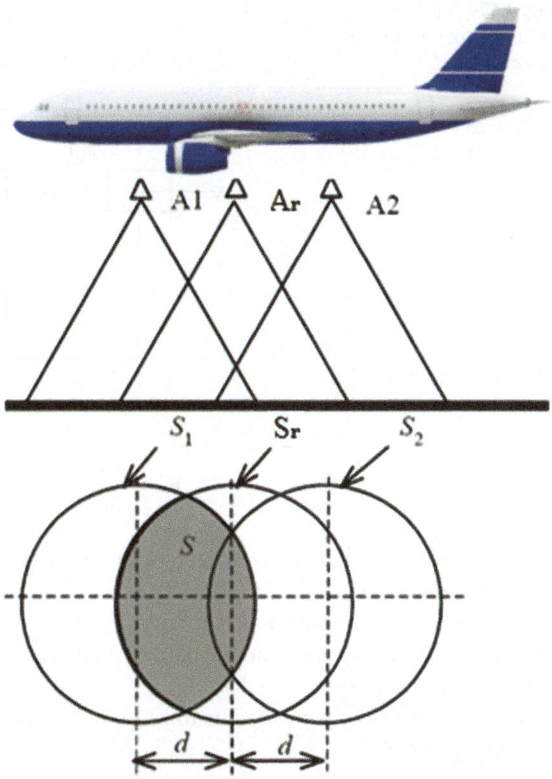

Fig. 4.16 Correlation velocity measurement method of an object

and two receiving antennas A_1 and A_2 are installed at equal distance l from each other along an object construction line (Fig. 4.16). Directional patterns of antennas are the same and directed to ground surface. Sounding area of antennas S_r and S_1 and S_2 correspondingly.

At some moment of time t_1 a signal e_1 in antenna A_1 is generated by an area $S = S_{и} \cap S_1$. In a time $\Delta t = d/V$, an antenna A_r will be at position of A_1 antenna and A_2 antenna will be at position of A_r antenna. Due to reciprocity theorem at a time moment $t_2 = t_1 + \Delta t$, a signal e_2 in antenna e_2 will be equal to signal e_1 in antenna A_1 at time moment t_1. Due to different reasons (random deviations of an object from a moving trajectory, receivers noise, etc.), there will not be an accurate signals $e_1(t_1)$ and $e_2(t_2)$ coincidence. However, if we examine a correlation function:

$$K(\tau) = \int e_1(t)e_2(t + \tau)\mathrm{d}t, \tag{4.30}$$

then its maximum will be observed in point $\tau_{\max} = \Delta t = d/V$. So that by measuring a position of function maximum (4.30) an object moving velocity can be determined.

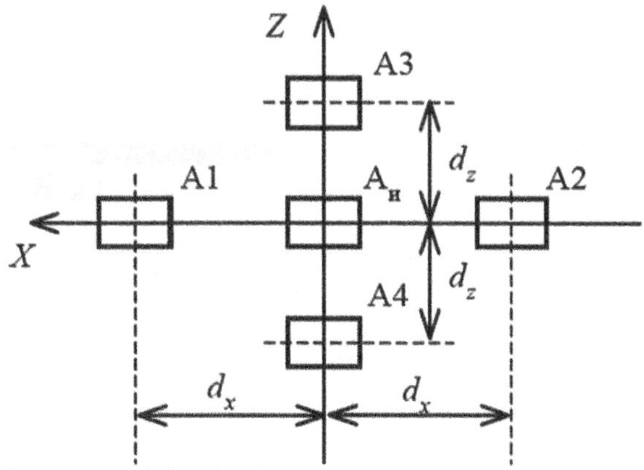

Fig. 4.17 Antennas arrangement onboard an object

$$V = \frac{d}{\tau_{\max}}, \tag{4.31}$$

To determine a full velocity vector, an additional pair of antennas is mounted onboard an object (Fig. 4.17).

Longitudinal component of velocity $V_x = \frac{d_x}{\tau_{x\max}}$ is measured using a pair of antennas A_1 and A_2, and using a pair A_3 and A_4 a lateral component of velocity is measured $V_z = \frac{d_z}{\tau_{z\max}}$.

Correlation velocity meters due to its high accuracy, simplicity and reliability are very prospective.

4.4 Object Position Finding Methods

Determination methods of an object position can be divided into three groups:

surveillance matching (visual reference (orientation); matching of television, radar and other terrain images with corresponding maps; correlation-extreme navigation on Earth physical field);

positional, i.e., position (surfaces) lines method using radio-engineering systems;

dead-reckoning methods (Doppler, inertial and its combinations).

Surveillance-matching methods are based on structure determination of some physical field typical for the terrain and on parameters comparison of this field with input parameters in memory units of navigational systems. The systems implementing a surveillance-matching method are correlation-extreme navigational systems using an information on heights field of terrain features. As for aerial vehicles in simplest case, a surveillance-matching method of aircraft or helicopter positioning is based on comparison of terrain image on map or navigational screen with actual view of ground surface observed by a crew visually or using technical aids. Visual reference is one of the oldest methods of aerial vehicle positioning. As practice shows, mean-square error of aerial vehicle position determination at visual orientation comprises 1–3 km at offset 5–15 km from reference points and 0.1–0.3 km during flying over a reference point. Herewith, the navigation reference points is meant by natural or artificial well-prominent at common landscape objects (populated point, river, road, factory chimney-stalk, etc.) with precisely known coordinates or position. Orientation accuracy of aerial vehicles can be extremely increased through use of surveillance radars or trackers of other types.

The main disadvantage of orientation as a method of an object positioning is impossibility in some cases to determine its position in bad meteorological conditions and on featureless terrain (sea, desert etc.).

Correlation-extreme navigation has a high potential accuracy and characterized by immunity to jamming and absence of accumulating errors. However, a practical realization of these systems is connected with a necessity to have large amounts of priori information on used physical fields parameters en route of an objects movement.

Besides, it possesses a low accuracy which is decreased with increasing of altitude and velocity of an object. However, considering a success in development of correlation-extreme navigation systems we can note that surveillance-matching method is a prospective one due to absence of accumulating errors.

Positional method is based on position determination of an object via intersection points representing point of two or more position lines (surfaces) intersection relatively to the known reference points. Position (surface) line is called a geometric locus of points on surface (in space), corresponding to some single value of navigational parameter. In radar navigation and radar location, the most widespread is for the following position lines.

Equal space line (ESL). ESL represents a position line all points of which are equally spaced from some fixed point. ESL represents a circle.

Equal bearing lines (EqBL). EBL represents a position in each point of which a direction to an object comprises some constant angle. EBL represents a straight line.

Equal differences in distance lines (EDDL) or loran line of position. EDDL represents a position line in each point of which a distance difference to two points of ground surface is a constant value. EDDL represents a hyperbola.

Equal sum in distance lines (ESDL). ESDL represents a position in each point of which a sum of distances to two points on ground surface is a constant value. ESDL represents an ellipse the focuses of which are the points relatively to which a sum of distances is determined.

Depending on what position lines are used for determination of an object position there are rho-rho fixing, theta-theta, range-difference (loran), rho-theta and sum-range distinguished methods.

During application of positional method the radio-navigational stations (radio beacons) located in points with the known coordinates (navigational guide-points (NGP)) are used as reference points. Radio-navigational stations can be mounted on moving platforms (e.g., artificial Earth satellites), if its movement law is known or can be predicted with prescribed accuracy.

An advantage of positional method consists in that determination of an object position is carried out without consideration or even knowledge of an earlier traveled distance, it has a quite high accuracy. Positional methods have a wide spread in maritime and aerial navigation. The disadvantages are discretion in fixing of an object position and possibility of an object coordinates determination only in operational zone of radio-navigational system, i.e., not an autonomous method.

Dead reckoning is called a method of an object position coordinates determination based on calculation of traveled distance relatively to the known initial (reference) position of an object. For dead reckoning, it is necessary to have information on movement direction of an object and its acceleration or movement velocity relatively to Earth. Herewith, a traveled distance by an object is calculated by double integration of acceleration or by single integration of velocity in time.

The method advantage is a possibility of permanent receiving of information on object position, an autonomous method. The main disadvantage of the dead-reckoning method concludes in errors increasing of an object positioning in time.

In practice, all three examined methods of object positioning are frequently used in combination.

Chapter 5
Performance Characteristics of Radar Location and Radio Navigation Systems

Radar location and radio navigation systems specification can be divided into tactical, technical and operational. This chapter will examine more precisely tactical characteristics since they at the most degree define a purpose and capabilities of practical use of radar location and radio navigation systems.

5.1 Operational Range of Radar Location and Radio Navigation Systems

Operational range of radar location and radio navigation systems is determined by a target detection range (aerial vehicle, maritime vessel or other object). **Target detection range** means a greatest distance $D_{\max}$, by which a decision on the presence of a target in conditions of specified probabilities, correct solution and false alarm is made. Here we should pay attention that in language for general purposes the detection range means a "maximum distance" at which a target can be detected. At such definition, a maximum range—is some fixed value (e.g., 50 km), above which a target is impossible to detect (e.g., 51 km). Further, throughout a material discussion, a probabilistic nature of the most generally accepted definitions will be underlined.

Any object, falling in radar coverage zone, is inside a radio beam of some limited time. Let a total energy, radiated by a radar in all directions in a time of one target illumination, be a value E_{Σ}. At uniform distribution of this energy in space around a radar, an energy flux density near a target, positioned at D distance, comprises a value:

$$\Pi_t = \frac{E_{\Sigma}}{4\pi D^2}. \tag{5.1}$$

D. A. Akmaykin et al., *Theoretical Foundations of Radar Location and Radio Navigation*, Springer Aerospace Technology,
https://doi.org/10.1007/978-981-33-6514-8_5

Application of directional transmitting antenna with directional gain (factor) G_1 permits to increase energy flux density, while a directional pattern beam of radar antenna will coincide with a directed one to a target. Then, an energy flux density near a target will be as follows:

$$\Pi_t = \frac{E_\Sigma G_1}{4\pi D^2}.$$

Energies, waves, scattered by a target in direction of receiving antenna, will be: $\Pi_{t1} = \frac{E_\Sigma G_1}{4\pi D^2} A$, where A—RCS of a target. In this case, an energy flux density, scattered by a target wave in vicinity of receiving antenna, will have a value $\Pi_{t1} = \frac{E_\Sigma G_1 A}{4\pi D^2} \cdot \frac{1}{4\pi D^2}$, and, consequently, an energy E_{rcv} at its output comprises a value:

$$E_{\text{rcv}} = \frac{E_\Sigma G_1 A}{4\pi D^2} \cdot \frac{1}{4\pi D^2} \cdot S_{\text{eff}}, \tag{5.2}$$

where S_{eff}—antenna effective area, which in, for example, ultra-high frequency band consists about 0.5–0.6 from geometric area of its aperture.

The antenna theory sets a relation between directional gain G of antenna and its effective area.

$$G = \frac{4\pi S_{\text{eff}}}{\lambda^2}. \tag{5.3}$$

In this case (5.2), formula can be represented as follows:

$$E_{\text{rcv}} = \frac{E_\Sigma G_1 G_2 \lambda^2 A}{(4\pi)^3 D^4}. \tag{5.4}$$

where G_2—directional gain of receiving antenna.

Conversely, by substituting directional gain factors of antennas by its effective areas in accordance with (5.3) formula, we obtain:

$$E_{\text{rcv}} = \frac{E_\Sigma S_{\text{eff1}} S_{\text{eff2}} A}{4\pi \lambda^2 D^4}, \tag{5.5}$$

where S_{eff2}—effective area of receiving antenna.

Formulas (5.4) and (5.5) are correct at any range of "radar–target" at radar surveillance in empty (free) space.

It is resulted from received relations that an issue of "maximum range" is certainly connected with a problem of lowest energy, which can be fixed by a receiver and interpreted that its source is a wave reflected from a target.

Whenever a receiver is perfect, to avoid noise signals on its output considering up-to-date science and technology development level is impossible in principle. Change of these signals in time, carrying a principally random character, creates a constant, so-called, noise background. The abovementioned explains why to determine some fixed minimum level E_{rcv}, serving as a boundary of dilemma: "Signal exists–no signal," is principally impossible. One can speak about conditional concept of a minimum value $E_{rcv\ min}$, designated as sensitivity of receiving unit, in which a signal reflected from a target can be detected with a probability accepted by an information recipient.

An obtained in such way value $E_{rcv\ min}$, determines a maximum detection range D_{max}, the values of which, for the case when transmitting and receiving is conducted by the same antenna, i.e., for the case when $G_1 = G_2$ и $S_{eff1} = S_{eff2}$, are of the form of:

$$D = \sqrt[4]{\frac{E_\Sigma G^2 \lambda^2 A}{E(4\pi)^3_{rcv\ min}}} = \sqrt[4]{\frac{E_\Sigma S^2_{eff} A}{E2_{rcv\ min}}}. \tag{5.6}$$

Correlation (5.6) bears a name of **range equation in free space**.

As we can see, radar detection—is a reciprocal process depending both on RCS of *A* target and on parameters of the radar itself. For pulse radar, we can obtain a more convenient form of range equation by introducing a frequently used power value in pulse *P*. This is simple to do, if we consider that intervals of radiation and receiving time are equal between each other. By dividing a transmitting and receiving energy in (5.4) and (5.5) formulas into the same time, i.e., radiation time of one target, we obtain:

$$D = \sqrt[4]{\frac{E_\Sigma G^2 \lambda^2 A}{E(4\pi)^3_{rcv\ min}}} = \sqrt[4]{\frac{P_\Sigma S^2_{eff} A}{PP2_{min}}}. \tag{5.7}$$

From Eq. (5.7), we can make a false conclusion that a range of radar detection depends not on sounding and reflected energy, but on corresponding powers. Here we should pay attention that correlation (5.5), basing on which a second Eq. (5.7) was obtained, guided by energy relations, thus in general case it is necessary to use equation $E_\Sigma = \int_0^T P(t)dt$, which only at constant in time power $P(t) = \text{const}$ derives to a second equation in (5.7) formula. Local power increase, without time considering of this period, can set forth nothing.

Let us assume the written above equation in the following form: $E_\Sigma = \int_0^T P(t)dt = P_{avr}T$, where P_{avr}—is an average radiation power in T time period. Increase of an average power will provide an increase in range.

As we can see, at radar detection, as in the most physical phenomena, for achievement of desirable result it is necessary to consume some certain energy. Specific processing technique of receiving signal in radar terminal unit requires an energy presence, larger than its some lowest value.

Receiver sensitivity $E_{rcv\ min}$ finally plays a crucial role in determining of radar operational range. Value of lowest required threshold signal energy $E_{rcv\ min}$ depends, from one side, on the presence and intensity of different random processes following a radar detection, and from another side—on effectiveness of signals processing applied in radar intended for filtration of random processes.

Random process constantly presented in a radar is internal electrical noises of a receiver. Energy or spectral density of internal noise in linear (up to detector) part of a receiver is expressed as follows:

$$E_n = K_n k T_0, \tag{5.8}$$

where K_n—noise coefficient of radar receiver, estimating a degradation of signal-to-noise ratio (in power or energy) in comparison to receiver input; T_0—absolute temperature, at which a noise coefficient is measured. Usually, a noise coefficient of receiver is measured at $T_0 = 290$ K, and sometimes, a corresponding recalculation of a noise coefficient is required at high deviation of input temperature of radar from a standard one. In most cases, the deviation of real temperature from a standard one is neglected. Boltzmann constant $k = 1.38 \times 10^{-23}$ in this case gives an increment value of spectral noise density at temperature increase in one degree. Spectral noise density in the mentioned above conditions with an idealized receiver will comprise a value $kT_0 = 4 \times 10^{-21}$ J or W/Hz. A transition to a real receiver requires a multiplication of this spectral density into a noise coefficient. Formula (5.8) expresses that physical fact that a reason, inducing electrical noises, preventing a detection process is a heat motion of charged particles in stages of a receiver. As long as amplification coefficient is not included in formula but coefficient K_n exists considering noises of the whole receiver, so a value E_n represents noises of a receiver recalculated by its input. Herewith, it is more convenient to compare them with a desired (useful) signal, usually measured at the output. Finally, a relation of energy (power) of noise gives a real signal-to-noise ratio at output of receiver linear part (e.g., at output of intermediate-frequency amplifier). By using an expression (5.8), let us refer all noises of receiver to some fictive active impedance (resistance) standing at receiver output and heated up to $K_n T_0$ temperature; herewith, a receiver itself can be assumed as noiseless. It will be recalled that a noise power of active impedance does not depend on resistance value. Since a spectral noise density E_n is a noise power coming on a frequency band of 1 Hz, a total power of receiver noise, correlated to input (arriving) signal, will have a value:

$$P_n = E_n \Delta f = K_n k T_0 \Delta f. \tag{5.9}$$

where Δf—pass band of receiver linear part.

Now then, (5.8) and (5.9) formulas give a possibility to consider one fluctuation process, limiting a detection possibility, internal noises of receiver. To consider another fluctuation processes, external noises, target RCS fluctuations and to consider an effectiveness of a certain method of signal processing at the background of all fluctuations processes, a recognition (discrimination) coefficient is introduced:

$$K_r = \frac{E_{\text{rcv min}}}{E_n} = \frac{P_{\text{rcv min}}}{P_n}. \tag{5.10}$$

As we can see from (5.10) formula, recognition coefficient represents a relation of signal to noise at output of receiver linear part, minimally necessary for signal detection from a target. All operations for extraction of desired signal and filtering of random processes, which are executed in receiver and terminal unit of radar, are considered by recognition coefficient. Reducing of recognition coefficient testifies to a more effective processing in radar and leads to reducing of baseline minimum of threshold signal $E_{\text{rcv min}}$ $(P_{\text{rcv min}})$.

From formulas (5.8)–(5.10), we have:

$$\begin{cases} E_{\text{rcv min}} = K_r E_n = K_r K_n k T_0, \\ P_{\text{rcv min}} = K_r P_n = K_r K_n k T_0 \Delta f. \end{cases} \tag{5.11}$$

Formula (5.11) shows that target detection effectiveness depends on relation between signal and noise (K_r).

Obtained relations give a possibility to get another form of range equation by introducing $E_{\text{rcv min}}$ and $P_{\text{rcv min}}$ values:

$$D_{\max} = \sqrt[4]{\frac{E_\Sigma G^2 \lambda^2 A}{(4\pi)^3 k T_0 K_n K_r}} = \sqrt[4]{\frac{E_\Sigma S_{\text{eff}}^2 A}{4\pi \lambda^2 k T_0 K_n K_r}}. \tag{5.12}$$

or

$$D_{\max} = \sqrt[4]{\frac{P_\Sigma G^2 \lambda^2 A}{(4\pi)^3 k T_0 K_n K_r \Delta f}} = \sqrt[4]{\frac{P_\Sigma S_{\text{eff}}^2 A}{4\pi \lambda^2 k T_0 K_n K_r \Delta f}}. \tag{5.13}$$

In favor of best filtering of receiver internal noises a band of intermediate-frequency amplifier (IFA) of receiver is coincided with duration of sounding (and receiving pulse from point targets) pulse $\Delta \cong 1/\tau$ in the most pulse radars. In this case, (5.12) and (5.13) formulas will be as follows:

$$D_{\max} = \sqrt[4]{\frac{P_\Sigma G^2 \lambda^2 A \tau}{(4\pi)^3 k T_0 K_n K_r}} = \sqrt[4]{\frac{P_\Sigma S_{\text{eff}}^2 A \tau}{4\pi \lambda^2 k T_0 K_n K_r}}. \tag{5.14}$$

At calculations, the most difficulties are at calculation of recognition coefficient value K_r.

Calculation of recognition coefficient if necessary is a probabilistic, since a determinated (non-probabilistic) value K_r should "embody" all fluctuations processes preventing to detection, and also all measures for filtering in radar of these processes are in favor of desired signal extraction—reflected target signal. Recognition coefficient value depends on probability distribution law of a target RCS value and rate of its fluctuations, on intensity and rate of fluctuations of additional external and internal noises matching along with common internal noises to a desired signal, on signal processing principle in radar, particularly on degree of non-ideality of signal

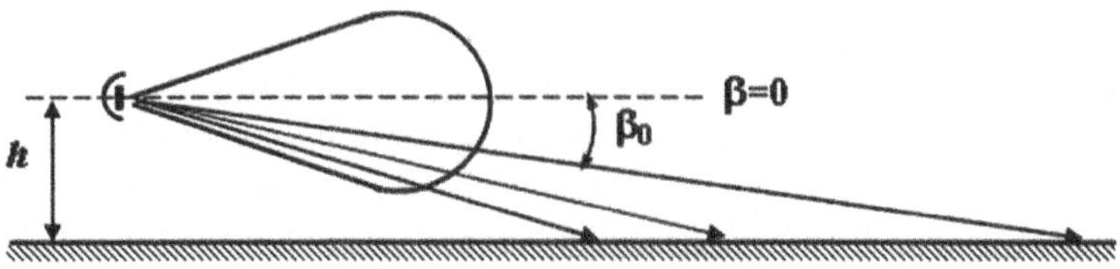

Fig. 5.1 Influence of ground surface on range of radar detection

accumulation (noises filtering) in processing devices. In one word, a recognition coefficient considers a probability character of target radar detection process.

Influence of radio waves reflection from ground surface on operational range of ground-based radar. Earth effects on radar detection of a target in those cases when a directional pattern of radar antenna intersects a ground surface (Fig. 5.1).

Radiated within directional pattern energy of radio waves reaches a ground surface in different directions, which correspond to negative angles of elevation.

In this case, a voltage in radar receiver is generated due to interference of direct ray directly reflected by a target, and a ray, consequently reflected by a ground and a target (Fig. 5.2). A result of direct beam interference and radio waves reflected by ground is a change of directional pattern of antenna: Real directional pattern differs from those that could radar have in a free space. To determine a resulting directional pattern of radar, let us examine a flat, smooth, perfectly reflecting surface, above which on height h a radar antenna is positioned with directional pattern in a free space $F(\beta)$. Radio waves are mirror reflected from smooth surface. So, a received by radar, reflected from a target, signal contains in this case two components propagated in different paths. Resulting signal value depends on amplitudes and phase difference of the mentioned components.

Let us assume also that a target is located at long distance relatively to a target height $D \gg H$, and that, consequently, elevation angle of a target β_t is small.

In the mentioned conditions, propagation paths of both components (direct and reflected from surface) are so close to each other that amplitudes of two signals are determined only by directional pattern. A phase difference between a direct and reflected from a surface waves is left to determine.

Fig. 5.2 Influence of ground surface on detection range

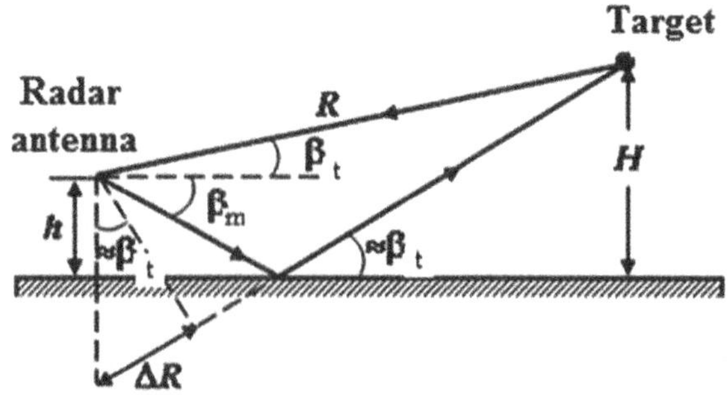

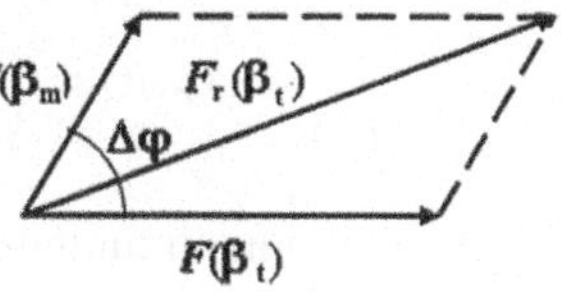

Fig. 5.3 Formation of resulting field in vicinity of radar target

Difference of travel path of two interfering rays, in accordance with Fig. 5.2 and with a condition $D \gg H$, will comprise:

$$\Delta D \cong 2h \ \sin \beta_t. \tag{5.15}$$

This travel path difference corresponds to a phase shift

$$\Delta\varphi^* = \frac{2\pi}{\lambda}\Delta D = \frac{2\pi}{\lambda}2h\sin\beta_t. \tag{5.16}$$

At reflection of radio wave, which we will calculate as horizontally polarized, from a surface, an additional phase shift in π is taken place. Then, a common phase shift between two beams:

$$\Delta\varphi = \pi + \Delta\varphi^* = \left(1 + 4\frac{h}{\lambda}\right)\sin\beta_t. \tag{5.17}$$

In accordance with Fig. 5.3, a resultant of electric field strength (intensity) in receiving end, representing a value of resulting directional pattern in direction β_t, is a vector sum of two amplitudes:

$$F_r(\beta_t) = \sqrt{F^2(\beta_t) + F^2(\beta_m) + 2F(\beta_t)F(\beta_t)\cos\Delta\varphi}. \tag{5.18}$$

Since values $F(\beta_t)$ and $F(\beta_m)$ are related with each other by the common reference (quiescent) directional pattern $F(\beta)$ and, consequently, one of it can be expressed through another, then we would not break commonality of discussion, if one of these value is assumed equal to one (we consider a directional pattern as normalized):

$$F_r(\beta_t) = \sqrt{1 + F^2(\beta_t) + 2F(\beta_t)\cos\Delta\varphi}. \tag{5.19}$$

Relation (5.19) remains (within earlier discussed constrains) for different elevation angles, which means and for different directions β in directional pattern:

$$F_r(\beta_t) = \sqrt{1 + F^2(\beta_t) + 2F(\beta_t)\cos\Delta\varphi}. \tag{5.20}$$

By introducing a formula (5.20) a phase shift value from (5.17), we obtain:

$$F_r(\beta) = \sqrt{1 + F^2(\beta_t) - 2F(\beta)\cos\left(\frac{4\pi h}{\lambda}\sin\beta\right)}. \tag{5.21}$$

Maximum and minimum (nulls) in directional pattern (5.21) will exist in those directions β, where cosine assumes a value ±1, that gives:

- condition for maximum

$$\sin\beta_{\max} = \frac{\lambda}{4h}(2n-1), \quad n = \pm1, \pm2, \pm3, \ldots \tag{5.22}$$

- condition for minimum

$$\sin\beta_{\min} = \frac{\lambda}{4h}n, \quad n = 0, \pm1, \pm2, \pm3, \ldots \tag{5.23}$$

The less relation λ/h, the more frequently maximum and minimum are alternated by resulting multi-beam directional pattern, and, consequently, the narrower lobes are become. In other words, a multi-lobe condition is increased with wave shortening and increasing of antenna lift above a surface (if directional pattern intersects a reflected surface). From (5.22) and (5.23) formulas follow that at $\lambda/h = 1/5$ a minimum of directional pattern are located at $\beta_0 = 0°;6°;11{,}5°;17{,}5°$ angles, and lobes width (between minimum) comprises of about 6° (Fig. 5.4).

At the presence of multi-lobe directional pattern, a detection range becomes a function of target elevation angle. In formula (5.21), a function value $F(\beta)$ changes on VHF waves (metric waves) band at β angle changing more slowly and then depending on relation h/λ cosinusoidal multiplier (cofactor).

In the mentioned above example (Fig. 5.4), this multiplier has been changed with 6° period, and at such change of angle, a directional pattern on metric waves is changing more slowly. Hence, it is reasonable to consider $F(\beta)$ a constant value equals to one for all directions. Instead of (5.21) for such initial isotropic emitter (radiator), we obtain:

Fig. 5.4 Multi-lobe directional pattern of antenna

$$F_r(\beta) = \sqrt{2\left[1 - \cos\left(\frac{4\pi h}{\lambda}\sin\beta\right)\right]} = 2\left|\sin\left(\frac{2\pi h}{\lambda}\sin\beta\right)\right|; \quad F(\beta) = 1. \quad (5.24)$$

Change of directional pattern (5.24) as opposed to initial pattern $F(\beta) = 1$ will lead to change of antenna directivity factor (gain), proportional to square of strengths:

$$\frac{G_r}{G} = 4\sin^2\left(\frac{2\pi h}{\lambda}\sin\beta\right). \quad (5.25)$$

By substituting a directivity factor G_r of multi-beam pattern into a range equation (instead of initial G), we obtain a range equation of VHF ground-based radar with respect to reflections influence from ground in the following form:

$$D_{\max} = \sqrt[4]{\frac{E_\Sigma \lambda^2 S_0 G^2 16 \sin^4\left(\frac{2\pi h}{\lambda}\sin\beta\right)}{(4\pi)^3 E_{\text{rcv min}}}}, \quad (5.26)$$

where G—initial directional factor.

Detection of low-flying targets problem has been acquired a particular importance. In this case:

$$\sin\beta \cong \beta \cong \frac{H}{R}, \quad (5.27)$$

$$\sin\left(\frac{2\pi h}{\lambda}\sin\beta\right) \cong \frac{2\pi h}{\lambda}\cdot\frac{H}{R}. \quad (5.28)$$

By substituting these values into (5.26), we obtain

$$D_{\max} = \sqrt[4]{\frac{E_\Sigma \lambda^2 S_0 \cdot 4\pi \cdot G^2}{E_{\text{rcv min}} \lambda^2}\,\frac{hH}{R_{\max}}}. \quad (5.29)$$

Radar detection range of low-flying target can be finally defined using the following relation:

$$D_{\max} = \sqrt[8]{\frac{E_\Sigma \lambda^2 S_0 \cdot 4\pi \cdot G^2 (hH)^4}{E_{\text{rcv min}} \lambda^2}}. \quad (5.30)$$

Principal differences of obtained correlation from that, which is correct for radar detection in a free space, are concluded in the following:

(a) detection range is more weaker and depends on standard radar parameters. Dependency in form of octic root is experimentally confirmed for low-flying and maritime targets;

(b) detection range due to multi-beam directional pattern depends not on product *G*, but on its relation G/λ;
(c) dependency of detection range is expressed from lift height of radar antenna and object altitude.

Horizontal radiation polarization, for which the abovementioned formulas were derived, should be used when it is necessary to get a longer detection range at small angles of elevation. For a full-space coverage in vertical plane, a vertical radiation polarization is preferable when a multi-lobe condition of a pattern is less expressed.

Reducing of radar operational range due to attenuation of radio waves energy in atmosphere. In radar location a very high frequency (VHF), ultra-high frequency (UHF, L-band), centimeter wave (S-band) and millimeter wave (extremely high frequency) bands of ultra-short waves are widely used. Radio waves of all mentioned bands experience an absorption and scattering by oxygen molecules and water vapors, hydrometeors (rain, fog, snow), dust particles and other atmospheric inhomogeneity. The more attenuation of waves energy, basically, the shorter a wavelength, however a monotonicity of attenuation in some parts of spectrum is breaking due to resonance between operational wavelength and sizes of absorbing particles. Wave attenuation coefficient experiences sizeable fluctuations due to resonance with oxygen and water molecules that exist in millimeter waves (Fig. 5.5). Value of attenuation of very high frequency and ultra-high frequency waves in Earth atmosphere is so modest that its influence on range of radar surveillance is neglected. From other side, for waves shorter than 10 cm, an attenuation record represents significant corrections in calculation of radar operational range (coverage).

As basic characteristic of radio wave attenuation, an attenuation coefficient is used, showing an attenuation of electromagnetic energy at wave shift at distance of 1 km. An attenuation coefficient is determined experimentally for different absorbent materials and different wavelengths.

Attenuation coefficient is more convenient to measure decibels:

$$\gamma = 10\,lg\frac{Э_\Sigma}{Э} = 20\,lg\frac{P_\Sigma}{P}\left[\frac{dB}{km}\right], \tag{5.31}$$

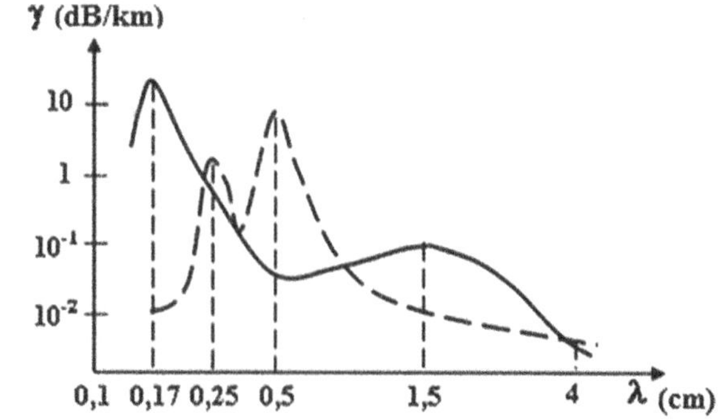

Fig. 5.5 Dependence of attenuation coefficient on wavelength

where $E_\Sigma(P_\Sigma)$—radiated energy (power); $E(P)$—energy (power), remained after passing by wave a segment of 1 km.

Then, overall attenuation at path "radar–target–radar" comprises a value:

$$\Delta = 2\gamma D[dB]. \tag{5.32}$$

In accordance with the range equation, a radiated energy E_Σ and E, remained after attenuation at path, needs to be compared to each other under root in fourth power (at other equal conditions):

$$\frac{D_{0\max}}{D_{\max}} = \sqrt[4]{\left(\frac{E_\Sigma}{E}\right)^{2D_{\max}}} = \sqrt[4]{\left(\frac{P_\Sigma}{P}\right)^{4D_{\max}}}. \tag{5.33}$$

where $D_{0\max}$—radar operational range in a free space; $D_{\max}$—real distance considering an attenuation.

By substituting a relation of energies, we obtain:

$$\frac{D_{0\max}}{D_{\max}} = \sqrt[4]{10^{0.2\gamma D_{\max}}} = 10^{0.05\gamma D_{\max}} = e^{0.115\gamma D_{\max}}. \tag{5.34}$$

Expression (5.34) permits to determine a real distance of radar coverage zone with regard to radio waves attenuation in atmosphere.

Besides notion of maximum operational range, there is another notion of minimum operational range of the system. **Minimum range** is a smallest distance, at which objects can be detected by the radar. Minimum range is defined basically by a duration of sounding pulse. In point 4.1, the relations were presented, permitting to estimate a minimum operational range of the radar.

5.2 Accuracy of Radar Location and Radio Navigation Systems

Accuracy of radar location and radio navigation systems depending on its operational modes can be characterized by different criteria. For object coordinates and moving parameters measuring mode, the most common parameter, featuring an accuracy, is a measuring error. If a measured value—p, and measuring result (or evaluation)—$\hat{p}$, then its difference $\Delta p = p - \hat{p}$ represents a measuring error (evaluation error).

Measuring errors have a random character and in general form conclude non-random and fluctuation component. Measurement precision index is an average value of squared error:

$$\varepsilon^2 = M\{\Delta p^2\} = \sigma^2 + \Delta_p^2(p), \tag{5.35}$$

where: σ^2—measuring error dispersion (variance); $\Delta_p(p)$—non-random (response) value, called fixed (systematic) error.

In practice, it should be achieved that $\Delta_p(p) = 0$, then total and mean-root-square errors are matched $\varepsilon = \sigma$.

In radar location systems, fixed errors can be eliminated through radar calibration, and errors arisen by random factors are known only at an average, for a large number of measurements. For radar location, the most important random processes are: electric noises and noises in radar receiver (self-generated and received by antenna); signal fluctuations, reflected from a target, stipulated by target movements and other reasons; unstable operation of different electric and mechanical measuring circuits; instability of radio wave propagation in continuously variable atmosphere. At reradiating of sounding signal by responder, mounted on object, target fluctuations stop influence on range measurement accuracy, but an error appears, connected with instability of responder actuation time. As we can see, random changes in real conditions of radar operation are confirmed:

- measuring object itself—reflected signal, transforming due to electric noises influence into "signal + noise" value, not identical in shape of envelope to probing radiation;
- radar measuring circuits;
- radio waves propagation path.

At measuring of radar targets coordinates in noise conditions a shape of pulse envelope acquires importance. If we compare as an example rectangular and triangular envelopes, then it will be clear that accuracy of range readout in noise conditions is connected with steepness of a front. In consequence, in radars, where a high measuring accuracy is required, a receiver band path is expanded by increasing thereby a steepness of reading fronts. In illustrated example (Fig. 5.6), a noise of the same intensity and shape influences on those points of pulses by which readings are taken (front of rectangular and peak of triangular pulse). Changing of pulse envelope shape reduces range measurement accuracy; e.g., peak bluntness of triangular pulse due to noise inputs an error in range measuring result in band $(a - b)$.

At noises absence or at very strong signal, there is no principal difference between pulses of different shapes. There is no difference as well in choice of specific reading

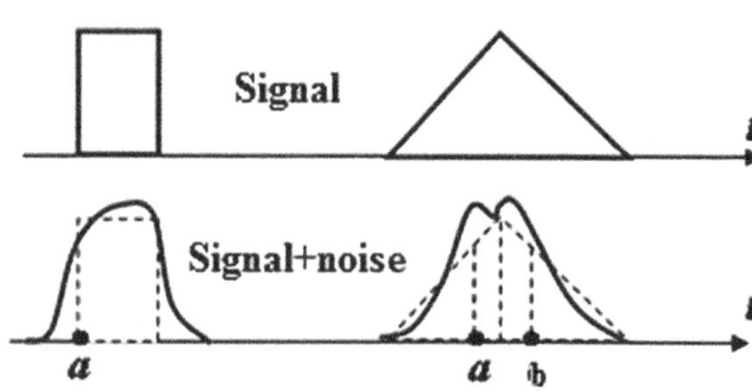

Fig. 5.6 Influence of noise signal on measuring accuracy

point (base) on pulses envelope. Particularly, at noises absence, a range measurement can be conducted with equal success as both for leading edge and trailing edge of pulses.

Concentrate in more detail on accuracy of radio navigational systems, where examine separately an accuracy of positioning systems and dead reckoning systems.

In positioning systems, a detection accuracy of position lines (surfaces) and detection accuracy of object position should be distinguished.

Detection accuracy of position lines (surfaces). Detection error of surface (line) position equals to normal distance between two surfaces (lines) of position corresponding to true and measured values of navigational parameter. Equation of navigational parameter in arbitrary Cartesian coordinates has a form $p = p(x, y, z)$—in space, $p = p(x, y)$—on surface. Each space (plane) point can be associated with a certain numeric value of reduced function. In such a way, there is a three-(two-) dimensional scalar field of navigational parameter p. Surface (line) position is a surface (line) of level. Within limits of radio navigation system operating space (coverage), the navigational parameter p represents a continuous and differentiable function, and then, changing of scalar field of parameter p is convenient to describe it by gradient. If unit vector $\vec{l}$ represents a normal vector to a surface (line) position directed to rise of navigational parameter p, then an expression, determining a gradient, has the following form:

$$\frac{\partial p}{\partial l} = \vec{l}\,\mathrm{grad}\,p, \tag{5.36}$$

where gradient modulus

$$|\mathrm{grad}\,p| = \frac{\partial p}{\partial l}. \tag{5.37}$$

Gradient $\mathrm{grad}\,p$ is a vector variable, showing a direction of quickest rise of navigational parameter p, where gradient modulus $|\mathrm{grad}\,p|$ characterizes a change rate of scalar function p.

Gradient gives a possibility to connect an error of measuring navigational parameter Δp with an error of surfaces (lines) position finding Δl. By transitioning in (2.2) to finite decrements, we obtain:

$$\Delta l = \frac{\Delta p}{|\mathrm{grad}\,p|}. \tag{5.38}$$

From (5.38), it is obvious that to reduce errors in detection of surfaces and lines of position, a field gradient of navigational parameter and its measuring accuracy should be increased.

If function p is given analytically in arbitrary Cartesian reference system, then a gradient modulus for surface and line of position correspondingly has the following form:

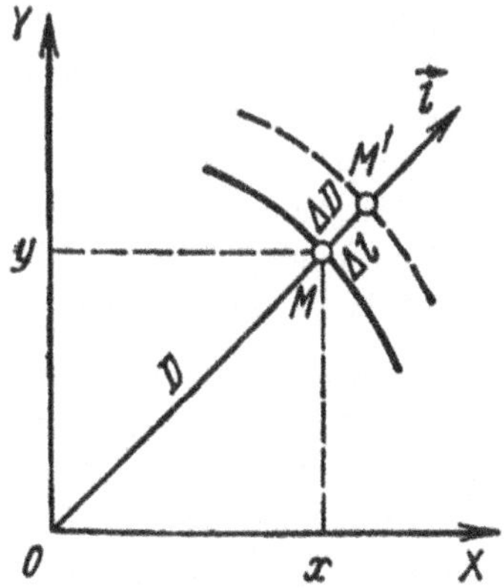

Fig. 5.7 Clarification to (5.41) formula

$$|\mathrm{grad}\,p| = \sqrt{\left(\frac{\partial p}{x}\right)^2 + \left(\frac{\partial p}{y}\right)^2 + \left(\frac{\partial p}{z}\right)^2}, \tag{5.39}$$

$$|\mathrm{grad}\,p| = \sqrt{\left(\frac{\partial p}{x}\right)^2 + \left(\frac{\partial p}{y}\right)^2} \tag{5.40}$$

Let us determine a detection accuracy of position lines for range-finding, goniometric and differential range-finding measuring techniques of navigational parameters.

Range-finding radio navigational systems (RNS) define a navigational parameter representing either a range in case of radio range finder without responder or doubled range in case of radio range finder with responder. To range-finding radio navigational systems refers a radio-altimeter as well, determining a parameter $p = 2H$, where H—flight altitude. Position line of range-finding RNS has a circle shape (Fig. 5.7).

Let us find a mean-square value of detection error of position line σ_l, stipulated by measurement error of distance *D*. In Fig. 5.7, *M* is the true position of an object, offset (shifted) position of an object, determined based on measuring of navigational parameters; *D* is the distance from navigational guide point NGP (radio beacon) up to an object; *O* is the positioning of NGP (responder); Δl is the linear offset (shift) of position line, caused by range measurement error. At selected system of coordinates, $D = \sqrt{x^2 + y^2}$ according to (5.40) $|\mathrm{grad}D| = 1$, $\Delta l = \Delta D$.

Consequently,

$$\sigma_l = \sigma_D = \frac{c}{2}\sigma_\tau, \tag{5.41}$$

where σ_D—mean-square value of measuring error ΔD; σ_τ—mean-square value of measuring error of signal delay time interval; *c*—adio wave propagation velocity.

In goniometric RNS, a measured navigational parameter is an angle α (Fig. 5.8), which is measured with an error $\Delta\alpha$. In Fig. 5.8, the same notations are admitted as

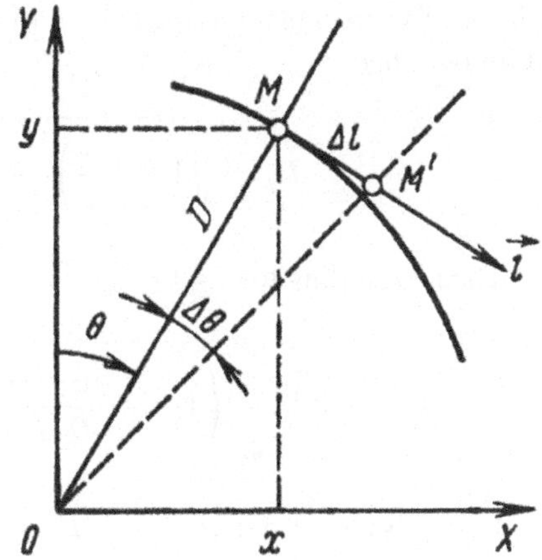

Fig. 5.8 Clarification to (5.42) formula

in Fig. 5.7. Position line represents a straight line OM. Let us find a root-mean-square error of position line finding σ_l. From Fig. 5.8, we can see that $\alpha = \text{arctg}\left(\frac{x}{y}\right)$. Then, according to (5.40) $|\text{grad}\alpha| = \frac{1}{\sqrt{x^2+y^2}} = \frac{1}{D}$, $\Delta l = D\Delta\alpha$.

Consequently,

$$\sigma_l = D\sigma_\alpha, \tag{5.42}$$

where σ_α—mean-square value of an error $\Delta\alpha$. Clearly, a value σ_l at constant value σ_α is proportional to distance D from NGP to an object.

Differential range-finding RNS determines a navigational parameter $p == D_p = D_A - D_B$—range difference of an object to two NGP (Fig. 5.9), where D_A and D_B—object ranges to two NGP. Position line of such RNS represents hyperbolic curve. In Fig. 5.9, d is the base of radio-navigational system; Ψ is the angle, at which a base is

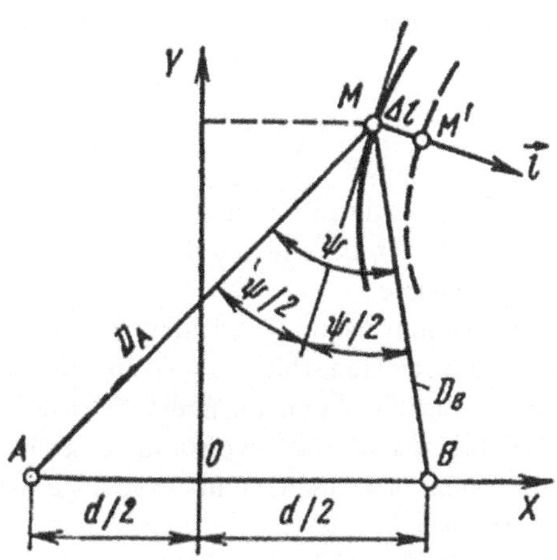

Fig. 5.9 Clarification to (5.44) formula

visible. The rest notations are similar to introduces earlier. After figure examination it follows that:

$$D_p = \sqrt{\left(x + \frac{d}{2}\right)^2 + y^2} - \sqrt{\left(x - \frac{d}{2}\right)^2 + y^2}.$$

Then, according to (5.40)

$$\left|\text{grad } D_p\right| = \sqrt{2\left(1 - \frac{x^2 + y^2 - \frac{d^2}{4}}{D_A D_B}\right)} = \sqrt{2\left(1 - \frac{D_A^2 + D_B^2 - d^2}{2D_A D_B}\right)}.$$

But, as is clear from triangle *AMB*, $D_A^2 + D_B^2 - d^2 = 2D_A D_B \cos\Psi$, from where we obtain:

$$\left|\text{grad } D_p\right| = \sqrt{2(1 - \cos\Psi)} = 2\sin\left(\frac{\Psi}{2}\right).$$

Then, linear offset of position line (hyperbolic curve), caused by measurement error of distance difference, equals to:

$$\Delta l = \frac{\Delta\Delta_p}{\left(2\sin\frac{\Psi}{2}\right)}.$$

Consequently, mean-square error value of hyperbolic curve finding σ_l is given by a relation:

$$\sigma_l = \frac{\sigma_{D_p}}{\left(2\sin\frac{\Psi}{2}\right)}. \tag{5.43}$$

where σ_{D_p}—mean-square value of measurement error ΔD_p. Considering that $\sigma_{D_p} = c\sigma_\tau$, we obtain:

$$\sigma_l = \frac{c\sigma_\tau}{\left(2\sin\frac{\Psi}{2}\right)} \tag{5.44}$$

Therefore, in differential range-finding systems, a mean-square error value of linear offset σ_l depends on measuring error of interval σ_τ and angle Ψ, at which a base is visible.

Maximum accuracy is achieved at $\Psi = \pi$, i.e., when object is directly above the system base. Herewith $\sigma_l = \frac{c\sigma_\tau}{2}$, a value σ_l is minimal and equals to mean-square error value of position line finding in case of range-finding system. At object removal from the system base, an error increases in inverse proportion to $\sin\left(\frac{\Psi}{2}\right)$. The larger the system base, the less an error, since with increase of base dimensions an angle Ψ increases.

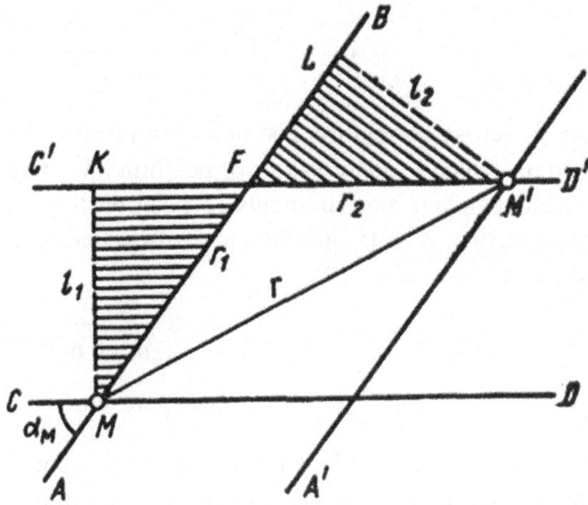

Fig. 5.10 Error of AV position finding

Accuracy of object position finding through positional method. At use of positional method, an object position is determined as an intersection point of two or more position lines. We consider as object locates at large distances from ground-based stations, relatively to which position lines are determined, and that errors of position lines finding are much lesser than these distances. Then, a family of position lines, corresponding to different values of measured navigational parameters near object position, is located in *M* point of intersection point of *AB* and *CD* position lines, corresponding to true values of measured navigational parameters (Fig. 5.10). Straights $A'B'$ and $C'D'$ defines position lines with regard to measuring errors, M'—calculated with errors object location.

Position lines *AB* and *CD* are intersected at angle α_M. Points *M* and M' are offset relatively to each other at distance *r*, which is called a radial error. Values l_1 and l_2 represent position lines *AB* and *CD* finding errors. As it clear from Fig. 5.10, $KM \perp C'D'$, $LM' \perp AB$. By denoting $r_1 = M'F$ и $r_2 = MF$, we have $r_1 = \frac{l_1}{\sin\,\alpha_M}$ and $r_2 = \frac{l_2}{\sin\,\alpha_M}$. From consideration of triangle $MM'F$, we find that:

$$r^2 = r_1^2 + r_2^2 + 2r_1 r_2 \cos\,\alpha_M.$$

Then, radial error of object position for a single measuring:

$$r = \frac{\sqrt{l_1^2 + l_2^2 + 2l_1 l_2 \cos\,\alpha_M}}{\sin\,\alpha_M}. \tag{5.45}$$

As errors l_1 and l_2 represent random values, then a radial error of object position finding is random as well. By considering a relation (5.45), at $\alpha_M = \text{const}$ we will obtain an expression for mean-square value of radial error:

$$\sigma_r = \frac{\sqrt{\sigma_{l_1}^2 + \sigma_{l_2}^2 + 2\sigma_{l_1}\sigma_{l_2}\rho\cos\alpha_M}}{\sin\alpha_M} \tag{5.46}$$

where ρ—cross-correlation coefficient of position lines AB and CD finding errors; σ_{l_1} and σ_{l_2}—mean-square values of position lines finding errors.

At independent measurements of position lines AB and CD, that is frequently met in practice, $\rho = 0$, and then a position-finding error is defined by the simpler formula:

$$\sigma_r = \frac{\sqrt{\sigma_{l_1}^2 + \sigma_{l_2}^2}}{\sin\alpha_M} \tag{5.47}$$

From (5.46) and (5.47) is clear that mean-square value of radial error of object position finding σ_r depends on errors dispersion of position lines AB and CD measuring and on angle, at which these lines are intersected and on joint correlation function as well.

Maximum accuracy at given σ_{l_1}, and σ_{l_2} will be achieved, when position lines are intersected at 90° angle. Note that a radial error r of object position finding is not distributed by Gaussian law even when errors l_1 and l_2 represent Gaussian random variables. In some cases at navigational calculations, an approximate estimate of position-finding errors of an object based on mean-square value of radial error is incomplete. Herewith, more complete statistical characteristics of RNS errors are used, and particularly, ellipse of errors is examined.

Let us estimate a probability of that expected value object position is in a certain area, surrounding its true position. Assume that random errors l_1 and l_2 of each position line finding are independent and are governed by Gaussian distribution law. Its probability densities will be as follows:

$$w(l_1) = \frac{1}{\sigma_{l_1}\sqrt{2\pi}}\exp\left(-\frac{l_1^2}{2\sigma_{l_1}^2}\right),$$

$$w(l_2) = \frac{1}{\sigma_{l_2}\sqrt{2\pi}}\exp\left(-\frac{l_2^2}{2\sigma_{l_2}^2}\right),$$

Joint probability density of errors l_1 and l_2 here equals to:

$$w(l_1, l_2) = w(l_1)w(l_2) = \frac{1}{2\pi\sigma_{l_1}\sigma_{l_2}}\exp\left[-\frac{1}{2}\left(\frac{l_1^2}{\sigma_{l_1}^2} + \frac{l_2^2}{\sigma_{l_2}^2}\right)\right]. \tag{5.48}$$

By equating an index of a power of (5.48) expression to constant number, we obtain a line equation, at which probability density $w(l_1, l_2)$, characterizing object position-finding error, is equal, i.e.:

$$\frac{l_1^2}{\sigma_{l_1}^2} + \frac{l_2^2}{\sigma_{l_2}^2} = c_1^2. \tag{5.49}$$

where c_1 is some constant value.

From (5.49), it is clear that line of constant probability density $w(l_1, l_2)$ represents an ellipse in oblique coordinate system, axes of which are coincided with normal to position lines, and origin—with true object position. Ellipse, defined by relation (5.49), is called an ellipse of errors or dispersion ellipse. At examining of ellipse of errors, it is usually turned to Cartesian (rectangular) coordinate system with an origin in the same point, where the new system axes are matching with ellipse axes. Define semi-axes a and b of ellipse of errors and rotation angle of its axes relatively to position lines. Let, as it was before, a family of object position lines around its position can be substituted with segments of parallel straight lines independently on shape of position lines. As it is performed in Fig. 5.11, position lines AB and CD are intersected at angle α_M $\left(\alpha_M \le \frac{\pi}{2}\right)$. True position is in point O. Line FE represents a bisector (bisectrix) of angle α_M. At independent l_1 and l_2, a direction of major semi-axis a of ellipse of errors comprises with line FE an angle ν, which is determined by a relation:

$$\mathrm{tg}2\nu = \frac{\sigma_{l_1}^2 - \sigma_{l_2}^2}{\sigma_{l_1}^2 + \sigma_{l_2}^2}\mathrm{tg}\alpha_M \tag{5.50}$$

From (5.50), we can see, ellipse of errors orientation depends on errors dispersions of position lines measurement and on angle α_M.

At any value of variables $\sigma_{l_1} \ne 0$ and $\sigma_{l_2} \ne 0$, we have that $\frac{\left|\sigma_{l_1}^2 - \sigma_{l_2}^2\right|}{\sigma_{l_1}^2 + \sigma_{l_2}^2} < 1$, and consequently, $\mathrm{tg}2\nu < \mathrm{tg}\alpha_M$; i.e., an angle ν is less than angle $\frac{\alpha_M}{2}$. At $\sigma_{l_1} = 0$ or $\sigma_{l_2} = 0$, we find that $|\nu| = \frac{\alpha_M}{2}$. At $\sigma_{l_1} = \sigma_{l_2} = \sigma_l$ major axis of an ellipse of errors coincides with FE line, i.e., angle $\nu = 0$.

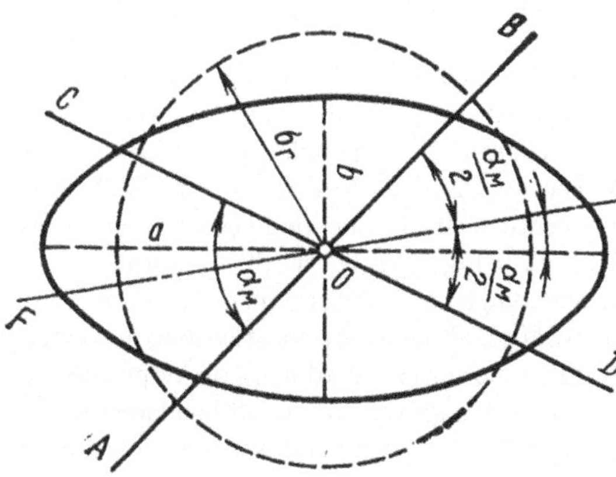

Fig. 5.11 Ellipse of errors

In practice, there is one case of interest, when accuracy of position lines finding AB and CD is similar (which is encountered, particularly, at changes of single-type equipment) and changes of navigational parameters are independent. In this case (at $\alpha_M \leq \frac{\pi}{2}$), an ellipse of errors semi-axis sizes is equal:

$$a = \sigma_l c_1 \sqrt{\frac{1}{1 - \cos\,\alpha_M}},\ b = \sigma_l c_1 \sqrt{\frac{1}{1 + \cos\,\alpha_M}}. \qquad (5.51)$$

From (5.51), it is clear, that at $\sigma_{l_1} = \sigma_{l_2} = \sigma_l$ and $\alpha_M = \frac{\pi}{2}$, an ellipse turns into a circle, as $a = b = \sigma_l c_1$. At $\alpha_M = 0$, an ellipse is regenerated into two parallel straight lines. Relations (5.49)–(5.51) permit to determine sizes and orientation of ellipse of errors.

Accuracy of object position finding through dead reckoning method. We conduct an estimation of dead-reckoning accuracy in relation to the most specific for radio-navigational systems technique of vector determination of full path object velocity using Doppler velocity meter; i.e., we will find object position-finding errors at Doppler reckoning.

To provide dead-reckoning coordinates for moving object, it is necessary to find its vector components of a full path velocity in axes of that system, in which a dead-reckoning is carried out. Usually, left-handed orthodromic (great circle) coordinate system is used as such system. Equations of dead reckoning in orthodromic coordinate system have the following form:

$$x(t) = x_0 + \int\limits_{t_o}^{t} W_x dt,\ y(t) = y_0 + \int\limits_{t_0}^{t} \frac{W_y}{\cos\left(\frac{x}{R}\right)} dt, \qquad (5.52)$$

where $x_0 = x(t_0)$ and $y_0 = y(t_0)$—initial values of AV position coordinates, corresponding to a moment t_0; W_x and W_y—vector projections of full path velocity on orthodromic coordinate system axis OXY (Fig. 5.12); $\cos\left(\frac{x}{R}\right)$—latitude correction; R—Earth's radius.

Based on measurements, vector components of a full path velocity $\vec{W}$ are determined in axes of coordinate system, connected with antenna of Doppler meter. After readout of corresponding corrections, a vector component $\vec{W}$ is calculated in connected Cartesian coordinate system $CX_cY_cZ_c$, the coordinate axes of which are matched with construction lines of an object (to be definite, we examine AV as an object). In Fig. 5.12, W_x' and W_y' is projections of vector $\vec{W}$ on CX_c and CY_c axes correspondingly; C is object center of mass (inertia); axes CX_1 and CY_1 are parallel to axes of orthodromic coordinate system; Ψ_0 is great-circle course of an object; $x(t)$ and $y(t)$ is the current coordinates of AV position. Due to agreed notations, components of a full path velocity in left-handed orthodromic coordinate system W_x and W_y, necessary for object position calculation, are defined by the following relations:

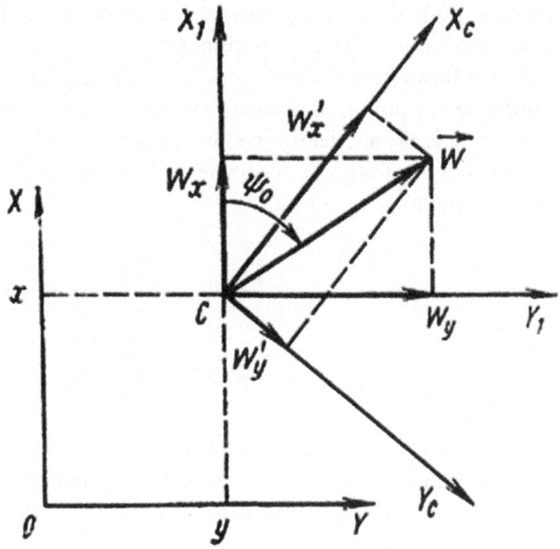

Fig. 5.12 Clarification to accuracy of object position finding through dead reckoning method

$$\begin{aligned} W_x &= W_x' \cos \Psi_0 - W_y' \sin \Psi_0, \\ W_y &= W_x' \sin \Psi_0 - W_y' \cos \Psi_0. \end{aligned} \tag{5.53}$$

Herewith, equations of Doppler dead reckoning (5.52) become a form:

$$\begin{aligned} x(t) &= x_0 + \int_{t_0}^{t} \left(W_x' \cos \Psi_0 - W_y' \sin \Psi_0\right) dt, \\ y(t) &= y_0 + \int_{t}^{t} \frac{W_x' \sin \Psi_0 - W_y' \cos \Psi_0}{\cos\left(\frac{x}{R}\right)} dt. \end{aligned} \tag{5.54}$$

Accuracy of Doppler dead reckoning is usually characterized by mean-square value σ_r of radial error of object position finding, which at independent values $x(t)$ and $y(t)$ equals:

$$\sigma_r(t) = \sqrt{\sigma_x^2(t) + \sigma_y^2(t)}, \tag{5.55}$$

where $\sigma_x^2(t)$ and $\sigma_y^2(t)$ are the dispersions of object coordinates finding errors in axes OX and OY.

In each time moment (point) t, instantaneous values of coordinates calculation errors of an object are determined by equations:

$$\Delta x(t) = x_m(t) - x(t), \Delta y(t) = y_m(t) - y(t), \tag{5.56}$$

where $x_m(t)$ and $y_m(t)$—measured (calculated) coordinates of an object; $x(t)$ and $y(t)$—true values of object coordinates.

As it follows from (5.56), errors $\Delta x(t)$ and $\Delta y(t)$ appeared by a reason that initial coordinates x_0 and y_0, comprising a path velocity W_x' and W_y', and orthodromic (great circle) course Ψ_0 are determined with errors as well.

Dispersions of object coordinates finding errors in axes *OX* and *OY* correspondingly are equal:

$$\sigma_x^2(t) = M\left[\left\{\overset{\circ}{\Delta} x(t)\right\}^2\right],$$

$$\sigma_y^2(t) = M\left[\left\{\overset{\circ}{\Delta} y(t)\right\}^2\right],$$

where index «°» means a centralized value of random value.

By defining with required accuracy with mathematical models of initial coordinates errors x_0 and y_0, comprising a vector of a full path velocity W_x' and W_y' and course Ψ_0, according to (5.56), the values $\sigma_x^2(t)$ and $\sigma_y^2(t)$ are calculated. Then, based on (5.55), a mean-square value of radial error of object position finding $\sigma_r(t)$ is determined.

5.3 Operating Space of Radar Location and Radio Navigation Systems

To estimate RNS possibilities and comparison them between each other a definition of operating space (functional area) is used. Operating space (area) is defined by operational range of the system, admitted error of object position finding and directional patterns of antennas. For different RNS types, operating areas have different sizes and contours.

Plotting of RNS operating space resolves into determination (for assigned position points of ground-based stations) of curves of admitted object position errors calculation; curves, limiting an operational range, and sectors, defined by directional patterns of antennas. Further, for simplicity, we examine RNS operating zones, corresponding to multi-directional operation mode of ground-based stations. Herewith, mathematically operating area (zone) is defined by the following in equations:

$$D \leq D\text{max}, \sigma_r \leq \sigma_{r\ \text{dr}}, \tag{5.57}$$

where $\sigma_{r\ \text{dr}}$—admissible limit value of mean-square value of object position finding. Value σ_r is calculated according to (5.46). Operating zones of applied in practice RNS of different types will be examined bellow. Outer boundary of RNS operating area is generated after plotting of curves on a map corresponding to $D = D\text{max}$ and $\sigma_r = \sigma_{r\ \text{dr}}$, as a boundary of general part of surface area, surrounded by these curves.

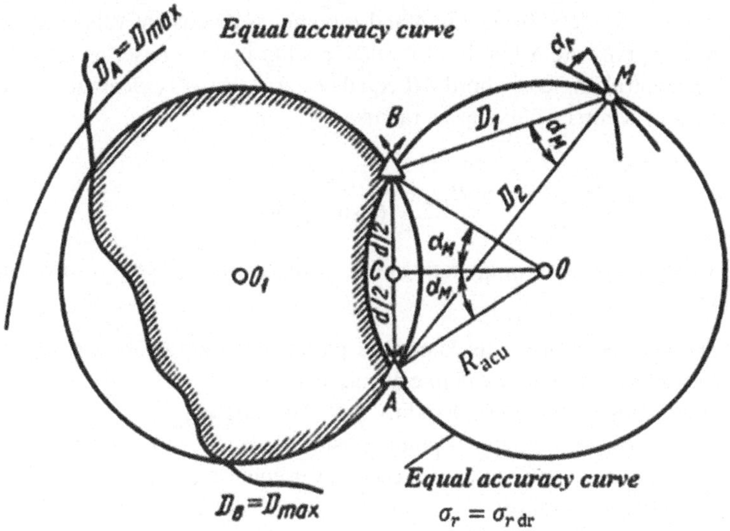

Fig. 5.13 Plotting technique of operating zone of rho-rho radio navigation system

Inner boundaries of operating zone are defined by radii of non-operational areas around ground-based stations which due to technical characteristics are considered as the known.

Plotting of operating zone will be examined on example of rho-rho radio navigation system (Fig. 5.13), which includes onboard interrogator and two ground responders located relatively to each other at d distance.

Considering a relation (5.41), mean-square value of object position lines finding errors can be written as follows: $\sigma_{l_1} = \sigma_{D_1} = \frac{c}{2}\sigma_{\tau_1}$ and $\sigma_{l_2} = \sigma_{D_2} = \frac{c}{2}\sigma_{\tau_2}$. At plotting of operating zone, we assume that measuring accuracy of time interval using both radio range finders is equal: $\sigma_{\tau_1} = \sigma_{\tau_2} = \sigma_\tau$. Then, for independent ranges measurement, using (5.47) formula, we obtain:

$$\sigma_r = \frac{\sigma_D\sqrt{2}}{\sin\,\alpha_M} = \frac{c\sigma_\tau\sqrt{2}}{2\sin\,\alpha_M}. \tag{5.58}$$

Equal accuracy curve, necessary for plotting of outer boundary of operating zone is defined in accordance with (5.58), defining $\sigma_r = \sigma_{r\,\mathrm{dr}}$ and calculating an angle α_M:

$$\sin\,\alpha_M \frac{\sqrt{2}\sigma_D}{\sigma_{r\,\mathrm{dr}}} = \mathrm{const}. \tag{5.59}$$

Angle α_M, representing an intersection angle of two position lines, equals to an angle, at which radii AM and BM are intersected (Fig. 5.10), from where it is clear, that equal accuracy curves are represent lines, in each point of which an angle between directions to responders, located in A and B points, is a constant value according to (5.59). Such curves are a family of circles, resting upon d base, as upon a chord (bisecant). Central angle of chord AB equals to $2\alpha_M$. As it is clear from a triangle AOC, an equal accuracy curve has a radius:

$$R_{\text{acu}} = \frac{d}{(2\sin\ \alpha_M)}. \tag{5.60}$$

Consequently, plotting of operating zone of range-finding RNS resolves into the following:

(1) responders position points A and B are plotted on a map and base d is measured;
(2) by given σ_D and $\sigma_{r\ \text{dr}}$ we define an angle α_M;
(3) in accordance with (5.60), a radius R_{acu} is calculated;
(4) knowing R_{acu} and A and B points position, centers of circles O and O_i are defined on a map and circles, representing an equal accuracy curves, are plotted;
(5) RNS operational ranges $D_{A\max}$ and $D_{B\max}$ are calculated and plotted on a map.

An area, located within an equal accuracy curve and maximum operational range, represents an operating zone of range-measuring RNS. Operating zone is located by both sides from a base of selected responders. Let us examine how RNS accuracy inside an operating zone will change.

Value σ_r is minimal on circle 1 (Fig. 5.14), corresponding to angle value $\alpha_M = \alpha_{M1} = \frac{2}{\pi}$, where sin $\alpha_{M1} = 1$ and $\sigma\sqrt{2}_{D_r\ \min}$. For circle 1, a base d is a diameter. At object offset from a base outside a circle 1 limits, an angle α_M is reduced that

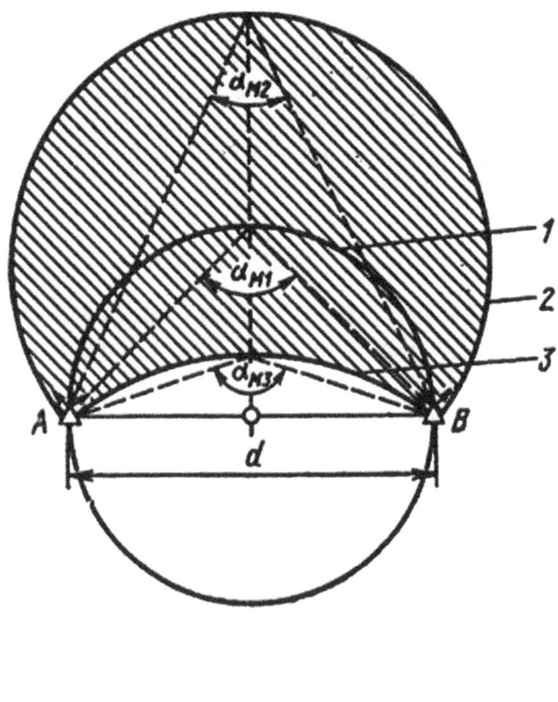

Fig. 5.14 Operating zone of range-measuring radio navigation system

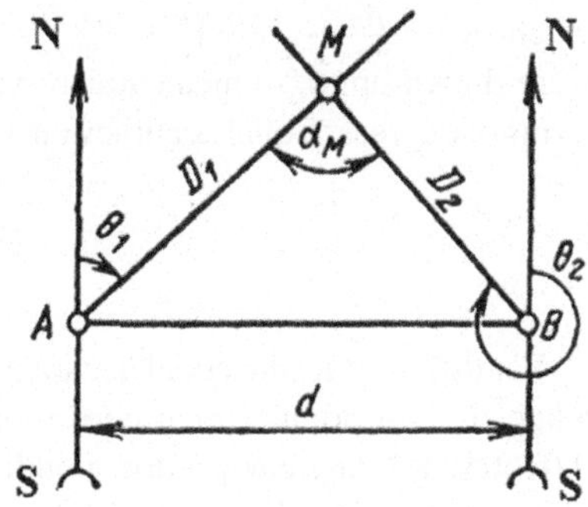

Fig. 5.15 Operating zone calculation procedure of angle measuring radio navigation system

leads to σ_r increase. At approaching to a base, beginning from a circle 1, an angle α_M increases. Herewith, $\alpha_M > \frac{\pi}{2}$. Such an angle increase leads to sin α_M reducing that stipulates a rise of values σ_r. In Fig. 5.14, circles 2 and 3 are depicted, which correspond to that for them $\sigma_r = 2\sigma_{r\min}$. If we define, that $\sigma_{r\text{ dr}} = 2\sigma_{r\min}$, then operating zone will be located between circles *2* and *3*. For this case, a half of operating zone is depicted in a figure. According to (5.59), the circles 2 and 3 correspond to value $\sin\ \alpha_M = \frac{1}{2}$ at angles $\alpha_M = \alpha_{M2} = 30°$ and $\alpha_M = \alpha_{M3} = 150°$.

As it is clear from (5.59), an error increase σ_D at given value $\sigma_{r\text{ dr}}$ leads to reduction of range-measuring RNS operating zone.

We will examine a plotting of angle measuring (theta-theta) radio navigation system (AMRNS) operating zone as an example of onboard radio direction finder, operating with two ground-based radar stations *A* and *B*, which are located in navigational guide point NGP/radio beacon, spaced at value *d* (Fig. 5.15). In a figure, *d*—a system base; θ_1 and θ_2 object azimuths relatively to ground stations *A* and *B*; D_1 and D_2—distances to stations *A* and *B*; α_M—angle, at which position lines are intersected; *M*—object position.

It is proposed that meridians in points *M*, *A* and *B* are parallel. Considering that $\sigma_{l1} = D_1\sigma_{\theta 1}$ and $\sigma_{l2} = D_2\sigma_{\theta 2}$, and assuming that azimuths measuring is conducted using the same onboard radio range finder and in the same conditions; i.e., assuming that $\alpha_{\theta 1} = \alpha_{\theta 2} = \alpha_\theta$, we obtain:

$$\sigma_r = \frac{1}{\sin\ \alpha_M}\sqrt{D_1^2\sigma_\theta^2 + D_2^2\sigma_\theta^2}. \tag{5.61}$$

To determine boundaries of AMRNS operating zone, it is necessary to plot an equal accuracy curve according to (5.61), in each point of which $\sigma_r = \sigma_{r\text{ dr}}$. With this purpose, we reduce (5.61) to a more convenient at calculations form:

$$\sigma_r = Kd\sigma_\theta^\circ, \tag{5.62}$$

where $K = \frac{0.117}{\sin \alpha_M}\sqrt{\left(\frac{D_1}{d}\right)^2 + \left(\frac{D_2}{d}\right)^2}$—coefficient, for finding of which special tables were drawn up; σ_θ°—mean-square value of azimuth measuring error, expressed in terms of degrees. Equal accuracy curve is defined according to the following formula:

$$K = \frac{\sigma_{r\ \mathrm{dr}}}{\left(d\sigma_\theta^\circ\right)}.$$

For different K, the equal accuracy curves of AMRNS are performed in Fig. 5.16, where C is a point, which corresponds to minimum value σ_r, equals to $\sigma_{r\min} = 0.01605 d\sigma_\theta^\circ$. Angle α_M, corresponding to point C, equals to 109°28', and segment OC equals to $\frac{d}{2\sqrt{3}}$. Equal accuracy curves of AMRNS have a complex shape. They, in particular, differ from circles, leant upon a base as on a chord. Curves, depicted in Fig. 5.16, characterize a half of operating zone. Another part is symmetric with respect to a base of the system. Note that surface area, limited by curves of maximum operational range of AMRNS, as a rule, is bigger than its operational zone; i.e., for AMRNS operational zone, an equation $\sigma_r \leq \sigma_{r\ \mathrm{dr}}$ is more non-slack than equation $D \leq D_{\max}$.

The specific features of operating zones of bearing-distance (rho-theta) RNS are: difference of position lines (one of which—great-circle course (orthodromy), another—equal ranges line), joining of ground stations of bearing and distance measuring channels of the system in one point and constancy of angle, at which position lines are intersected. These properties relate to different types of bearing-distance measurement systems: radar location radars of landing system, radio-beacon landing systems and short-range radio-technical navigation system.

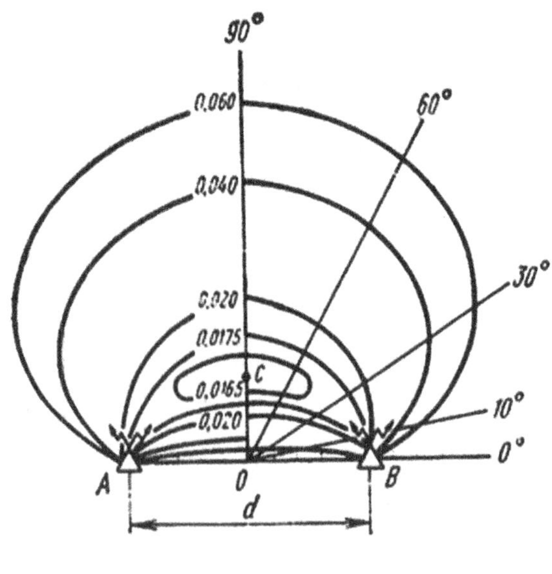

Fig. 5.16 Equal accuracy curves of AMRNS

Almost all in-service present bearing and distance RNS operate in VHF band, and hence, a calculation of $D_{\max}$ should be carried out with respect to object altitude and terrain features between object and ground-based station.

At plotting of equal accuracy curves, considering that $\sigma_{l1} = D\sigma_\theta, \sigma_{l2} = \sigma_D$ and $\alpha_M = \frac{\pi}{2}$, according to (5.49), we obtain:

$$\sigma_r = \sqrt{(D\sigma_\theta)^2 + \sigma_D^2}. \tag{5.63}$$

Based on expression (5.63), giving $\sigma_r = \sigma_{r\ \mathrm{dr}}$, a curve of equal accuracy of bearing and distance measuring system can be defined:

$$R_{\mathrm{tr}} = \frac{\sqrt{\sigma_{r\ \mathrm{dr}}^2 - \sigma_D^2}}{\sigma_\theta} = \text{const}, \tag{5.64}$$

which represents a circle of radius R_{tr}. A center of this circle coincides with a position of ground-based station.

Outer boundary of operating zone of bearing and distance measuring RNS (BDRNS) is formed after curves were plotting on a map, corresponding to $D_{\max}$ and R_{tr}, as a boundary total areas part, surrounding by these curves. Inner boundary of operating zone is defined by a radius of non-operating zone around a ground-based station, stipulated by its technical characteristics (e.g., by directional patterns).

Operating zones of hyperbolic navigation systems (HNS) have a more complex shape. An object position using HNS is defined as a point of position lines intersection, each of which represents a hyperbolic curve (Fig. 5.17), from two pairs of ground-based stations.

Considering a relation (5.44), for mean-square values of position lines finding errors, formed by each pair of RNS stations, we can write:

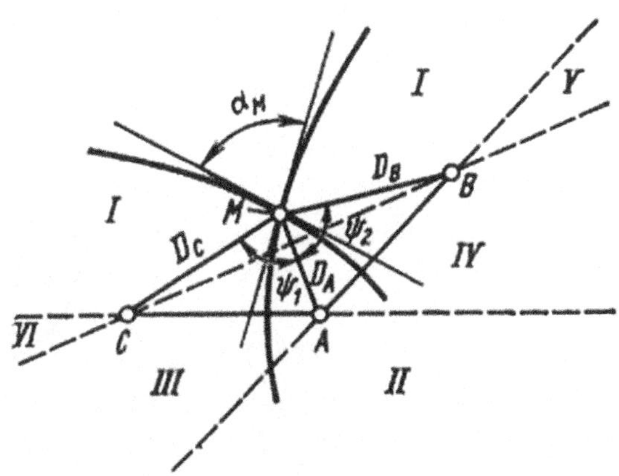

Fig. 5.17 Operating zone plotting technique of hyperbolic radio navigation system

$$\sigma_{l_1} = \frac{c\sigma_{\tau_1}}{2\sin\left(\frac{\Psi_1}{2}\right)} \text{ and } \sigma_{l_2} = \frac{c\sigma_{\tau_2}}{2\sin\left(\frac{\Psi_2}{2}\right)}.$$

where σ_{τ_1} and σ_{τ_2}—mean-square values of navigational parameters finding errors τ_1 and τ_2, representing a difference of signals arrival moments correspondingly for the first and for the second pair of stations; Ψ_1 and Ψ_2—angles, at which the first and the second bases of the system are visible.

At independent errors of position lines finding according to (5.47) for mean-square value of radial calculation error of object position, we obtain:

$$\sigma_r = \frac{c\sqrt{\sigma_{\tau_1}^2 \sin^2\left(\frac{\Psi_1}{2}\right) + \sigma_{\tau_2}^2 \sin^2\left(\frac{\Psi_2}{2}\right)}}{2\sin\ \alpha_M \sin\left(\frac{\Psi_1}{2}\right)\sin\left(\frac{\Psi_2}{2}\right)}. \tag{5.65}$$

In some cases, measuring errors of time intervals τ_1 and τ_2 are equal, i.e., $\sigma_{\tau_1} = \sigma_{\tau_2} = \sigma_\tau$, then:

$$\sigma_r = \frac{c\sigma_\tau}{2\sin\ \alpha_M \sin\left(\frac{\Psi_1}{2}\right)\sin\left(\frac{\Psi_2}{2}\right)}\sqrt{\sin^2\left(\frac{\Psi_1}{2}\right) + \sin^2\left(\frac{\Psi_2}{2}\right)}. \tag{5.66}$$

In the case when HNS consists of three stations (Fig. 5.18), a relation (5.66) for accuracy evaluation of object position finding is simplified a little, since an angle α_M can be expressed via Ψ_1 and Ψ_2. Taking into account the fact that hyperbolic curves bisect angles Ψ_1 and Ψ_2, from Fig. 5.18, we find:

in I and II zones $\alpha_M = \frac{(\Psi_1+\Psi_2)}{2}$;
in III and VI zones $\alpha_M = \frac{(\Psi_1-\Psi_2)}{2}$;
in IV and V zones $\alpha_M = \frac{(\Psi_2-\Psi_1)}{2}$.

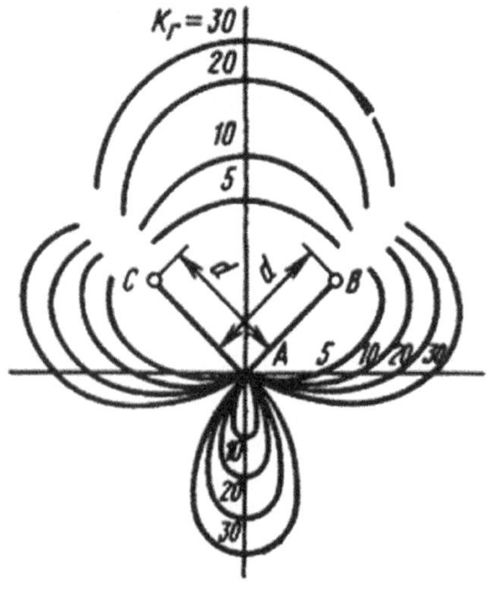

Fig. 5.18 Equal accuracy lines of hyperbolic radio navigation system

Therefore, for I and II zones, which are primary for HNS, formula (5.66) becomes a form:

$$\sigma_r = \frac{c\sigma_\tau}{2\sin\frac{\Psi_1}{2}\sin\frac{\Psi_2}{2}\sin\frac{\Psi_1+\Psi_2}{2}}\sqrt{\sin^2\frac{\Psi_1}{2}+\sin^2\frac{\Psi_2}{2}}. \tag{5.67}$$

At defining of operating zone boundaries of HNS it is necessary to plot an equal accuracy curve according to (5.65), (5.66) or (5.67), in any point of which $\sigma_r = \sigma_{r\,\mathrm{tr}}$. For this, in cases of expression (5.66) or (5.67) result in more convenient form during calculations:

$$\sigma_r = Kc\sigma_\tau$$

where K—tabular coefficient. Coefficient value is calculated by formulas, in case of (5.67) the coefficient is calculated according to formula:

$$K = \frac{\sqrt{\sin^2\frac{\Psi_1}{2}+\sin^2\frac{\Psi_2}{2}}}{2\sin\frac{\Psi_1}{2}\sin\frac{\Psi_2}{2}\sin\frac{\Psi_1+\Psi_2}{2}}.$$

HNS equal accuracy curve is plotted on the base of the following relation:

$$K_g = \frac{\sigma_{r\,\mathrm{dr}}}{(c\sigma_\tau)}.$$

For different values of K HNS, equal accuracy curves are depicted in Fig. 5.18.

As we can see from the above figure and given above formulas, the maximum accuracy of object position finding is achieved on bases of the system. At given σ_{r_1} and σ_{r_2}, an object position-finding accuracy worsens with offset from station due to rise of calculation errors of position lines and angle α_M reducing. Due to low accuracy, HNS cannot be used for object position finding in directions, being continuation of stations bases, for which bases angles are equal: $\Psi_1 = 0°$ or $360°$, either $\Psi_2 = 0°$ or $360°$, and in areas, where position lines are travel in a parallel way, i.e., at $\alpha_M = 0°$ or $180°$.

To obtain an operating zone of HNS from an area (square), limited by a curve equals to an accuracy, the segments are excluded in which signals receiving is not provided by one or several stations of the system; i.e., such areas are excluded that fall outside operational range of ground-based stations.

5.3.1 Radar Surveillance and Its Characteristics

The most modern radars have a directional pattern width comprised of units of degrees. Since there is a requirement to radars to provide targets surveillance in a significant wider angular zone, measured by tenths and hundreds of degrees, a necessity arises to shift a directional pattern in one or two angular coordinates. Such shift of radio beam, providing by mechanical movement of antenna, electronic beam scanning or antenna relocation due to movement of radar platform, provides a **radar air surveillance**.

Observing of a given range within limits of $D_{\min} - D_{\max}$ is carried out in period of time, incomparable shorter, then an air scan time in angles within limits of $\alpha_{\min} - \alpha_{\max}$ and $\beta_{\min} - \beta_{\max}$. For example, at observing of a target at distance 150 km and scan time of assigned air sectors in 1 s, single scan of the whole distance is conducted in time $\frac{2D_{\max}}{c} = 10^{-3}$s, i.e., to thousand times faster, then at angle. It is practically possible to consider a scan of assigned range as an instantaneous one, in consequence of which a term of radar surveillance is applied in a greater degree for air observing in angular coordinates.

Methods of radar surveillance are divided into single-beam, multi-beam, instantaneous (parallel) and sequential. There is also mixed-mode method of radar surveillance.

Instantaneous surveillance is used in radars by one fixed beam, measuring only a range and, consequently, have no use for beam scanning. For instance, airborne radio altimeters, range-finding radio-beacon stations. Instantaneous surveillance is possible at measuring of angular coordinates, but for this purpose, a directional pattern (DP) of radar antenna should be multi-beam and instantly cover the whole surveillance zone of radar. Such instantaneous multi-beam surveillance is otherwise called a parallel surveillance.

In multi-beam systems of instantaneous surveillance, the antenna beams remain stationary, but a number of simultaneously operating receiving beams should provide an overlapping of the whole surveillance zone. In these conditions, a number of beams equals to a number of angularly positioned elements, containing in radar surveillance zone. Each of beams is connected with its receiving unit, due to which a target angular coordinate is defined by a number of a receiver, at output of which a maximum signal is fixed. Range is still defined by reflected signal dwell time relatively to a sounding one. All target coordinates are measured simultaneously and almost instantaneously. Main disadvantage of sequential method, a longtime of surveillance, collapses. However, multi-element aerials, multi-channel receiving units, complex information-processing devices, necessary for multi-beam radars, make them a complicate in manufacturing and in operation.

Single-beam surveillance with scanning is called a sequential surveillance, since it requires a sequential illumination of all zone elements. There is a variety of sequential surveillance.

If one of two angular coordinates is measured, e.g., in aviation onboard surface surveillance radars or ground-based surveillance radars, then an antenna beam is imparted with wide width in a plane of non-measured angular coordinate in order to overlap the whole surveillance zone per one rotation (one sweep) of antenna. Such surveillance is called a circular, or sector surveillance. In Fig. 5.19, variants of circular and sector radar surveillance are illustrated.

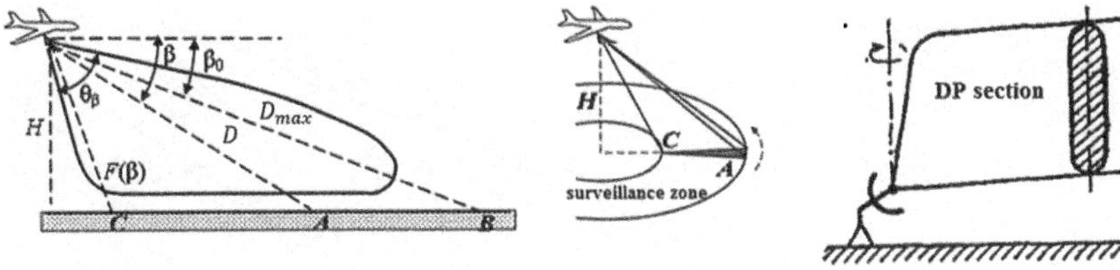

Fig. 5.19 Circular and sector radar surveillance

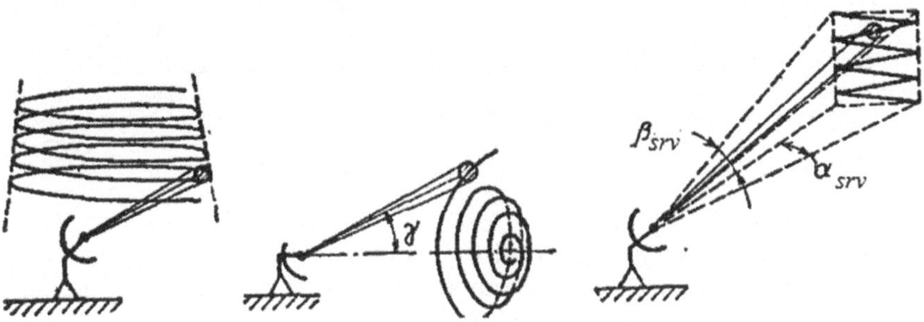

Fig. 5.20 Types of radar surveillance in two planes

If for precise measuring of both angular coordinates a single-beam surveillance is used, then scanning is provided by pencil beam in line by line, along helical curve, spiral, zigzagging or in another trajectory (Fig. 5.20).

Helical scan consists of circular rotation (revolution) in azimuthal plane and slowly change of a beam position in elevation angle. *Spiral scan* means that beam projection to a plane, perpendicular to rotation axis, has a shape of spiral. *Conical scan* can be examined as a special case of spiral scan, when angle γ between rotation axis and beam axis is not changed. This angle at spiral scan is less than 45°, and at helical scan can be considerably larger. Helical and spiral scans are used in target search mode, and conical scan is carried out in target automatic search mode. *Zigzag* (or *line by line*) *scan* means an oscillatory beam motion (scanning) in azimuth or in elevation angle with gradual change of its position by another angular coordinate. Such a beam trajectory is beneficial that permits to set surveillance sectors independently in azimuth and elevation angle.

There are radars, in which beam travel in horizontal plane is provided due to translational object motion, e.g., aircraft motion (Fig. 5.21).

Such surveillance method has a name of side-looking (lateral) radar scan. Besides sequential and parallel scanning, there is a mixed scan, at which a several beams instantly cover a surveillance zone by one angular coordinate, and by another, a sequential circular scanning is carried out as in panoramic radars. Mixed scan is a certain trade-off between complexity of multi-beam radars and long scan period of radar with scanning beam. For example, in Chap. 4, we have already examined a

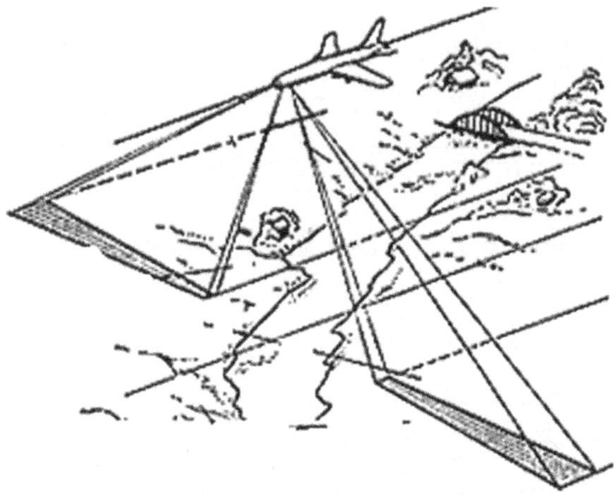

Fig. 5.21 Side-looking (lateral) radar scan

method of stacked beam patterns, realized in some ground-based radars and representing a combination of a single-beam sequential scan in horizontal plane with multi-beam instantaneous scan in vertical plane.

Radar scan is characterized by the following parameters:

Radar surveillance zone—space area, within limits of which illumination, receiving and processing of reflected signals is conducted, including targets detection with given characteristics. This zone (Fig. 5.22) is limited in range with limits $D_{\min} - D_{\max}$, in azimuth with value α_{srv} and in elevation angle with value β_{srv}.

Scanning time—a time necessary for single scan of a given operating zone of radar location or radio navigation system. Scanning time T_{srv}, is the most important parameter of radar surveillance. At selection of scanning time T_{srv}, a set of factors should be considered: special features of targets movement, necessity of targets illumination with enough portion of electromagnetic oscillations energy and other. If a target per scan time T_{srv} can travel (fly over) the whole surveillance zone in angular coordinates, then a target miss is possible and scan will be breaking, not continuous (full).

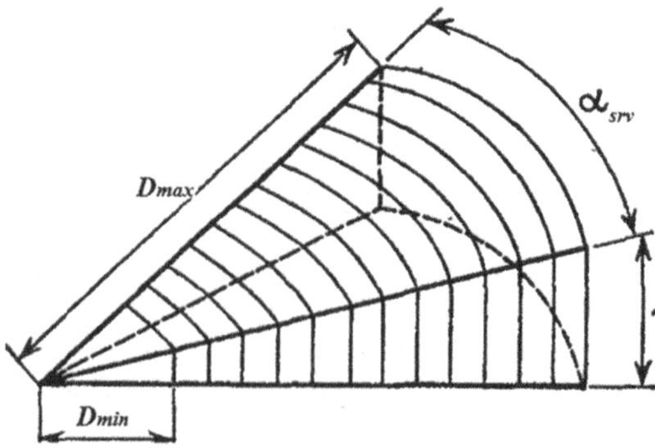

Fig. 5.22 Surveillance zone

We assume that a target fly over time of this surveillance zone $t_{t.fly}$ is more than scan time (scanning period) T_{srv}:

$$t_{t.fly} \geq T_{srv}. \tag{5.68}$$

Target fly over time $t_{t.fly}$ should be selected as a least of two values, corresponding to a target fly over of surveillance zone as both in radial and in tangential directions. Fly over time in tangential direction (e.g., in azimuth):

$$t_{t.fly} = \frac{\psi_{a3} D}{V_{t.tng}}, \tag{5.69}$$

where ψ_{az}—admissible sector of target fly over in azimuth; D—distance to a target. This distance should be selected by the least from boundary values. Fly over time in radial direction $t_{t.fly.r}$ is defined by a value:

$$t_{t.fly.rd} = \frac{\Delta D_{fly}}{V_{t.rd}}, \tag{5.70}$$

where ΔD_{fly}—admissible distance of target fly over in radial direction (e.g., resolution element size in range).

Characteristics of radar surveillance at the most degree depend on type directional pattern of radar antenna.

Radar surveillance method, applied in radar of a certain type, is imposed by tactical purpose of the system. Radar structural design should correspond to a set of requirements to radar air surveillance, the most important of which are:

1. Scan should cover the completely assigned space, for which an illumination and target signals receiving should be provided, positioned in any point of surveillance zone in range and angular coordinates. As single and multiple radio beams pass through a target (depending on surveillance method), the last one should receive an amount of sounding energy sufficient for its detection at maximum range.
2. During scanning, an assigned measuring accuracy of target coordinates and assigned resolution capability on its coordinates should be provided, that is also determined by efficient radar energy management during beam pass through a target.
3. Duration of each scanning cycle (period) should be quite short (repetition scan (sweep) rate—long enough), in order to have a possibility of identification of the same target, observed in different scan cycles. At fulfillment of the mentioned condition, a continuous observing (tracking) of a target will be provided, despite to high approaching (offset) speed of a target and radar.
4. Predetermined operation reliability, specified mass and dimensions of antenna unit, engaging in realization of radar surveillance should be provided.

Listed requirements in some cases are contradict one another. At instantaneous scanning, the indicated contradictions are removed, but great difficulties arise in satisfaction of the last requirement—equipment becomes a multichannel and more sophisticated.

5.4 Resolution Capability of Radar Location and Radio Navigation Systems

One of the main radar and RNS characteristics is its resolution capability, i.e., a possibility of separate detection, coordinates measuring or movement parameters of several objects located in surveillance zone. In some cases, it differs a target resolution in range, velocity (in Doppler frequency) or in angular coordinates.

It is admitted to evaluate a resolution capability in minimum interval value between two targets by corresponding coordinate or targets movement parameter. It is assumed that resolution value by one or another coordinate is defined at equal values of other coordinates or targets movement parameters.

Targets resolution means a possibility at radar surveillance to answer the following questions: What is the relative position of targets in space, what are its relative velocities, how much targets are observed and what its specific features (e.g., RCS)?

Radar resolution capability at the most defines a possibility of detection and target coordinates measuring at the background of returns (clutters), passive and active jamming of different origins. Consequently, a resolution capability is important enough at evaluation of radar interference immunity and at development of countermeasures systems. Particularly, passive jamming and different interfering reflecting surfaces can be examined as aggregate of targets which are of no interest at observing of some assigned target, but using resolution capabilities, it is possible to select this assigned target, provide its detection and determine its coordinates and moving velocity.

As it mentioned before, a resolution standard in the examined case can be such interval between input signals for two targets, at which it is possible to observe independently the separate maximums, and a "gap" ("null") between them does not exceed some specified value relatively to a maximum value of output signal.

According to *Rayleigh criterion* (introduced for the first time in optics), output responses in a form of function $(\sin x)/x$ are examined. As defined in this criterion, two equal output signals are differ, if a maximum of one signal coincides with a minimum (zero) value of another (Fig. 5.23).

Herewith, a null is originated of not more than 20% from a maximum (peak) value. In this case, an interval between maximums numerically equals to response width of output signal (for a point object of observing). Hence, according to Rayleigh criteria, two targets are resolved (in corresponding coordinate), if an interval between them $\delta\alpha$ is not less than response width of output signal α_{out}

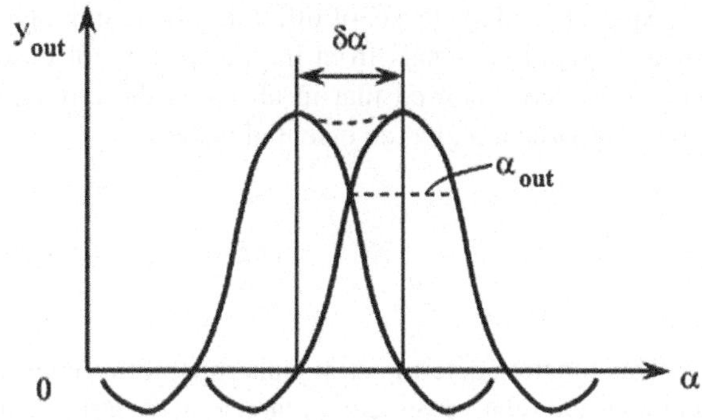

Fig. 5.23 Rayleigh criterion

$$\delta\alpha \geq \alpha_{out}. \quad (5.71)$$

where $\delta\alpha$—resolution in corresponding coordinate axis α, and α_{out} a response width 0.5 level from a maximum.

More general criteria of a resolution (*Sparrow criterion*), which is applicable for output signals of almost any shape, defines that resolution is achieved only when a "null" between maximums is reduced up to a zero. Herewith, an interval between maximum values is as follows that output signal responses are intersected in inflection point of output signal envelope $y_{out}(\alpha)$; i.e., when equation is complied (Fig. 5.24):

$$\left.\frac{d^2 y_{out}(\alpha)}{d\alpha^2}\right|_{\alpha = 0} = 0. \quad (5.72)$$

In case application of Sparrow criterion, two targets differ, if interval between maximums of output signals $\delta\alpha$ is not less than a response width α_{out}. Thus, independently from a selected criterion, targets resolution is evaluated by a width of output response at 0.5 level from a maximum in corresponding coordinate axis.

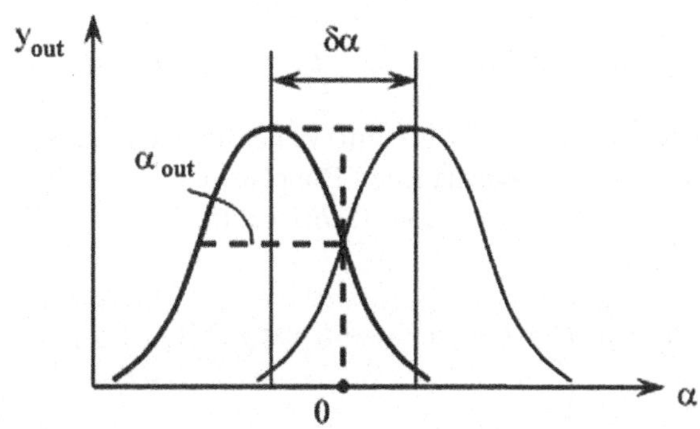

Fig. 5.24 Sparrow criterion

Analysis and synthesis of different processing algorithms of radar location and radio navigation signals from the perspective of ensuring the required resolution capability have showed that in all cases the cross-correlation signals function—simple exponential, has an essential value:

$$\dot{\Psi}_s = \int_{-\infty}^{\infty} S\big(t - t_{\partial}, f_0 + f_{Dp}\big) S^*(t - t_c, f_0 + f_c)dt. \quad (5.73)$$

This function determines signals capabilities in ensuring of resolution capability in range and relative velocity of targets. Research of this function properties permits to discover boundary capabilities and resolution parameters of targets depending on signals used.

Let us write down a radiated signal in the following form:

$$s(t) = \dot{S}(t)\exp(j2\pi f_0 t), \quad (5.74)$$

where $\dot{S}(t)$—complex amplitude (envelope) of used signal, in which all possible types of modulation are considered:

$$\dot{S}(t) = A(t)\exp[j\phi_s(t)], \quad (5.75)$$

where $A(t)$ describes an applied amplitude modulation (e.g., pulsed), and function $\phi_s(t)$—all types of phase and frequency modulation.

Reflected signal differs from a radiated one (as it was mentioned earlier) by amplitude attenuation (damping), considered by coefficient a, time delay on range t_{d} and frequency shift on Doppler frequency value F_d. As a result, write down a received signal as follows:

$$\dot{S}_{\text{rcv}}(t) = a\dot{S}(t - t_{\partial}, f_0 + F_d) = aS(t - t_{\partial})\exp[j2\pi(f_0 + F_d)(t - t_d)]. \quad (5.76)$$

Reference signal in processing device we will correspondingly write down as:

$$\dot{S}_{\text{ref}}(t) = \dot{S}(t - t_s, f_0 + f_s) = S(t - t_s)\exp[j2\pi(f_0 + f_s)(t - t_s)]. \quad (5.77)$$

where t_{s} and f_{s}—offset (shift) values in signal, characterizing a change of delay value on range (t_c) and change of Doppler frequency (f_{s}).

Substitute expressions (5.76) and (5.77) into (5.73) formula:

$$\dot{\Psi}_s = a\int_{-\infty}^{\infty} \dot{S}(t - t_d)\dot{S}^*(t - t_s)\exp j2\pi[(f_0 + F_d)(t - t_d) - (f_0 + f_s)(t - t_s)]dt. \quad (5.78)$$

Introduce new relative shift variables of reference signal and received one, reflected from a target:

$$\tau = t_s - t_d, \tag{5.79}$$

$$\nu = F_d - f_s. \tag{5.80}$$

We introduce a variable of integration as well:

$$t' = t - t_d. \tag{5.81}$$

Due to evident transformations, we obtain (omitting a stroke in notation t'):

$$\psi_s = a\left\{\int_{-\infty}^{\infty} \dot{S}(t)\dot{S}^*(t-\tau)e^{j2\pi\nu\tau}dt\right\} \exp\, j\ 2\pi[(f_0 + f_s)\tau]dt. \tag{5.82}$$

Function in curved brackets in (5.82) formula, we designate as:

$$\Psi_s(\tau, \nu) = \int_{-\infty z}^{\infty} \dot{S}(t)\dot{S}^*(t-\tau)e^{j2\pi\nu\tau}dt. \tag{5.83}$$

Function Ψ_s was named as **uncertainty function** of a signal. Parameters τ and ν are connected with range differences and radial velocities by relations:

$$\tau = \frac{2(D - D_s)}{c};\ \nu = \frac{2(V - V_s)}{\lambda}, \tag{5.84}$$

where D and V—distance to a target and relative radial velocity correspondingly. Values D_s и V_s—distance and velocity, on which signals processing system is adjusted (matched filter or correlation device). Consequently, a function ψ_s can be written as follows:

$$\psi_s(\tau, \nu) = a \underset{s}{\dot{\Psi}}(\tau, \nu)e^{j\theta(\tau,\nu)}, \tag{5.85}$$

where $\theta(\tau, \nu)$—phase of complex-valued function ψ_s, equals to:

$$\theta(\tau, \nu) = 2\pi[(f_0 + f_c)\tau] = 2\pi[f_0 + F_d - \nu]\tau. \tag{5.86}$$

In Fig. 5.25, the coordinate systems of variables are depicted t_s и F_d, and τ and ν for a certain point of target position (TP_1) $t_s = t_{d1}$ и $f_s = F_d$, where $\tau = 0$, $\nu = 0$. At evaluation of resolution capability, a position of second-order chain relatively to the first one corresponds to τ and ν shifts.

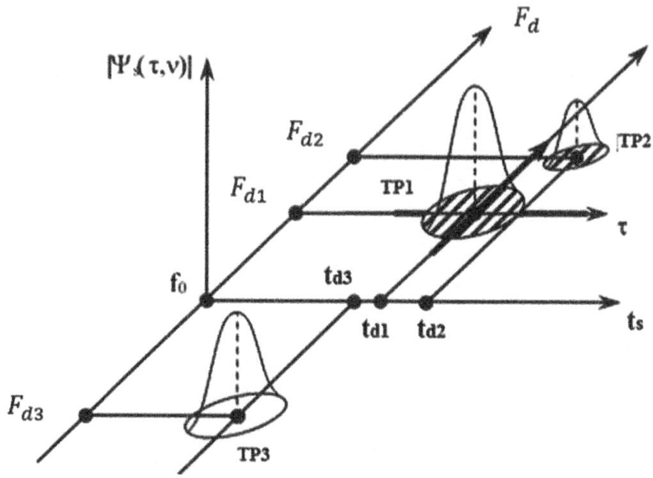

Fig. 5.25 For clarification of uncertainty function

Introduction of uncertainty function term is justified by the reason that in area, close to $\tau = 0$ and $\nu = 0$, a function $|\Psi_s|$ a little differs from its maximum value. If a second target is located in this area, then an uncertainty arises in decision making on position of two targets and on possibility of its resolution. Volume, limited by uncertainty function (UF) modulus on plane τ and ν, is often called uncertainty function body. Almost all signals properties can be determined by exploring UF structure or UF function body.

One of the most important UF properties concludes in the following. Change of form and signal modulation parameters can change a UF body shape, but its volume would not change. Attempts to obtain a narrow UF peak in area of coordinates origin, will lead to UF volume redistribution, to appearing of other maximums or to increase of side-lobes level of UF. UF width reducing in any axis inevitably leads to UF extension in any other direction.

Exploration of UF structure in its volume representation is quite difficult. By this reason, contour lines method is frequently used for UF representation on a plane τ and ν. Such figures are called *ambiguity diagrams*.

A knowledge of UF form permits to evaluate the radar systems properties: resolution capability, coordinates measuring accuracy, and observing possibilities of targets at the background of clutters (returns).

In Fig. 5.26, a complete solid figure of UF and image of plane sections, parallel to a plane (τ, ν), in a form of contour lines are depicted. The analogue sections for ground profile imaging are applied in topography.

UF section view at level of not less than 0,5 from a maximum permits to clarify a sense of "*uncertainty function*" term once again. In the limits of such sections (ellipse view), it is almost impossible to separately determine positions (coordinates) of several (at least two) targets. Its UF jointly forms a complicated solid (spatial) figure,

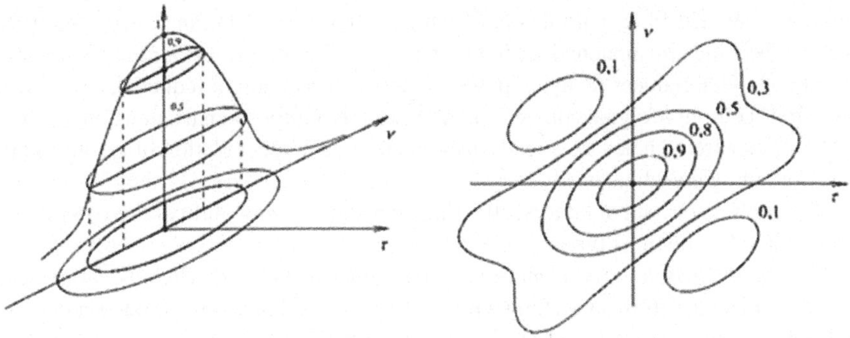

Fig. 5.26 Ambiguity diagram

and ambiguity diagram is formed as some plane figure, in which it is impossible to determine centers position (peak points) of targets locations. Consequently, an ambiguity arises in targets resolution and in coordinates finding of observed targets. At the presence of internal noises in radar, an ambiguity also arises within UF section in position finding of even one target, as a section form and its center are random and depend on relation of signal energy and noise. At sufficiently large relation of signal energy and noises spectral density, the measured target position is located within ambiguity ellipse for this ambiguity diagram with a center, corresponding to true target coordinates.

By analogy with time signals, *spatial uncertainty function* Ψ_x is introduced, which describes properties of spatial signals transformation.

Therefore, in-depth analysis of signals uncertainty function permits in conditions of specified to radars requirements to select correctly a type of radar signal.

Now, we define resolution capabilities in coordinates and targets moving parameters and examine what effects on radar resolution capability.

Range resolution capability is estimated by minimum distance $\delta D_{\min}$ between two presented in one direction targets, at which these targets are observed separately.

Angular coordinates resolution capability is a possibility of a separate surveillance of two and more targets, located at the same distance from a radar, but positioned in different angular directions. Targets resolution in angular coordinates is admitted to measure by a minimum angle between directions to separate targets, at which its separate detection and coordinates measuring are provided.

Velocity resolution capability is determined by a minimum difference of radial velocities of two targets with the same coordinates, at which these targets are still observed separately.

In real conditions of radar operation its resolution capability differs from a frontier one. Degradation of resolution capability (resolution loss) is stipulated by influence of many disturbing factors, which appear in some radar components, by imperfect design of functional elements, wave distortions in radar paths and during process of signals propagation in space. Mismatched signals and characteristics of its processing units lead to degradation of resolution capability. Particularly, if reflected signals come to matched filter, parameters of which differ from a nominal, then this filter already becomes mismatched with all consequent particularities: a minimum value of output signal reduces (a signal power to noise ratio at output reduces), a response function at output of signals processing unit extends. Output signal extension in delay and Doppler frequencies axes leads, as it is obvious, to degradation of resolution in corresponding coordinates.

Concentrate on a set of factors, effecting on resolution probability of radar systems and typical for the most type of radar systems.

Signals distortion at its forming in radar transmission path. Signals parameters change will lead to its mismatching with processing unit that was mentioned above. At forming of high-frequency pulses, for example, random changes of carrier frequency are possible. Additional random modulation appears and spectrum structure changes. These factors will lead to degradation of targets resolution both in range and in Doppler frequencies. If frequency modulation, for example, is used in signals, then distortions in transmitter will lead to FM deviation from a given modulation law that in its turn also will lead to distortion of output signal at processing of signals and, consequently, to degradation of resolution capability.

Errors at arrangement of antenna systems. Different manufacturing faults in antenna structure, e.g., in scatterer profile generation, lead to distortions of antenna DP, to main beam width spread and to side-lobes level increase. All these errors will lead to degradation of radar resolution capability in angular coordinates.

Huge effect on radar resolution capability can be done due particularities arising at *signals propagation in space.* For instance, atmospheric inhomogeneity can lead, at radio waves travel, to appearance of signals phase fluctuations and as a result—to wave fronts structure distortion, that in its turn will change antenna DP at receiving and degrade targets resolution in angular coordinates.

Irregularities in target moving or radar platform (in case of onboard radars) create distortions in a form of signal, in its spectrum structure, in change of modulation laws. In all cases, it is arguably that to limiting value of resolution magnitude $\delta\xi_{\text{lim}}$ in some coordinate ξ some "increment" is added $\Delta\xi$, which defines a degradation of targets resolution. Real resolution capability, evaluating by a value $\delta\xi_r$, can be written as follows:

$$\delta\xi_r = \delta\xi_{lim} + \Delta\xi.$$

Resolution degradation is numerically can be evaluated by resolution degradation coefficient:

$$\alpha_r = \frac{\delta\xi_r}{\delta\xi_{\text{lim}}} = 1 + \frac{\delta\xi_{\text{lim}}}{\delta\xi_{\text{lim}}}. \tag{5.87}$$

Influence of indicator (displaying) units parameters on resolution capability of radar location systems. Radar indicator units have a largest effect on targets resolution quality. This refers to almost all types of radars.

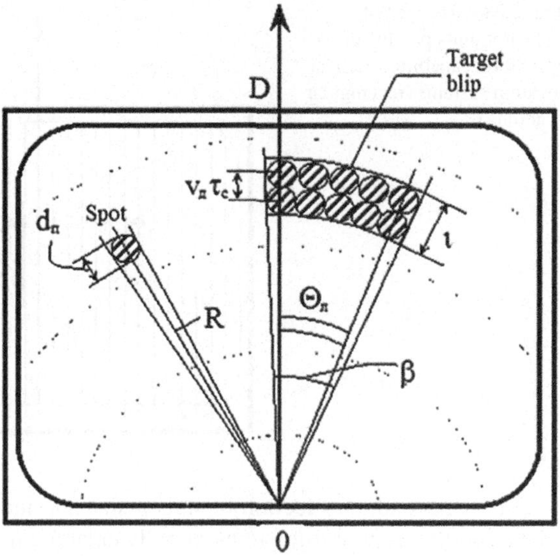

Fig. 5.27 Influence of indicator units parameters on resolution capability of radar location systems (radial-sector scanning)

On display of any indicator (multifunctional indicators of AV crew, flight dispatcher and engineer specialists indicators), a target blip of some aspect should be generated (formed). Size of such blip for a point target on a screen defined a possible resolution magnitude of targets. For simplicity, we consider that intensity (brightness) per blip area is constant as per intensity. Certain surface segment is highlighted on a screen—so-called spot is forming. Different images of target blips are composed from a set of such "blips." As an example, examine a formation of a target blip on indicator screen in "range-azimuth" mode with radial-sector scanning (sweep) (Fig. 5.27).

Resolution degradation in azimuth at radial-sector scanning equals to:

$$\delta\beta_{az.r} = \frac{d_n}{R}, \tag{5.88}$$

and for rectangular screen (scanning):

$$\delta\beta_{az.r} = d_n M_n. \tag{5.89}$$

Introduced correlations for a real resolution capability in range and in angular coordinates permit to point a set of improvement directions for indicator units in order to reduce influence of mentioned factors.

For the first, it is reasonable to reduce blip diameter on cathode-ray tube display. This could be achieved by quality improvement of focusing or via transfer to another displaying methods of indicating information, e.g., by application of high-resolution matrix displays (Fig. 5.28).

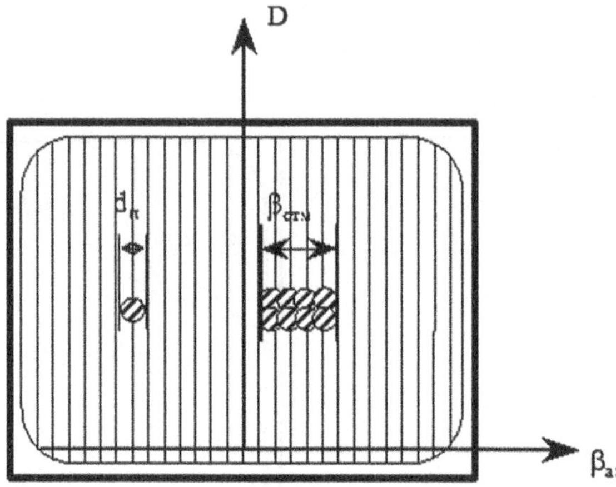

Fig. 5.28 Influence of indicator units parameters on resolution capability of radar location systems (rectangular scanning)

For the second, at selection of screen size and indicator type, it is necessary to create conditions permitting to maximally enlarge image scale in range and angular coordinates and in some cases in Doppler frequencies.

In modern radars, a target indication on display is usually appeared after special processing of radar data. For instance, a special target blip is forming. In these conditions, the limiting factor is possibility to observe a minimum image element (feature).

5.5 Bandwidth Capacity of Radar Location and Radio Navigation Systems

The radar location and radio navigation systems bandwidth capacity is one of the most important tactical characteristics and is relevant for systems with interrogation and response channels.

Response channel capacity is determined by a maximum number of responders which an interrogator can simultaneously successfully operate with. Limiting of response channel capacity is carried out due to of its saturation, when a number of responders increases to an extent that normal operation of responder is disrupted. Such mode for the most radars and RNS is not common since in each response channel, as a rule, one responder operates (e.g., aerial vehicle usually operates with one ground-based radio beacon (RB).

Interrogation channel capacity is determined by a maximum number of interrogators, which can simultaneously successfully operate this responder. Limiting of interrogation channel capacity is carried out due to decrease of responses which is given by this responder to each interrogator. A number of responses reduces as a responder locking device switches off a receiver on safety time t_s after each response. By this

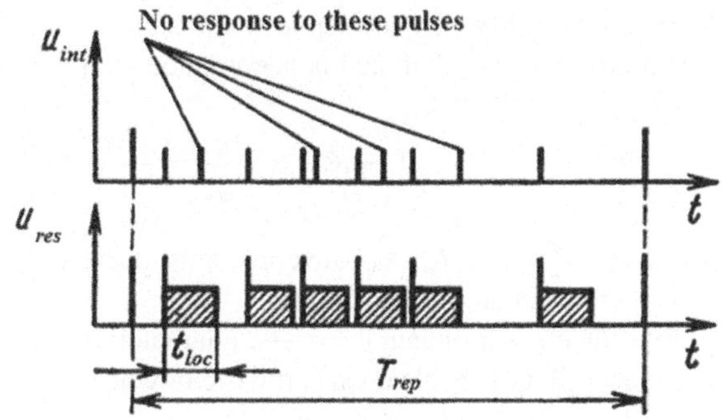

Fig. 5.29 For clarification of capacity

means, a protection of transmitter responder is provided from energy overloading at increasing of a number of interrogations.

Let us examine dependence degree of a response receiving probability by each separate interrogator from a number of functioning interrogators. Suppose that radio beacon operates simultaneously with N equal interrogators, radiating pulse signals with a repetition period T_{rep}. After radiating of each reply pulse, an interrogator locks for a time t_{loc}, equals $t_{\text{loc}} = t_r + t_{\text{ant}}$, where t_r—recovery time of normal equipment functioning, t_{ant}—anti-overloading interval, selected from a condition of overloading absence of transmitter responder.

Part of interrogators will not receive a response to its interrogation signals $u_i(t)$ (Fig. 5.29), and hence with increase of a number of interrogators, a probability of response receiving by each separate interrogator is reducing. This will lead to increase of range measurement errors and loss of tracking at using of follow-up tracker. Consequently, a capacity is determined by energy performance of transmitter responder at its operation with a large amount of interrogators.

Responder servicing quality of interrogators is usually estimated by response probability (response ratio):

$$K_{\text{res}} = \frac{N_{\text{res}}}{N_{\text{int}}} \tag{5.90}$$

where N_{res} and N_{int} — correspondingly an average number of response an effective (capable to cause actuation of responder) response pulses in a time T_{rep} in noiseless condition $N_{\text{int}} = N$.

For a minimum admitted coefficient of responses, $K_{\text{res min}}$ value is assumed, at which a necessary operational range and normal operation of interrogators is provided.

Let us define a response probability, when $N_{\text{int}} = N$, interrogations during T_{ar} are arriving independently of each other and arrival probability of each within an assigned time interval depends not on this interval position in a period T_{ar}, but only on its duration; $T_{\text{ar}} \gg \tau_{\text{pls}}, t_{\text{int}}$, where τ_{pls}—duration of interrogation (response) pulses. For the mentioned above conditions, a probability of response receiving by any one of N interrogators is determined per one interrogation by the relation (5.90).

Then, a possibility of non-response equals $1 - K_{res}$. Non-response probability by this interrogator is calculated in accordance with the following expression:

$$1 - K_{res} = \frac{t_{\Sigma}}{T_{arv}} = \frac{(N-1)K_{res}t_{int}}{T_{arv}}, \tag{5.91}$$

where $t_{\Sigma} = (N-1)K_{res}t_{int}$—average time, during which a responder replies to other $(N-1)$ interrogators.

Introduce a parameter $\nu = \frac{t_{int}}{T_{arv}}$, characterizing a relative value of safety time. Then, according to (5.91), we can write down:

$$K_{res} = \frac{1}{1 + (N-1)\nu}. \tag{5.92}$$

For practice, an inverse problem also is of interest, when by specified probability of response receiving $K_{res\ int}$ and at known value ν, it is required to define a maximum number of interrogators N_{max}, which this responder can operate. With regard to (5.92), we have:

$$N_{max} = \frac{[1 - K_{res\ int}(1-\nu)]}{K_{res\ int}}. \tag{5.93}$$

For modern radar location and radio navigation systems, $N_{max} \leq 100$.

One of the possible techniques of bandwidth capacity increase of radar location and radio navigation systems with interrogator is reducing of interrogated pulses repetition frequency in tracking mode. However, in this case, there are more severe requirements to distance measuring unit since it should operate normally at less number of response pulses in a time unit.

5.6 Interference Immunity of Radar Location and Radio Navigation Systems

From the technical point of view, noise influence on radar and RNS can be in a form of the following effect:

- desired (valid) signal waveform distortion and change of its main parameters;
- desired signal suppression (signal-to-noise ratio reducing) at signals and noises traveling through nonlinear elements;
- overloading of receiver, its separate elements (amplifiers, transducers etc.), signals and data processors, and indicator.

From the tactical point of view, interference effect on radar and RNS can lead to screening of desired signal (target) by noises, or to imitation of desired signals by forming of false signals (decoy targets).

From the point of view of signals transmission and transformation theory, interference influence leads to linear and non-linear effects. Linear effects, as a rule, are observed at low noise level and concludes in that nuisance (interfering) vibrations in form independent additive components are added into receiving channels. In these cases output interference effect does not depend on signal action and can be examined independently. Shape distortion of desired signal due to additive composition with a noise refers to linear effects. Other forms of noise influence are connected with nonlinear effects. Nonlinear effects appear at large level of noises when, for example, the nonlinearity of amplitude characteristics of different receiver elements is implied.

Operation of radars and RNS in noise conditions of different type is normal, i.e., normal mode of its functioning. The system developed without considering of possible interference effect, cannot be admitted as satisfactory since it is not compatible with up-to-date requirements.

For quantitative anti-jamming performance, different criteria (index) are applied. Among them, we can underline the following main criterion groups: tactical; energy; informational; accuracy, generalized and combined.

The most important tactical criterion of interference immunity is reducing degree of maximum target detection range at noises effect $D_{\max n}$ comparing to maximum target detection range at no noises $D_{\max}$, which is defined by *distance reduction coefficient* K_d:

$$K_d = \frac{D_{\max n}}{D_{\max}}. \tag{5.94}$$

Energy indicators are frequently used as interference immunity criteria, among which the most widespread is *suppression coefficient*:

$$K_{\text{sup}} = \left(\frac{P_n}{P_s}\right)_{\text{inp rcv min}}, \tag{5.95}$$

where P_n—noise power at input of a receiver; P_s—power of desired signal at input of a receiver.

Suppression coefficient is determined as minimum signal-to-noise ratio in power at input of a receiver, at which the system suppression takes place; e.g., target correct detection probability becomes less than 0.5, and false alarm probability is more than 0.001.

Based on suppression coefficient K_{sup}, the effectiveness of one or another techniques in interference immunity improvement can be evaluated by the following criterion:

$$B = \frac{K_{\text{sup}}}{K_{\text{sup}0}}, \tag{5.96}$$

where K_{sup0}—suppression coefficient of radar system without additional techniques (units, algorithms) of noise protection; K_{sup}—radar suppression coefficient with additional anti-jamming measures.

To estimate the effectiveness of spatial-time signals processing operations, another energy criterion of interference immunity is used sometimes—*energy efficiency coefficient*.

$$K_{en} = \frac{q}{q_0}, \tag{5.97}$$

where q—signal-to-/(noise + internal noise) ratio in power at output of processing system; q_0—the same ratio, but with no noise.

In some cases, the effectiveness of interference eliminators is evaluated by *interference immunity coefficient*, which is determined by the following expression:

$$K_{nn} = \left.\frac{P_{n3}}{P_n}\right|_{\mathrm{P_{no}} = \mathrm{const}}, \tag{5.98}$$

where P_{np}—minimal noise power at input of the same radar system with interference protecting (algorithm) unit; P_n—minimal noise power at input of the same system without interference protecting (algorithm) unit, at which a specified degree of system suppression is achieved; P_{cd}—probability of correct detection of a target. Advantage of this interference immunity criterion concludes in that by its using it is possible to characterize the effectiveness of different interference protection units and algorithms. Disadvantage concludes in impossibility of estimation of the system as a whole since it does not consider such characteristics of radar system as radiating power, antenna gain factor, etc., at the most determining the system interference immunity.

More complete criterion, permitting to consider both effectiveness of interference protection units and algorithms and technical parameters of the system, effecting on its interference immunity, is an *overall coefficient (index)* K_{ov}, calculated in decibels:

$$K_{\text{ov}} = K_{\text{key}} + K_{\text{add}}, \tag{5.99}$$

where K_{key} and K_{add}—key and additional criteria interference immunity.

Key indicator K_{key} is determined by transmitter power, resolution capability in angular coordinates, in range and approaching (rendezvous) velocity and can be introduced, for example, as follows:

$$K_{\text{key}} = 10\, lg(P_{\text{rad}} \tau \Delta f_s G_a), \tag{5.100}$$

where P_{rad}—radiation power; Δf_s—signal bandwidth; τ—target observing time; G_a—absolute gain of antenna. Values P_{rad}, τ and Δf_s in expression (5.100) are dimensionless due to its normalizing to 1 W, 1 s and 1 Hz, correspondingly.

Key criterion of the system interference immunity considers contribution of its most important criteria in ensuring of interference protection. Additional criterion of interference immunity, measured in decibels, serves for revealing and contribution registration of special interference protection units and algorithms.

5.7 Specifications and Performance of Radar Location and Radio Navigation Systems

Technical characteristics of radars and RNS, providing assigned tactical characteristics, are diverse enough. The main among them are:

- frequency band of emitted oscillations and carrier frequency;
- pulses duration and its repetition frequency;
- signal shape and modulation law;
- shape and width of directional pattern;
- radiating power (average, pulsed, peak);
- sensitivity (response) of receiving device;
- pass bandwidth;
- noise ratio;
- system mass and dimensions.

Operational characteristics of radars and RNS include the system reliability criteria (single and complex), maintenance and repair burden (labor-intensive characteristic), etc.

Operational characteristics of radars and RNS are specified at the stage of systems. These characteristics should above all reflect requirements to a new developed or upgrading system from the position of ensuring assigned fit-for-purpose requirements, ease in technical maintenance (serviceability), increase of reliability and effectiveness of maintenance.

Chapter 6
Detection of Radio Signals and Its Parameters Measuring

6.1 Detection of Radio Signals

Issues on detection of radio signals and its parameters measurement will be examined in this chapter on example of detection of radar signals (detection of radar targets).

Decision making on target detection can be done based on threshold principle; i.e., a target is considered detected if output signal of optimum system y_{out} exceeds some specified level—threshold y_{thr}. In Fig. 6.1, an output signal of processing system y_{out} by module during measuring of reference signal delay t_s at presence of several targets and in noise conditions effect is depicted as an example.

It is obvious, that at no-noise case, any signal increase y_{out} above zero level will testify on target presence in observation zone. However, noises availability extremely complicates a situation since a noise component of output signal y_{out} makes difficult to answer for sure a question on a target presence. From Fig. 6.1, it follows that a lot of "bursts" of output signal is observed, which could be mistaken for a sign of target presence. In this case, it is reasonable to set some (threshold) y_{thr} level and consider for a sign of target presence only those bursts, which exceed a threshold. It is clear, that in Fig. 6.1 a false burst of "target" is available, which is stipulated by a noise signal. It is generally accepted to consider such burst as "false alarm." A question arises, how to select a y_{thr} level to provide a reliable detection of a target and decreasing of threshold probability increase by noise bursts or other types of noises? To solve this problem we use methods, developed in theory of statistical hypothesis testing.

At each element of delay in range t_s and in Doppler frequency f_s an output signal y_{out} can exceed thresholds or not exceed it. With this connection, in simplest case of binary selection of statistical events we adopt conventionally that in observation zone (at each t_s and f_s position) a target can be presented (first hypothesis) or a target is absent (second hypothesis). As an output signal y_{out} is a random, then four statistical events are possible at decision making on target detection.

D. A. Akmaykin et al., *Theoretical Foundations of Radar Location and Radio Navigation*, Springer Aerospace Technology,
https://doi.org/10.1007/978-981-33-6514-8_6

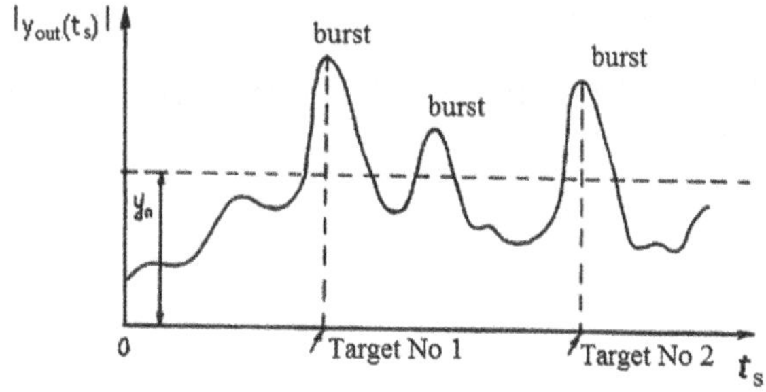

Fig. 6.1 Output signal realization of processing signal

Correct decision. Decision making on that a target at t_s position is detected at threshold exceed by y_{out} signal, is a correct in that case, when it is really known (first hypothesis) that a target in observation zone is presented. Due to noise effect, this event is a **probability of correct detection** P_{cd}.

False alarm. This corresponds to decision making on target presence in that case, when there is no target in observation zone (second hypothesis) and no signal, reflected from a target. Threshold is exceeded in this case by a noise at output of processing unit. This is false decision. Probability of false decision—**probability of false alarm**, we designate as P_{fa}.

Target miss (non-acquisition of available target). This is also a false decision, which originates in that case, when output signal y_{out} even at presence of reflection from a target, does not exceed a threshold y_{thr}. **Probability of target miss** we designate as P_{tm}. It is obvious, a probability P_{tm} can be determined by a probability of correct detection P_{cd}, since a target miss and correct detection comprise a full group of events and their total probability equals to one $P_{cd} + P_{tm} = 1$.

Correct non-acquisition. This is correct decision making that there is no target in observation zone in that case, when it is really absent (second hypothesis). It is clear, that in this case y_{thr} threshold is not exceeded by output signal y_{out}. We designate a probability of this event as P_{cn}. False alarm and correct non-acquisition also comprise a full group of events and total probability equals to $P_{fa} + P_{cn} = 1$.

Since an output signal y_{out} is random, then to find probabilities P_{cd} and P_{fa} it is preliminary necessary at testing of statistical hypothesis to find corresponding conditional probability density function (distribution) of output signal for a case of the mentioned two hypothesis: $S_t \neq 0$ and $S_t = 0$. To determine a false alarm probability P_{fa}, it is necessary to find $p(y_{out}|S_t \neq 0)$—conditional probability distribution density y_{out} at condition that at output there is only noise and no target $S_t = 0$. To find P_{cd}, it is necessary to find $p(y_{out}|S_t \neq 0)$ conditional probability distribution density y_{out} at mutual effect of noise and signal, reflected from a target. In Fig. 6.2, some realization of random output signal y_{out} and corresponding conditional probability distribution densities, describing statistical properties of read-out values y_{out} for each position on axis of delay t_s is depicted. Threshold level also is shown, which permits to evaluate conditional probabilities P_{cd} and P_{fa}. Actually, conditional probability of correct detection P_{cd} can be calculated by the known rules of probability theory as follows:

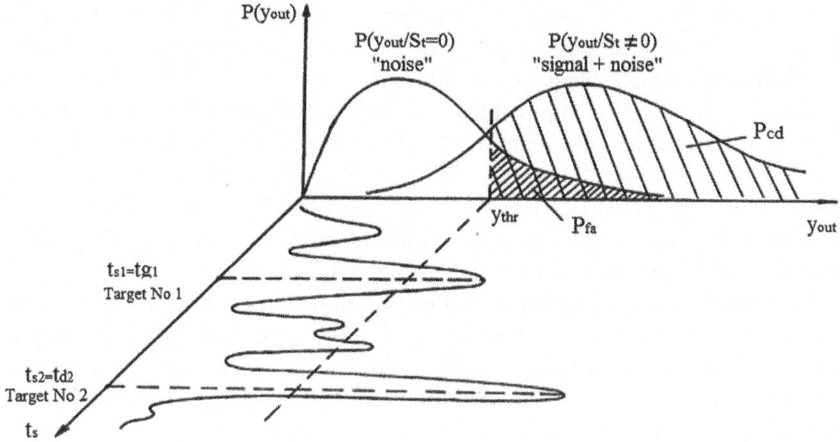

Fig. 6.2 Realization of random output signal y_{out} and corresponding conditional probability distribution density, describing statistical properties of read-out values y_{out}

$$P_{cd} = \int_{y_{thr}}^{\infty} p(y_{out}|S_t \neq 0)dy_{out}, \tag{6.1}$$

and conditional probability of false alarm as follows:

$$P_{fa} = \int_{y_{thr}}^{\infty} p(y_{out}|S_t = 0)dy_{out}, \tag{6.2}$$

In Fig. 6.2, the shaded areas of conditional distributions to the right of threshold y_{thr} correspond to probabilities P_{cd} and P_{fa}.

Threshold selection criterion. Probabilities P_{cd} and P_{fa} essentially depend on selection of threshold y_{thr}. Finally, certainty and reliability of target detection depend on correct selection of y_{thr} value. Threshold depends on criterion, which should be admitted at testing of statistical hypothesis. Nowadays, a several different approaches to threshold selection are known in mathematical statistics. However, not all the known criteria can be applied in target radar detection theory. This is explained that the most required parameters at criterion setting cannot be preset beforehand or defined during radar surveillance. Clarify this on example of **ideal observer criterion**. Based on this criterion, the threshold is selected in a way that to minimize average losses C_{ls} at decisions-making, which are stipulated by false decisions probabilities P_{fa} and P_{tm}. Average losses C_{ls} can be written down as follows:

$$C_{ls} = \beta_m P_m P_t + \beta_{fa} P_{fa}(1 - P_t), \tag{6.3}$$

where β_m and β_{fa}—corresponding coefficients of incorrect decisions "cost" (weight) at targets miss or false alarm; P_t—priori absolute (unconditional) probability at specified point and at a given moment of time.

However, it is almost impossible to define or set a probability P_t beforehand. Besides, it is difficult or even impossible to evaluate a cost of wrong decisions and, consequently, to set coefficients β_m and β_{fa}. It is impossible to set a conditional probability of target miss $P_{tm}(1 - P_{cd})$ for all possible values of RCS of radar targets, which are included as a parameter (in a form of amplitude of reflected signal) into conditional probabilities distribution densities $p(y_{out}|S_t \neq 0)$.

Neyman–Pearson criterion is the most commonly used in radar location. This criterion sets a probability of false alarm P_{fa} by a specified value, that permits ambivalently to determine a required threshold. There is no necessity to set any other parameters and coefficients. The most important, that a threshold value does not depend on signal presence, reflected from a target and its energy.

In should be noted that in the mentioned form the Neyman–Pearson criterion is used at condition of already found algorithm of optimum signals processing and optimum processing unit.

However, the Neyman–Pearson criterion can be used at selection of optimum signals processing unit or at comparison of different real non-optimal target detection units (at comparison, e.g., of optimum and none-optimum systems). In this case, the Neyman–Pearson criterion is formulated in a slightly different way. It is considered that signal energy, reflected from a target, is specified (e.g., signal amplitude is specified). Probability of false alarm P_{fa} is specified and probabilities of target miss P_{tm} or P_{cd} are evaluated. The more P_{cd} in these conditions, the closer this unit to an optimal. In this case, average losses $C_{ls} = P_{tm} + P_{fa}$ are also minimized.

Detection characteristics. Properties of radar systems at targets detection problems solution can fully be estimated through plotting of so-called detection characteristics, which represent a dependence of probability of correct detection P_{cd} from a value of signal energy relation, reflected from a target and spectral noise density ($R = 2E/N_o$) at specified as a parameter value of false alarm probability P_{fa} (Fig. 6.3).

It is possible another variation of detection characteristic, when probabilities P_{cd} and P_{fa} are specified in a form of parametric dependence, where parameter is a relation of signal energy and noise R (Fig. 6.4).

Detection characteristics are used, for example, at calculation of maximum range of targets detection. At specified probabilities P_{cd} and P_{fa}, the required value of relation R is found, i.e., relation of signal energy E at input of radar system to spectral noise density, at which obtaining of specified probabilities P_{cd} is provided at detection of a target. Often such a required value R is called a **detection parameter** and for an ideal (perfect) radar system without any deviations (without additional losses) and is designated as R_d (Fig. 6.3). If energy losses and deviations from ideal (perfect) operation conditions are observed in the system, then detection characteristics at specified value P_{fa} are lower than imperfect. Detection parameter in this case has a larger, than R_d value. Let us designate this parameter for imperfect (real) system

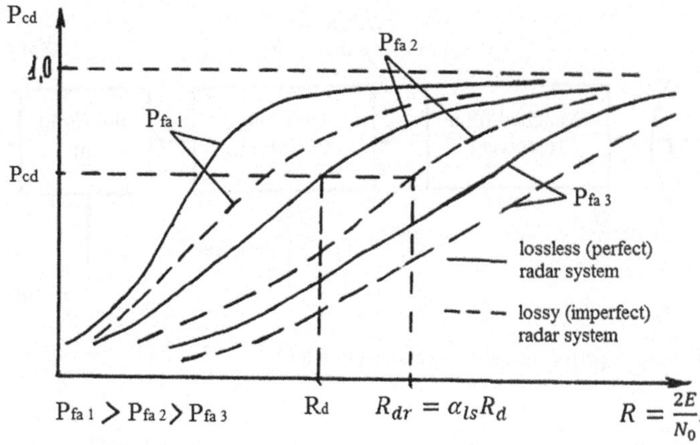

Fig. 6.3 Detection characteristics

Fig. 6.4 Detection parameter of perfect and imperfect radar

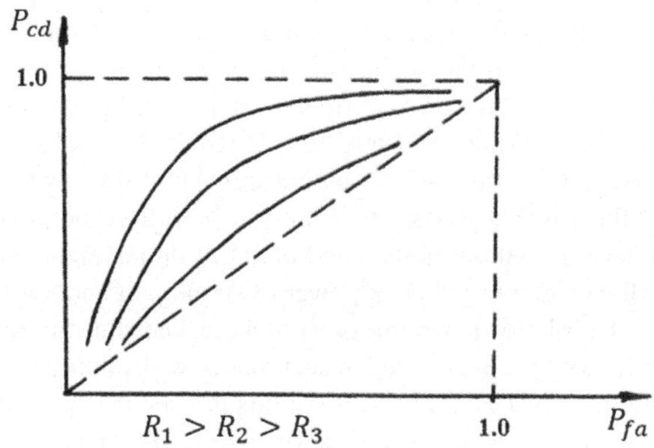

as R_{dr}. Consequently, to obtain the specified values of probabilities P_{cd} and P_{fa} in imperfect system, $R_{dr} > R_d$.

Relation R_{dr} and $R_d > 1$ and is called (α_{ls}) a loss coefficient in radar system

$$\frac{R_{dr}}{R_d} = \alpha_{ls}. \tag{6.4}$$

By this means, **loss coefficient** shows, by how much times it is necessary to increase a relation of signal energy to noise at radar input as compared to R_d parameter in order to provide a target detection with specified probabilities of correct detection P_{cd} and false alarm P_{fa}.

Let us examine simplified structure of a perfect radar at receiving of coherent signals with the known shape, when there are no various deviations from an optimal processing unit. Such system can be examined as a sample, by which real radars can

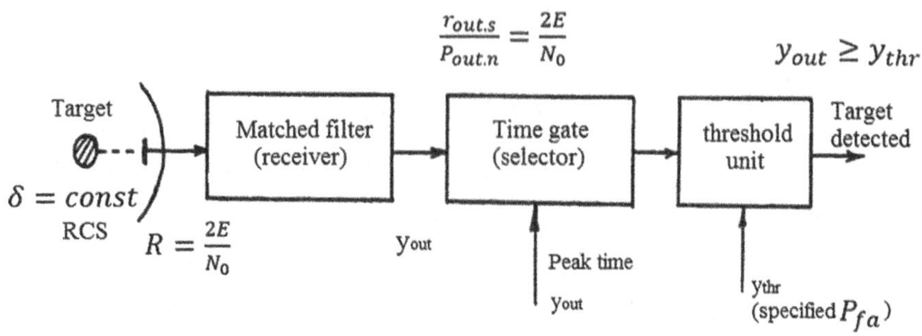

Fig. 6.5 Structure of perfect lossless target detection unit

be compared having energy losses and various deviations from perfect conditions of observation and signals processing.

In Fig. 6.5, a structure of perfect target detection unit is represented. It is clear, that a target has a constant value of RCS. At radar input, the energy of reflected from a target signal equals to E. At noises effect with spectral density $N_o/2$ (within limits $\pm\Delta f_n$ video frequency) relations of signal energy to noise equals to $R = \frac{2E}{N_0}$.

Received signal arrives to matched filter, the characteristic of which id fully corresponds to a shape and spectrum of radiated by radar signal. We consider that in this case shaping time of peak output signal is also known. Initial phase of reflected signal in this case is supposed to be known as well. At output of optimum signal processing system (at output of matched filter) an output signal y_{out} is shaped, at maximum of which a relation of peak power of signal part and noise equals to relation $R = \frac{2E}{N_0}$.

Threshold unit permits to make a decision on detection of a target. Level of threshold y_{thr} is selected in accordance with admitted acceptable value of false alarm probability P_{fa}. At increase of this threshold y_{thr} by signal from output of matched filter y_{out}, a decision is made on detection of a target.

The represented perfect unit for detection ensuring with specified probability P_{cd} requires the least value of reflected signal energy $E = E_{\min}$ and correspondingly the least relation $R_{\min} = 2E_{\min}/N_o$, which further acts as detection parameter $R_d = R_{\min}$.

In accordance with the abovementioned technique, for plotting and calculation of detection characteristics it is necessary at first to find conditional densities of probability distribution of output signal y_{out} for two main hypotheses, adopted at detection of targets.

Noise signal $n(t)$ is in effect at input of the system. We adopt that this noise process is Gaussian with normal probability distribution of read-out values $n(t)$. To simplify a process, we consider that noises are wideband ("white") with average value equals to zero $M[n(t)] = 0$, and correlation function in a form of delta-function:

$$M\{n(t_1)n^*(t_2)\} = \frac{N_o}{2}\delta(t_1 - t_2), \tag{6.5}$$

where $\frac{N_o}{2}$—spectral density of noise at input of radar system.

As is known, statistical properties of output signal y_{out} are fully determined by statistical properties of noise process $n(t)$. As process $n(t)$ is Gaussian, and signal processing system is a linear, then output signal $y_{out}(t)$ is Gaussian. Random read-out value y_{out} under Gaussian (normal) probability density distribution law:

$$p(y_{out}) = \frac{1}{\sqrt{2\pi\sigma^2_{y_{out}}}} \exp\left[-\frac{(y_{out} - \bar{y}_{out})}{2\sigma^2_{out}}\right] \tag{6.6}$$

where $\bar{y}_{out}$—average value of output signal y_{out} and $\sigma^2_{y_{out}}$—dispersion of output signal y_{out}.

To determine conditional distributed probability densities $p(y_{out}|S_t = 0)$ and $p(y_{out}|S_t \neq 0)$, it is necessary to find an average values $\bar{y}_{out}$ and dispersion $\sigma^2_{y_{out}}$ for two admitted hypothesis.

Examine a case, corresponding to the first hypothesis—no signal S_t at radar input $(S_t = 0)$. In this case, an output signal can be written as follows:

$$y_{out} = \int\limits_0^{t_n} n(t) S^*(t - t_s;\, f_d + f_s) dt \tag{6.7}$$

Average value of this random value y_{out} equals to zero, since $\bar{n} = 0$.

Signal dispersion y_{out} (assuming, that $\bar{y}_{out} = 0$):

$$\begin{aligned} \sigma^2_{y_{out}} &= M\{(y_{out} - \bar{y}_{out})^2\} = M\{y^2_{out}\} = M\{P_{n.out}\} = \overline{P}_{n.out} \\ &= M\left\{\int_0^{t_n} \int_0^{t_n} n(t_1) S(t_1 - t_s) S(t_2 - t_s) dt_1, dt_2\right\} \end{aligned} \tag{6.8}$$

Value of average noise power at output equals to signal y_{out} dispersion

$$\sigma^2_{y_{out}} = \overline{P}_{n.out} = \frac{N_0 E_0}{2} \tag{6.9}$$

As a result, conditional probability density distribution of output signal y_{out} for the hypothesis, no signal, reflected from a target, equals to the following:

$$p(y_{out}|S_t = 0) = \frac{1}{\sqrt{2\pi\sigma^2_{y_{out}}}} exp\left[-\frac{y^2_{out}}{2(\frac{N_0 E_0}{2})}\right] \tag{6.10}$$

Fine a conditional probability density distribution of output signal y_{out} for the second—target presence—and in receiving signal $y(t)$ a reflected signal is presented $S_t = 0$.

Maximum of output signal $\max\{y_{out}\}$ corresponds to coincidence of delays t_s and t_{dly} and Doppler frequencies $f_s = f_{Dp}$. Then for an output signal, we write down the following:

$$y_{out}(t_s;\ f_s) = a\int_0^{t_n} S\big(t - t_{dly};\ f_d + f_{Dp}\big)S^*(t - t_s;\ f_d + f_s)dt + \int_0^{t_n} n(t)S^*(t - t_s;\ f_d + f_s)dt \tag{6.11}$$

Determine statistical moments of conditional probabilities distribution for the second hypothesis. An average value $M\{y_{out}\}$ equals to (at maximum):

$$M\{\max(y_{out})\} = a\int_0^{t_n} S^2\big(t - t_{dly}\big)dt + \int_0^{t_n} \bar{n}(t)S(t - t_{dly};\ f_d + f_{Dp})dt \tag{6.12}$$

Considering that the first integral in (6.13) corresponds to signal "shape energy" E_0 and $\bar{n}(t) = 0$, then we obtain:

$$M\{\max(y_{out})\} = \bar{y}_{out} = aE_0 \tag{6.13}$$

Dispersion of output signal y_{out} for the second hypothesis equals to dispersion y_{out}, related to the first hypothesis (6.9), since a dispersion of random value sum in (6.11), represented by the second summand, and constant (nonrandom) value (first summand in (6.11)), equals to dispersion of a random value. Consequently, by substituting expressions (6.13) and (6.9) for an average value and dispersion of output signal into probabilities density distribution formula (6.6), we obtain a conditional density for the second hypothesis—presence of a target—in the following form:

$$p(y_{out}|S_t \neq 0) = \frac{1}{\sqrt{2\pi\sigma^2_{y_{out}}}}\exp\left[-\frac{(y_{out} - aE_0)}{2\left(\frac{N_0E_0}{2}\right)}\right] \tag{6.14}$$

In Fig. 6.6, distributions of (6.10) and (6.14) form are performed. Obviously, that they have a similar dispersion, however dispersion (6.14) is shifted along y_{out} axis at a value, proportional to signal amplitude *a*, consequently, at a value, proportional to signal energy $E = a(aE_0)$.

Find a probability of false alarm P_{fa}. For this purpose, integrate a conditional density (7.10) from a threshold level y_{thr} (Fig. 6.6) ad infinitum:

$$P_{fa} = \frac{1}{\sqrt{2\pi\sigma^2_{y_{out}}}}\exp\left[-\frac{y^2_{out}}{2\left(\frac{N_0E_0}{2}\right)}\right]dy_{out} \tag{6.15}$$

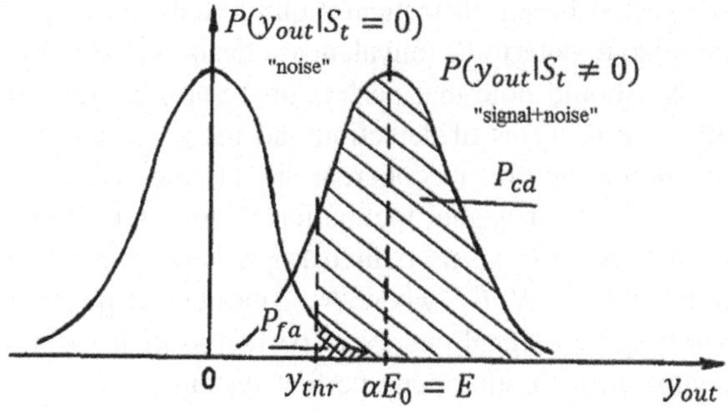

Fig. 6.6 Probabilities distribution at no target and target presence

Reduce this expression to a form of the known function—probability integral F(x):

$$F(x) = \frac{1}{\sqrt{2\pi}} \int_{-\infty}^{x} e^{-\frac{z^2}{2}} dz \tag{6.16}$$

Probability of false alarm:

$$P_{fa} = 1 - F\left(\frac{y_{thr}}{\sqrt{N_0 E_0/2}}\right) \tag{6.17}$$

Probability of correct detection P_{cd} is calculated in the same way:

$$P_{cd} = \frac{1}{\sqrt{2\pi\left(\frac{N_0 E_0}{2}\right)}} \int_{y_{thr}}^{\infty} \exp\left[-\frac{\left(y_{out-a E_0}\right)^2}{2\left(\frac{N_0 E_0}{2}\right)}\right] dz. \tag{6.18}$$

$$P_{cd} = 1 - F\left(\frac{y_{thr}}{\sqrt{\frac{N_0 E_0}{2}}} - \sqrt{\frac{2E}{N_0}}\right) = 1 - F\left(\frac{y_{thr}}{\sqrt{\frac{N_0 E_0}{2}}} - \sqrt{R}\right) \tag{6.19}$$

Weighty conclusions are followed from formulas analysis.

Firstly, for targets detection with specified probability it is necessary to have at radar input enough energy of receiving signal E, more precisely, it is necessary to have a required relation of signal energy and spectral density $R \geq R_o = 2E/N_o$. Here we should note, that P_{cd} probability does not depend neither signal shape, nor on width and shape of spectral characteristic of using signal, neither on modulation type, no on radiation wavelength, etc. и т.п. The single parameter of a signal, from which a probability of correct detection P_{cd} is depending on, is a signal energy E (more precisely, a value $R = 2E/N_o$).

Secondly, probability of false alarm depends on spectral noise density at input of the system and, of course, on threshold value y_{thr}. At specified value P_{fa}, which is selected based on tactical requirements and conditions of radar operation, it is possible to determine univalently a desired threshold value y_{thr}.

We should note that perfect processing system of coherent signals is the most effective in terms of detection and use of signal energy at radar input. As it was mentioned before, to coherent signals are referred all signals, in which a phase structure of HF signal inclination is definitely known. In particular, if a coherent pulse burst (train) is used, then a signal energy E equals to energies sum of all pulses in burst $E = N_p E_1$, where N_p—number of pulses and E_1—energy in one pulse. The required signal energy in perfect system for a target detection is minimal as compared with other none-perfect systems.

Detection characteristics are calculated and graphically plotted based on obtained formulas (6.15–6.19). The detection characteristics for an ideal case of target detection unit operation are performed on Fig. 6.7. The dependence of correct detection probability P_{cd} (expressed as percentage) as a function of R parameter (relation of signal energy to spectral noise density $R = 2E/N_o$) in decibels at specified values of false alarm probability P_{fa} is performed. R value in decibels, as normal, can be found according to the following formula:

$$R_{[dB]} = 10lgR. \tag{6.20}$$

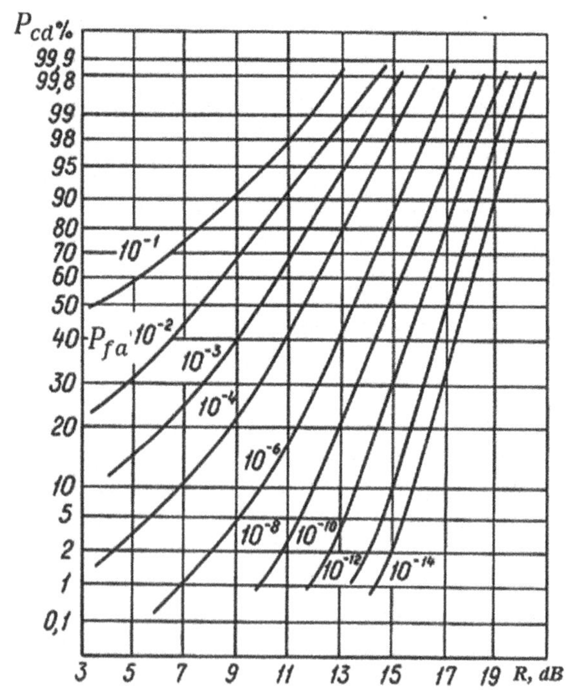

Fig. 6.7 Targets detection characteristics at use of coherent signals in lossless

Values of R parameter also numerically equal to a relation of maximum peak power of output signal of matched filter at input of threshold unit to average noise power at output of matched filter:

$$Q_{avn.out} = \left(\frac{P_{s}.\max}{P_n}\right)_{out} = R \tag{6.21}$$

If we evaluate an average power of signal $P_{m.f}$ at output of matched filter per period of HF oscillations, then, as is known:

$$P_{m.f} = \frac{P_{s}.\max}{2} \tag{6.22}$$

In this case, at input of threshold unit a signal-to-noise relation equals:

$$Q_{sn.av} = \frac{P_{s.av}}{P_n} = \frac{P_{s}.\max}{2P_n} = \frac{Q_{sn.out}}{2} = \frac{R}{2} = \frac{E}{N_0}. \tag{6.23}$$

Operation of real radar systems largely differs from those ideal conditions, at which targets detection is carried out at least energy of incoming from a target signal (least value $R = 2E/N_o$) to provide target detection with specified probability. At targets detection in real radar systems, in general case, a large signal energy is required, and consequently, it is necessary to have a large signal-to-noise ratio R at input of the system at similar required probabilities P_{cd} and P_{fa}.

At practical application of conclusions of the targets detection theory in radar location, the main objective is a determining of required for target detection energy of reflected from a target signal. To solve the mentioned task is possible by two following approaches.

Firstly, to compose a full signal (passing) and noises model in any given radar system and find density probabilities distribution of output signals for different hypothesis before threshold unit. By further plotting the real characteristics of detection, it is possible to determine a real value of relation R for ensuring of required probabilities P_{cd} and P_{fa}. However, the mentioned approach has an essential disadvantage in that there could be a huge amount of real radar model variants and it becomes very difficult and lengthy to solve this problem. Besides, in some cases it is difficult to formalize some nodes of radar model for rigorous solution of detection problem.

The second solution approach concludes in that an objective is divided into several stages. Firstly, decreasing level of signal-to-noise ratio at coming through any typical block or node of real radar structure is to be found, or we should determine at how many times the required signal energy will increase at radar input, reflected from a target, at considering of different deviations from ideal conditions of operation in the examined structure or node.

As it was mentioned before, to consider the pointed deviations the coefficient (factor) of losses in α_{ls} system is introduced. Coefficient of losses α_{ls} shows, at how

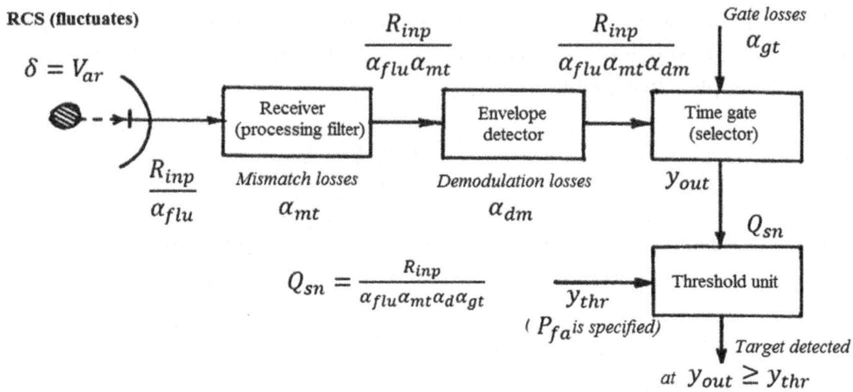

Fig. 6.8 Radar structure, differs from the perfect one $\alpha_{flu}\alpha_{mt}\alpha_{d}\alpha_{gt}$

many times a relation of signal energy and noise R in real radar operation conditions at target detection should be more, than a detection parameter R_d in perfect system for ensuring of required probability of correct detection P_{cd} at set value of false alarm probability P_{fa}. The resulting coefficient of losses α_{ls} can be performed in the form of product of partial (individual) coefficients of losses α_{ils}, characterizing the losses at coming through the separate typical nodes, or at considering of some typical operation conditions of the real radar.

$$\alpha_{ls} = \prod_{i=1}^{M} \alpha_{ils} \tag{6.24}$$

where M—a number of considered factors. Formula (6.24) in its structure is an analogue to formula of resulting gain coefficient of transducers (quadrupole) serial strings, which have partial gain coefficients.

As an example let us examine a structure of radar system (Fig. 6.8), differing from the perfect (Fig. 6.5) by a set of particularities, which lead to losses at targets detection.

In ideal case, the target RCS—a constant value, and in real conditions the targets have RCS, changing randomly. This leads to the fact that **RCS fluctuations loss** arise α_{flu}. Input signal-to-noise ratio R_{inp} reduces in α_{flu} times.

Receiving signal comes to processing unit (to receiver and matched filter). However, the processing unit structure (matched filter) in general case can be not completely matched with signal by some parameter. Consequently, at output of this unit a signal-to-noise ratio decreases as opposed to input. This fact is considered by introducing the **matching loss coefficient** α_{mt}.

After output of processing unit, the signal is a high frequency. Peaks of this signal are quite numerous and its position is random within a period of HF due to phase randomness in input signal or due to inexact knowledge of distance to a target. Consequently, an envelope extractor (some type of envelope detector) is usually used, which eliminates the signal dependence at processing unit output from a random phase. However, envelope detector leads to loss of some certain part of signal and this should be considered by introducing the **demodulation (detecting) loss coefficient** α_{dm}.

In those cases when a time gate (selector) is applied, gating a signal in a moment of signal maximum from processing unit output (or from output of envelope detector) one should consider a duration of time strobe. Maximum position y_{out} is unknown precisely and detection threshold exceed moment by a signal y_{out} can be within limits of some signal delay range in time (usually within time resolution interval). Time selector should be open at this time interval. Due to possible passing of additional noises, the losses arise, considered by **gate loss coefficient** α_{gt}.

Consequently, at target detection in the described non-perfect system (Fig. 6.8), under otherwise equal conditions, a signal-to-noise ratio at input of threshold unit Q_{sn} is happen to be considerably lower than required value R_d in perfect system. However, in this case a probability of detection P_{cd} also turns out lower of the specified (required), since:

$$Q_{sn} = \frac{R_{inp}}{\alpha_{flu}\alpha_{mt}\alpha_d\alpha_{gt}} \tag{6.25}$$

To obtain the specified value of target detection probability P_{cd}, it is necessary that value Q_{sn} should be less than R_d, and this requires a signal energy increase, reflected from a target, and correspondingly increase of input signal-to-noise ratio R_{inp}. Value R_{inp} should be more or equal to product of R_d to resulting losses coefficient $\alpha_{ls} = \alpha_{flu}\alpha_{mt}\alpha_d\alpha_{gt}$:

$$R_{inp} \geq R_d\alpha_{flu}\alpha_{mt}\alpha_d\alpha_{gt} \tag{6.26}$$

If we substitute R_{inp} into (6.26) formula, then we obtain $Q_{sn} \geq R_d$, that is required for detection with specified value P_{cd}. R_{inp} value in this case we designate as a detection parameter in non-perfect radar R_{dnr}. Increase of signal energy at radar input can be provided by different ways, e.g., by increasing of radiation power or by time increase of target illumination (dwelt time). Usually, an energy increase is achieved just by distance reducing to a target.

Certainly, at examining of detection conditions in real radar systems the losses can arise due to other reasons o factors (e.g., losses, connected with displaying or radar operator actions etc.). These losses record is possible at examining of certain types of radars.

6.2 Coordinates and Movement Parameters Measuring of Radar Targets

The previous chapters stated that at reflection from a target there is an essential modulation of radio signal. Reflected signal parameters carry information on object, its coordinates and movement parameters. This information can be placed in amplitude, frequency, phase, signal delay time, direction of arrival, etc. In order to obtain an information on coordinates and movement parameters of an object (target), it is necessary to measure values of the mentioned parameters, i.e., conduct its evaluation. Targets surveillance process is followed by noises of different nature, i.e., obtained during radar receiving evaluations of signal parameters are not necessarily precisely correspond to true values. Besides, signal parameters, which do not carry useful information, can much effect on measurements and thus are called **non-informative** parameters. For instance, the delay time carries information on distance to a target in radar signal, and amplitude and phase of signal have almost no information on distance and, consequently, amplitude and phase of radar signal in this case are non-informative parameters. Further, parameters of radar signal, which carry useful information on an object, we will call as **informative** parameters. Such parameters, for example, are: delay time t_{dly}, arrival angle of reflected wave γ, Doppler frequency shift f_{Dp} in reflected signal, since they permit to obtain information correspondingly on distance (range), angular coordinates and approaching velocity.

We will assume that desired (useful) signal $S(t)$ and noise $n(t)$ are additively included in receiving oscillation $y(t)$. Then receiving oscillation can be written as follows:

$$y(t) = S\big[t, \alpha(t), \alpha_{nip}(t)\big] + n(t); \quad t \in [0, t_{obs}] \tag{6.27}$$

where $\alpha(t)$—in general case a vector of informative parameters, the components of which are: t_d, f_{Dp}, γ etc.; $\alpha_{nip}(t)$—vector of non-informative parameters; $n(t)$—noise, the nature of which can be widely differing; t_{obs}—time of observation $y(t)$.

By analyzing the receiving oscillation (7.44), it is necessary to make a decision on what values are for informative parameters α of signal $S\big[t, \alpha(t), \alpha_{nip}(t)\big]$ in current time moment t. If during observation time t_{obs}, the components $\alpha_i(t)$ are constant ($\alpha_i(t) = const = \alpha; t \in [0, t_{obs}]$), then the described procedure is called **parameters evaluation**, when it is prohibited to disregard the signal parameter change $\alpha_i(t)$ per $t \in [0, t_{obs}]$ time and it is required the tracking of changing informative parameters values, then it is called as **parameters filtering** of radar signal.

Due to the fact that measuring process of informative parameters of radar signal is accompanied with noises, then random measuring errors are inherent to this process. Enumerate the primary reasons leading to errors in measurement of informative parameters $\alpha_i(t)$ of radar signal.

At first, availability of internal noises of radar receiver and external noises lead to random change of signal shape depending on measured parameter. Errors, stipulated by internal noises, are called noise errors and are evaluated by the mean-square value of measured parameter σ_{ns}.

Secondly, even in the absence of target movement relatively to radar due to noises, the delay time t_{dly}, Doppler frequency shift f_{Dp} and direction of arrival γ of reflected wave are changed, that leads to errors in determination of these parameters.

Thirdly, at target movement, its elementary reflectors (glittering dots) are rotating relatively the center of mass. This leads to random change of amplitude and field phase of reflected wave in radar deployment point, and consequently, leads to errors in coordinates and parameters finding of target movement. These errors are called fluctuation errors $\sigma_{fl}(\alpha)$.

Fourthly, propagation conditions of electromagnetic waves (EMW) during radar surveillance can change that, for example, leads to EMW propagation trajectory bending, and as a consequence, to measurement errors of received signal parameters $\alpha_i(t)$. These measurement errors are called propagation errors and are evaluated by $\sigma_{prop}(\alpha)$.

Since the mentioned factors, leading to measurement errors, act independently from each other, then summary error of informative parameter measurement α of radar signal can be represented as follows:

$$\sigma_\alpha = \sqrt{\sigma_{ns}^2(\alpha) + \sigma_{prop}^2(\alpha)} \tag{6.28}$$

Along with random errors, mentioned before, systematic errors also effects on radar change process, under which the difference is meant the following:

$$\Delta\alpha = \alpha_0 - \alpha_{avr} \tag{6.29}$$

where α_0—true value of measuring parameter; α_{avr}—average value of radar signal, obtained after multiple measurements.

The reasons of systematic errors, for example, can be stable defects of radar equipment (instrumental errors); errors, arising at approximation of calculating formulas (procedure errors); steady errors of operator. As opposed to random errors, the systematic errors can be corrected during adjustment and alignment of equipment. Hence, further we discuss only random errors, which follow the radar measurement.

Above all, we examine those random errors, which cannot be eliminated. For this reason at further examining of issues on radar measurement we will assume that a noise $n(t)$, including in receiving oscillation $y(t)$ (6.27), is an internal noise of radar receiver, which can be represented by wideband Gaussian noise.

Posterior (inverse) density of probability distribution permits fully argue on parameters of signals:

$$p_{ps}(\alpha_s) = kp_{pr}(\alpha_s)\exp\left(-\frac{E}{N_0}\right)\exp\left(-\frac{E}{N_0}\int_0^T y(t) - S(t, \alpha_s)dt\right) \tag{6.30}$$

From (6.30) expression, it follows that at the known priori density, the determining of posterior density of probability distribution is equivalent to find sufficient statistics, which is a function of unknown coordinates and parameters of targets movement.

$$y_{out}(\alpha_s) = \int_0^T y(t)S(t, \alpha_s)dt \tag{6.31}$$

Function $y_{out}(\alpha_s)$ defines those significant operation, which is necessary to fulfill with receiving oscillation $y(t)$ in order to get all available information on parameters α_s. In another words, $y_{out}(\alpha_s)$ function is also a sufficient statistics for evaluation of informative parameters α_{si} of radar signal.

It is known that there are several methods, in compliance of which an evaluation of useful signal parameters is sought.

Estimation from condition for minimum of posterior dispersion

$$\sigma_\alpha^2 = \int_{(\alpha)} (\alpha - \hat{\alpha}_s)^2 p_{ps}(\alpha)d\alpha, \tag{6.32}$$

that leads to evaluation of form of average value (first moment) from density $p_{ps}(\alpha_s)$:

$$\hat{\alpha}_s = \int_{(\alpha)}^{0} \alpha p_{ps}(\alpha)d\alpha. \tag{6.33}$$

Maximum-estimation posterior probability, when evaluation $\hat{\alpha}_s$ is taken as a value α_s, at which for the specified observation $y(t)$ a posterior density of probability distribution has an absolute maximum, i.e.,

$$\frac{dp_{ps}}{d\alpha}|_{\alpha=\hat{\alpha}} = 0 \tag{6.34}$$

Maximum likelihood estimation method, when an evaluation $\hat{\alpha}_s$ is taken as a value α_s, at which likelihood function $p(\alpha_s/y) = L(\alpha)$ reaches its maximum value, i.e.,

$$\frac{dL(\alpha)}{d\alpha}|_{\alpha=\hat{\alpha}_s} = 0 \tag{6.35}$$

Maximum method of posterior probability transfers to maximum likelihood estimation, when priori density is unknown and it is possible to consider it as quite uniformly distributed at interval of possible values of informative parameter of radar signal (e.g., rectangular or normal with high dispersion).

When prior density $p_{ps}(\alpha)$ has Gaussian distribution law, estimations obtained via mentioned methods are matched.

Criteria of maximum likelihood are widely used at radar measurements, since its application does not require of priori data during both at solution of detection tasks and radar signals resolution, and at estimation of its informative parameters. In estimation theory, it is proofed that in this case estimations, obtained via method of maximum likelihood, ensure an unbiasedness of estimations. Along with estimations obtaining of informative parameters of radar signal, it is important to know the effectiveness of these estimations, i.e., what is an error of measurements, what is a value of mean-square root error of estimation, how it is related with characteristics of the radar system itself and used radar signal. In statistical theory of estimation, an estimation error dispersion in accordance with Rao–Cramer bound is commonly used as quantitative characteristic of estimation effectiveness. As it was mentioned before, in parameters of reflected from targets electromagnetic oscillations, the coordinates and data on target movement are "encoded." Depending on solving tasks by radar system, the different assumptions on coordinates changing behavior and targets movement parameters can be examined.

In some cases, for example, during air space observation, it is reputed that during dwell time the movement coordinates and parameters are not changed in time, they are constant. In another case, at continuous target tracking, the targets coordinates are continuously changed, but even in this case during signals receiving at some period (e.g., pulse duration or pulse burst duration), it is also reputed that movement coordinates and parameters are constant. In all abovementioned cases, the coordinates measurement is reduced to estimation problem of constant signals parameters, constant values of coordinates and target movement parameters.

When assumption concerning coordinates constancy is prohibited to apply, an estimation of changing in time signals parameters is examined, i.e., filtering of parameters is performed. The two-stage estimation method of targets coordinates and parameters is used in radar location. At the first stage, the HF radar signals processing is carried out as well as estimation of "instantaneous" read-out values of movement coordinate and parameters for a time when assumption on its constancy is fulfilled. At the second stage, the processing of "radar data" is provided, i.e., an estimation of variables at time of read-out values of coordinates, the filtering of coordinates changing processes is fulfilled. This second stage is carried out in digital read-out values and, as a rule, corresponds to video-frequency or low-frequency spectral region. Usually, the filtering in this case is related to the linear type of filtering.

To determine a structure of optimum meter (discriminator/pulse filter), we examine the correlation integral $\Psi(\alpha_s)$, which for maximum determination is necessary to be differentiated in shift parameter α_s and equaled to zero:

$$\frac{d\Psi(\alpha_s)}{d\alpha_s} = 0, \text{ or} \frac{d}{d\alpha_s}\left[\int_{-\infty}^{\infty} y(t)S(t,\alpha_s)dt\right] = 0 \quad (6.36)$$

Solution of Eq. (6.36) permits to find a value of α_s parameter, at which the likelihood function $L(\alpha_s) = p(y/\alpha_s)$ reaches its maximum value. Consequently, the optimum measuring (estimation) unit of radar signal parameters is a unit of maximum positioner of correlation integral and finding of corresponding to it value of $\hat{\alpha}_s$ parameter. Often an optimum unit is called an optimum discriminator, underlying the meter capability to determine a deviation degree of true value of α_t parameter apart of its value $\hat{\alpha}$ in maximum (peak) point $\Psi(\alpha_s)$.

By performing of differentiation in α_s in (6.36) expression, we obtain an equation, which permits to find a value $\hat{\alpha}_s$, corresponding to maximum $\Psi(\alpha_s)$:

$$y_{out} = \int_{-\infty}^{\infty} y(t)\frac{d}{d\alpha_s}S(t,\alpha_s)\bigg|_{\alpha_s=\hat{\alpha}_s} dt = 0 \tag{6.37}$$

In accordance with (6.37), the functional chart of optimal meter can be represented as shown in Fig. 6.9.

It includes the differentiation circuit of reference signal $S(t, \alpha_s)$, multiplier and integrator, performing accumulation of signal for a time t_{acc} (existence time of desired signal). At parameter α_s change relatively to point $\hat{\alpha}$, a signal at integrator output is also changed proportionally to derivative of signal envelope $\Psi(\alpha_s)$ (Fig. 6.10a). At signal presence, reflected from a target, at input the zero value of output signal corresponds to estimation $\alpha_s = \hat{\alpha}_s$ of coordinate or parameter of target movement (Fig. 6.10b).

Input signal of discriminator is a function of mismatching $\alpha_s - \hat{\alpha}_s$. If we send a signal $(\alpha_s - \hat{\alpha}_s)$ to control circuit and close (lock) a feedback path, by supplying a

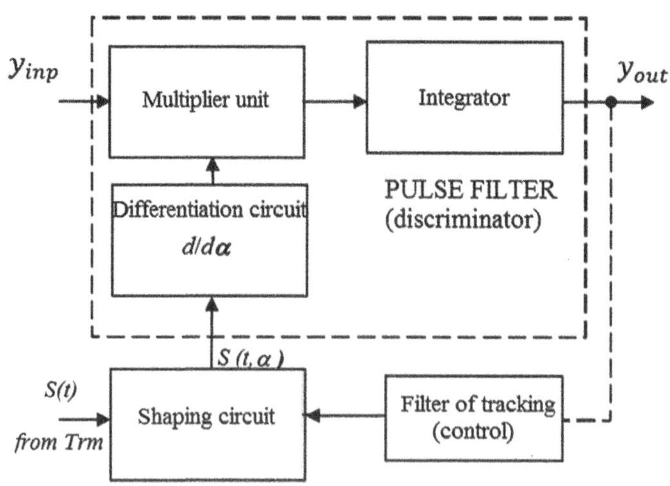

Fig. 6.9 Functional chart of optimal meter (discriminator)

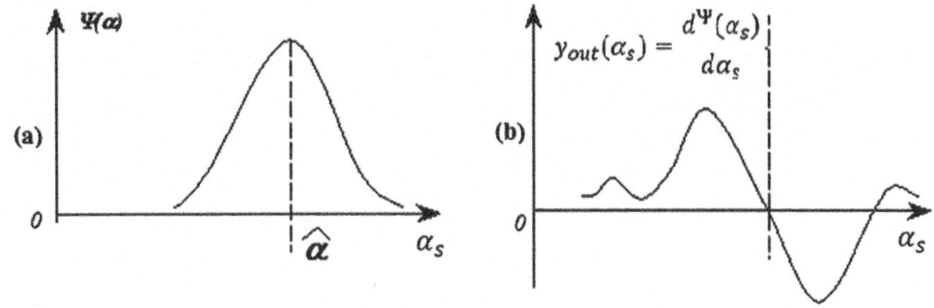

Fig. 6.10 Illustration of discriminator operation

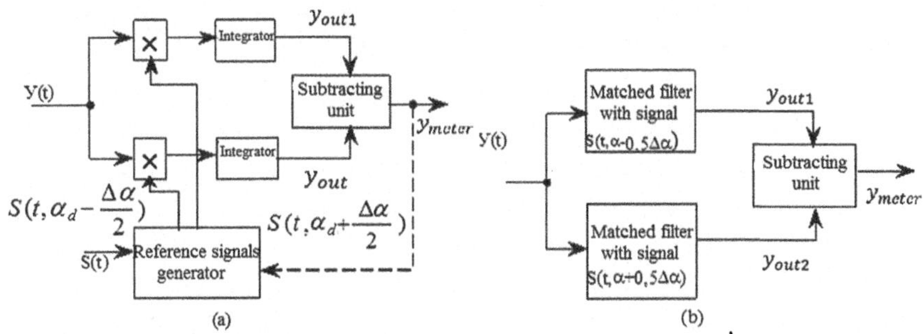

Fig. 6.11 Functional circuits of quasi-optimal meters

control action to discriminator, then we obtain the system, tracking the changing of parameter α_s (follow-up meter). In Fig. 6.11, the feedback circuit is shown dashed.

Functional circuits of quasi-optimal meters are represented in Fig. 6.11.

Circuit in Fig. 6.11a is realized using correlation units, and in Fig. 6.11b—with use of filters, matched with signals $S\left(t, \alpha_s + \frac{\Delta\alpha}{2}\right)$ and $S\left(t, \alpha_s - \frac{\Delta\alpha}{2}\right)$, which are mismatched relatively to each other at value $\Delta\alpha_s$. Consequently, maximums of output signals y_{out1} and y_{out2} are a bit shifted relatively to each other. After subtraction of these signals, we obtain a voltage, proportional to difference between parameter value $\hat{\alpha}_s$ and its reference value α_s. If we fix a value of output voltage of subtracting unit at specified $\hat{\alpha}_s$ and different values of α_s parameter, then we obtain a characteristic, analogue as shown in Fig. 6.11b. Equation to zero of a signal at output of subtracting unit corresponds to a moment, when $\alpha_s = \hat{\alpha}_s$.

Other variants of plotting of quasi-optimal meters are possible.

Part II
Radar Location Systems

A key moment at arrangement of radar systems is a selection of processing principle based on phase dependences of radiated and reflected radio signals. In this connection, the coherent and non-coherent radars are distinguished.

Oscillations are called coherent during a certain time interval, if there is a functional relation of some part of oscillation with its any another part. Otherwise, the oscillations are non-coherent. For two harmonic oscillations, coherence is achieved in case if phase difference between them on a certain interval remains constant.

As for pulse radars—if initial phase of radio pulses in a sequence is constant or is changed according to the known law, then such a sequence is coherent, and if a variation law is random, then a sequence is none-coherent.

Chapter 7
Non-coherent and Pseudo-coherent Radar Systems

Operation principle of non-coherent radar system (NCRS) is based on antenna radiation of electromagnetic waves into direction of an object of interest and receiving of reflected energy, its processing and information obtaining on parameters of this object, i.e., its coordinates—azimuth, elevation angle and range. NCRS structure is depicted in Fig. 7.1.

Electromagnetic waves are generated in transmitting device (TRD), channelized (constrained) in waveguide duct and through antenna switch (AS) directionally radiated by antenna (9A) into a space in direction to objects. Reflected from objects, radio waves through antenna-feeder unit come to a receiver (Rcv) and after transformation and information highlighting on object into terminal unit—display (indicator) unit (DU). Through matching unit (MU) in NCRS, the information is introduced from systems and complexes matched with it. Signals from object attitude sensors (AAS) are applied for controlling of antenna driving and stabilization mechanism (ADSM).

NCRS set composition and purpose of main components:

Antenna block provides:

- radiation into space of HF pulses coming from transmitter;
- receiving of signals, reflected from ground or aerial objects;
- beam traveling in azimuth;
- beam traveling in elevation angle;
- gyroscopic beam stabilization in horizontal or other specified plane.

Transmitting/receiving block provides:

- generation of powerful HF pulses and its transmission into the radar waveguide duct;
- processing of pulses, reflected from objects, into intermediate-frequency pulses, its amplification and detection;
- automatic frequency (tuning) control (AFC) of local heterodyne;
- temporary automatic gain (adjusting) control (AGC) of receiver;

D. A. Akmaykin et al., *Theoretical Foundations of Radar Location and Radio Navigation*, Springer Aerospace Technology,
https://doi.org/10.1007/978-981-33-6514-8_7

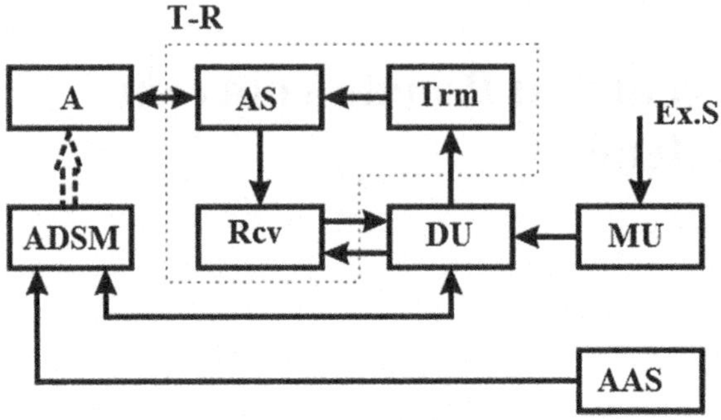

Fig. 7.1 NCRS structure. A—antenna; T-R—transceiver (transmitter/receiver); AS—antenna switch; Trm—transmitter; ADSM—antenna driving and stabilization mechanism; Rcv—receiver; DU—display unit; MU—matching unit; Ex.S—external systems; AAS—aircraft attitude sensor

- generation of start (initial) pulses for triggering of pulse circuits in other blocks of the station;
- supply of other blocks of the station with rectified stabilized (regulated) voltage.

Display block provides:

- acquisition of colored representation (displaying) of reflected signals on screen;
- radial-sector scanning on screen with different scales in range;
- objects azimuth and range determination;
- generation of necessary for station operation control (synchronization) pulsed signals, and calibration range markers;
- voltage amplification of video signals coming into block;
- video signals mixing with calibration range markers and screen intensifying pulse;
- display brightness and range marker control, extraction from general video signal of signals, reflected from objects of interest;
- station operation management;
- switch-on, operation modes and range marker spacing switching;
- manual control of antenna in elevation angle.

The transceiver is intended for formation of non-coherent sequence of radio pulses and receiving of reflected from target signals, its shifting at intermediate frequency and primary amplification.

Ultra-high-frequency generator (UHFG)/microwave emitter is used in non-coherent-pulsed radars as a transmitter, since it provides a high-pulsed power at wavelength of several centimeters. Repetition period of sounding pulses T_r is selected according to a condition of unambiguous distance finding to all targets located in radar surveillance zone: $T_r < 2D_{\max}/s$, and pulse duration determines a resolution in range. For continuity of average power, the following condition should be observed: ${}_r F_r = \text{const}$.

NCRS receiver (detector) refers to superheterodyne type and is usually with one frequency transformation (conversion). Due to high-frequency instability of oscillations formed by magnetron, to maintain the continuity of intermediate frequency (IFq) of receiver, the automatic frequency control (adjustment) system is used.

Construction particularities of NCRS devices are defined by parameters of input signals and characteristics of output equipment. Input signal, in general, represents a random process with broad amplitude range (up to 60 dB and higher). This stipulates a necessity to provide a large dynamic range of receiving path. Besides, a dynamic range of displaying units happens to be considerably lower than variation range of input signal amplitudes.

To solve this problem, the process of signal amplification in receiving device is divided between intermediate-frequency pre-amplifier (IFPA) and directly intermediate-frequency amplifier (IFA). There is a function of manual adjustment of amplification gain factor in NCRS, realized in IFPA, and IFA with logarithmic characteristic prevents limitation of powerful signals and permits to maintain its specifics.

Another distinctive feature of receivers is application of temporal automatic gain control (TAGC), which provides reducing of receiver amplification right after transmitter pulse sending and smooth rising of amplification with the passage of time. TAGC circuit is adjusted in a way that after 50–60 μs after pulse radiating of transmitter, an amplification reaches its ordinary value. Such adjustment provides reducing of signal amplification, reflected from close-spaced targets that prevents intense strobing of screen by these signals.

Considerable disadvantage of magnetron transmitters is non-coherence of forming pulses sequence; in each new radio pulse, the initial phase is a random value. Coherence ensuring is possible at signal processing. Burst processing coherence is ensured by memorizing of initial phases of sounding pulses at a period of its repetition.

In some radar application cases, it is necessary to solve a problem of "wind shear" and turbulence detection. Doppler principle is taken as a basis of wind shear detection, concluding in frequency difference of radiating and reflected from target pulses. To solve this problem, it is necessary to apply a coherent processing of radiated and received signals. It is impossible to solve the problem through the abovementioned NCRS.

NCRS composition, including relatively effective and at the same time inexpensive transmitting units, characterizing by high pulsed power, has led to application of pseudo-coherent method of signal processing in order to detect the most hazardous for flight moisture targets. The major advantage of this method is that it permits to extract signals with frequency increment due to Doppler effect at reflection of sounding signals from moving objects, which represent the hydrometeors, presenting at phenomenon of "wind shear" and turbulence.

Functional diagram of pseudo-coherent radar system (PCRS) is depicted in Fig. 7.2.

The main components of the chart are the following: antenna with stabilization and control unit, intended for formation in air space of specified directional pattern and control of its shift independently on AV trajectory irregularities. The purpose of

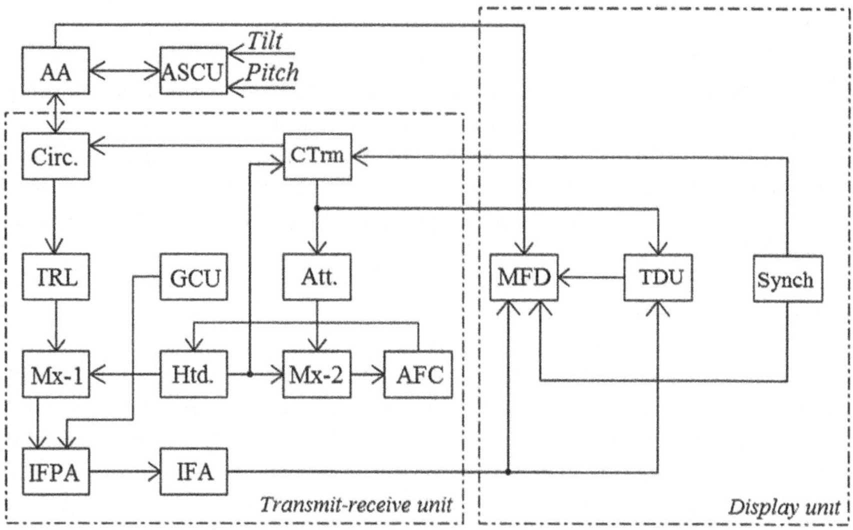

Fig. 7.2 PCRS functional chart. AA—antenna assembly (unit); ASCU—antenna stabilization and control unit; CTrm.—coherent transmitter; Circ.—circulator; TRL—TR limiter/spark-gap receiver; GCU—gain control unit; Att.—attenuator; Mx—mixers; Htd.—heterodyne; AFC—automatic frequency control unit; IFA—intermediate-frequency amplifier; IFPA—intermediate-frequency pre-amplifier; TRU—transmit/receiver unit; TDU—turbulence and "wind shear" detection unit; Synch.—synchronizing unit; MFD—multi-function display

transmitter/receiver (transceiver) is to form non-coherent sequence of radio pulses and receiving of reflected from target signals, ensuring of coherent processing for information transmission on hazardous meteorological phenomenon to display unit. Display unit is intended for visual information representation, and it also provides an operation synchronization of all radar components.

The transceiver has the following main components:

Superheterodyne type receiver with single frequency conversion is provided by a mixer 1 and heterodyne. Heterodyne frequency is adjusted in frequency of radiating sounding signal by AFC unit. The IFPA and main IFA are used in receiver. The last has a logarithmic-magnitude characteristic with linear initial section. Intermediate-frequency pre-amplifier is intended for amplification of reflected signals coming to its input with a possibility of gain factor change, since a radar signal is characterized by broad dynamic range.

Circulator is designed for antenna switching from receiving into transmitting channel. Reflected from radar target signals (microwave pulses), received by radar antenna assembly unit, are coming via waveguide duct through circulator, attenuator to the second mixer. AFC unit circuit serves for maintaining of constant frequency difference of magnetron and heterodyne (intermediate frequency) due to low stability of magnetron. Considering that intermediate frequency of radar should be constant and magnetron oscillator is a frequency-sensitive unit (admitting frequency offset within limits of ±1–3%, and this is about several megahertz), then a unit should ensure a function of frequency adjustment (control).

After conversion at mixer output, the intermediate-frequency pulses are generated, which come to AFC unit input, which generates a voltage proportional to deviation of intermediate frequency from its nominal value. This voltage, by influencing on control electrode of heterodyne, leads to its frequency change by reducing an intermediate-frequency deviation from a nominal value.

From transceiver IFA output, a signal is delivered to turbulence detection unit (TDU) of MFD. In TDU, a video signal is amplified and sent to liquid-crystal display of MFD unit.

TDU operation principle is based on Doppler effect application. The Doppler effect means waves frequency change registered by a receiver which happens due to its wave source movement and a receiver. Differential frequency allocation between radiated and reflected from a target signal is originated in TDU, and non-zero value of this parameter testifies on movability of reflection object.

MFU—graphical and telecommunication information imaging on color liquid-crystal display from onboard systems and sensors.

As it was mentioned before, the main requirement in order to ensure the coherent receiving consists in hard phasing of reference and radiating oscillations. For this purpose, a hard phasing of radar HF transmitter and reference oscillations generator is required. Different radars with inner coherence differ from each other by the following features:

- direction of phase synchronization (transmitter synchronizes a generator of reference oscillations or vice versa);
- synchronizing frequency (at what frequency a synchronization is performed: at high or intermediate);
- oscillation comparison frequency in phase detector (at what frequency a comparison is performed: at high or intermediate).

Let us examine an applied in centimeter wave radars functional pseudo-coherent diagram (Fig. 7.3) with phasing from radar transmitter and oscillations comparison are intermediate frequency of receiver. Time oscillograms in different points of diagram are represented in Fig. 7.4.

Pulses of modulator (1) control an operation of magnetron oscillator (generator). Part of the energy of sounding (probing) oscillation (2) of high frequency f_h is mixed with output voltage of a stable local (heterodyne) oscillator with frequency to obtain beats of intermediate frequency f_{hdy} at output of phasing mixer (3).

An intermediate-frequency pulse, phased relative to the radar generator, is used for phase synchronization of coherent local oscillator, operating at intermediate frequency. The oscillation phase of the coherent local oscillator, which acts as a reference voltage generator for phase detector, is coupled in the described manner with the phase of sounding pulse. Therefore, the output voltage of the coherent local oscillator (5) can be used at receiving of reflected signals, related to this sounding pulse. In each pulse repetition period T during the pulse duration, the coherent local

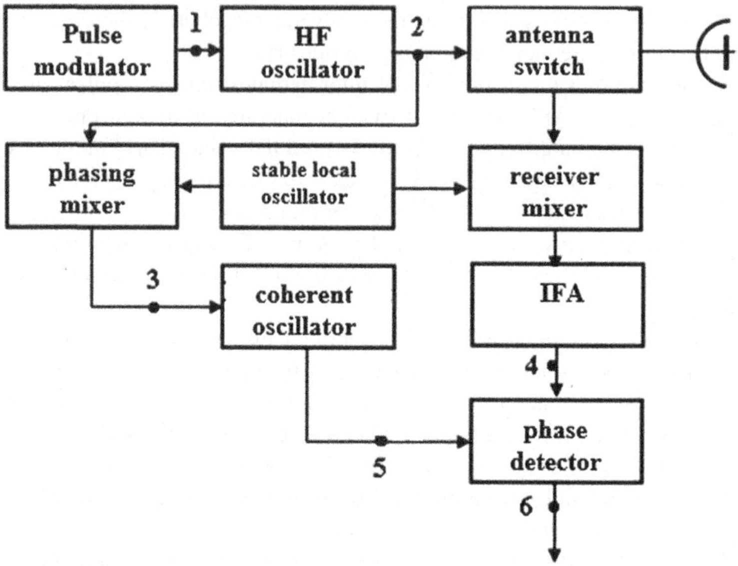

Fig. 7.3 Functional pseudo-coherent chart with transmitter phasing (IFA—intermediate-frequency amplifier)

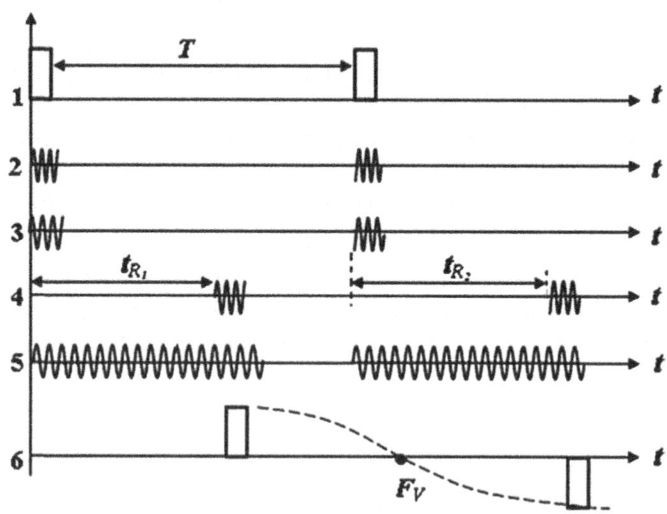

Fig. 7.4 Time waveforms (oscillograph traces) at various points in the circuit

oscillator is phased again through the phasing mixer by the transmitter, the pulses of which are incoherent with respect to each other.

The duration of the generation time of the coherent local oscillator exceeds the delay time t_R of reflected pulse arriving from the maximum range. Before arrival of the next phasing pulse, the coherent local oscillator is locked by a special control circuit and is unlocked a little later than the moment of phasing pulse arriving. In this case, the phasing pulse reliably imposes its phase and frequency on the coherent local oscillator.

By providing a phasing of coherent local oscillator at intermediate frequency, the stable local oscillator simultaneously serves as a conventional local oscillator of superheterodyne receiver. However, the continuous oscillations generated by it are characterized by increased frequency stability in order to ensure a constant phase shift between the voltage of coherent local oscillator (5) and receiving signal (4). The receiving signal of intermediate frequency and reference voltage (5) is delivered to a phase detector that multiplies these voltages. After filtering out high combination frequencies (by low-pass filter at output of phase detector), an output video signal represents a sequence of pulses that are modulated in amplitude (6) by Doppler frequencies, determined by velocities of targets movement, since the delay time in adjacent probing periods is different (t_{R1} and t_{R2}), and accordingly, the phase shift between signal of coherent local oscillator and the reflected signal has a different meaning.

In the overwhelming majority of cases, coherent equipment is used for moving target discrimination by suppressing (compensating) of reflections from stationary (e.g., ground) or low-moving (e.g., passive radar chaffs/dipoles) objects, or solving the problem of fast discrimination, detecting the type and degree of moveability of reflected objects ("wind shear," turbulence).

An automatic phase-locked-loop frequency control (PLL) loop can be used as a device that implements the function of memorizing and storing the value of initial phase of sounding pulse during operation period of the radar. The block diagram of a pseudo-coherent transmitter using a PLL is shown in Fig. 7.5.

From synchronizer output, the pulses are delivered to input of pseudo-coherent transmitting unit providing a formation of frequency repetition via subsequent division into FrD and further into M for ensuring of pulse generation of specified shape and duration. Further, a signal comes to PA where the main amplification is provided up to a value ensuring a magnetron operation. Magnetron is a microwave generator (oscillator) operating in intercrossed by magnet electrical fields, capable to shape signals of high pulsed power.

Shaped sounding signals through a circulator come to antenna unit and are radiated into an air space. Since the formed oscillations are non-coherent, it is possible to provide coherent processing via registration of initial phase value at one repetition period. This is possible by shifting into intermediate frequency in mixer and signal delivering to a unit, providing this value storage, representing a phase-locked-loop frequency control. Circuit includes VCO, which is tuned by a signal in such way that it corresponds to a signal on loop input. PD allocates a signal of phase difference of sounding signal and VCO. Integrator smooths random HF oscillations induced

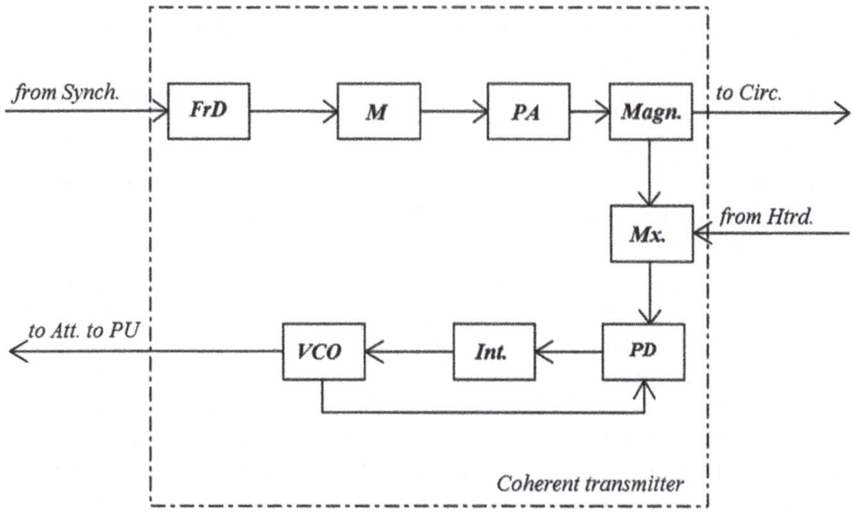

Fig. 7.5 Pseudo-coherent transmitter functional diagram (FrD—frequency divider; M—modulator; PA—power amplifier; Magn.—magnetron; Mx.—mixer; VCO—voltage-controlled oscillator; Int.—integrator; PD—phase detector (discriminator))

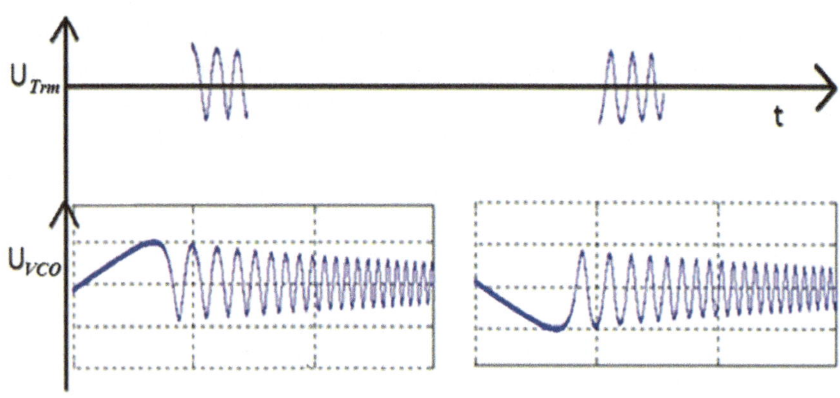

Fig. 7.6 Voltage-controlled oscillator tuning on transmitter signal

by equipment noise. This voltage tunes VCO circuits up to complete matching with mixer signal (Fig. 7.6). Then, a signal delivers to attenuator and processing unit (PU).

Chapter 8
Coherent Radar Systems

In NCRS systems examined above, target detection and its parameters measurement are performed based on envelope analysis of reflected oscillations, pick-out by envelope detector. High-frequency pulse filling is used as a carrier frequency that transports information on target to the radar receiver, and as a means of optimal energy storage of a single pulse in the resonant circuits of IFA. All phase and frequency information on target, which is contained in the sequence (burst) of pulses, reflected from a target, is destroyed.

Naturally, that incomplete use of possible information on target, contained in high-frequency filling, leads to energy losses in a signal and to decrease of radar observation efficiency. The signal-to-noise ratio in receiving path of the radar decreases, the minimum required input signal-to-noise ratio, i.e., the station coefficient of discrimination, increases. The main physical process in NCRS, which degrades a signal detection at background of noise, is a suppression of each of burst pulses with noise in a nonlinear element—an envelope detector. At weak single pulses, the signal-to-noise ratio at detector output decreases as compared to input.

Failure to use the information on target contained in the high-frequency filling leads to another defect of NCRS—its poor protection from highly correlated (compared to receiver noise) passive interference: reflections from local objects, reflections from hydrometeors, the sea surface, metallized tapes. Interfering reflections from stationary or slowly moving objects can significantly disrupt the normal operation of ground-based, air-borne, ship-borne and other radars.

The radar efficiency increases significantly upon the transition from non-coherent receiving to coherent receiving, which allows, to one degree or another, to extract additional information on target, contained in high-frequency filling of pulse burst.

To implement coherent receiving in a pulsed radar, it is fundamentally necessary to have three oscillations that are coherent with each other: a sounding oscillation, a receiving oscillation, a reference oscillation in receiver, intended for comparison in phase or in frequency with the received oscillation.

D. A. Akmaykin et al., *Theoretical Foundations of Radar Location and Radio Navigation*, Springer Aerospace Technology,
https://doi.org/10.1007/978-981-33-6514-8_8

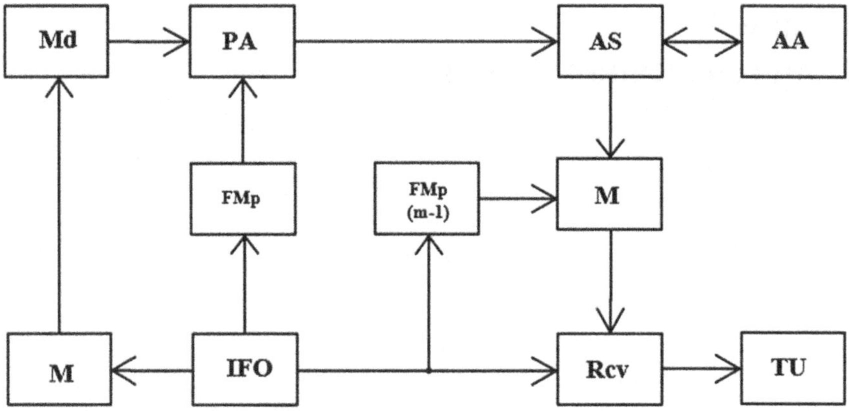

Fig. 8.1 Structure of coherent radar (Md—modulator; PA—power amplifier; AS–antenna switch; AA—antenna assembly; FMp—frequency multiplier; M—mixer; FrD—frequency divider; IFO—intermediate frequency oscillator; Rcv—receiver; TU—terminal unit)

The solution to this problem is possible by forming a coherent sounding sequence of pulses implemented in fully coherent radar (CRS), the structure of which is shown in Fig. 8.1. The coherence of sounding signals is determined by stability of intermediate frequency oscillator (generator) IFO, which forms a reference continuous oscillation. Its F_{rcv} is multiplied by a frequency multiplier by a factor of m and used as a carrier. The power amplifier PA amplifies the carrier wave and, using the modulator signals, forms a pulse sequence from it. The pulse repetition rate is set by the frequency divider FrD, associated with the IFO, the formed sequence of sounding pulses is coherent, since it represents a cut from a continuous harmonic oscillation at certain time intervals, set by the FrD, is emitted by the antenna assembly (AA) into space. The received by the antenna assembly (AA) signal through the antenna switch (AS) comes to the mixer, to the second input of which a signal at frequency $F_{rcv}(m-1)$ is delivered from frequency multiplier (FMp $(m-1)$). As a result of signal transformations in the mixer, the signal reflected from the target arrives at the receiver input at a frequency close to the intermediate one, which differs from it by the presence of a Doppler shift, stipulated by velocity of a target moving relatively to the radar. To the second input of receiver a signal from the IFO arrives, which is the reference one, for implementation of coherent processing. After the implementation of joint processing of reference signal and reflected from a target, the information goes to the terminal unit (TU) for indication or transmission of radar data to users.

The most important element of coherent radar receiver is a phase detector, which replaces the envelope detector of none-coherent radar. The phase detector compares (multiplies) the received and calibration (reference) voltages, as well as high-frequency filtering and the formation of a target video pulse. The output voltage of the phase detector after elimination of high-frequency components is proportional to the cosine of phases between received and reference oscillations:

$$U_{fd} = U_0 U_{0n} \cos \Delta\varphi. \tag{8.1}$$

Let us consider the structure of the high-frequency filling of oscillations arriving at input of phase detector. Let the radar radiates a wave

$$U_r = U_0^* \sin \omega_0 t. \tag{8.2}$$

When reflecting from a moving target, a signal arrives at the radar receiver

$$U_{sr} = U_0 \sin((\omega_0 \pm \Omega_V)t - \omega_0 t_R - \varphi_0), \tag{8.3}$$

where $\Delta\varphi_V$—phase incursion, caused by the presence of a radial velocity component with respect to the radar—V_r (Doppler effect); $\Delta\varphi_R$—phase incursion associated with the time of passage of the radio wave to a target and back; φ_0—jump-like phase change (phase jump) of reflected signal by some random value φ_0, caused by the presence of electrical conductivity in the surface layer of a target.

The numerical values of the listed quantities can be calculated using the following formulas:

$$\Delta\varphi_V = \pm\Omega_V t = \pm\omega_0 \frac{2V}{c} t, \tag{8.4}$$

(herewith the plus sign is taken when approaching the target, and the minus sign—when moving away from it)

$$\Delta\varphi_R = \omega_0 t_R = \omega_0 \frac{2R}{c}, \tag{8.5}$$

where R—is the distance from radar to a target; φ_0—usually considered as some random variable, uniformly distributed in the interval $0 \div 2\pi$.

For further consideration, the received oscillations (8.3) are reasonable to be represented in the following two forms in accordance with formulas (8.4) and (8.5):

$$U_{sr} = U_0 \sin\left(\left(\omega_0 \pm \frac{2V_r}{c}\right)t - \omega_0 \frac{2R}{c} - \varphi_0\right), \tag{8.6}$$

$$U_{sr} = U_0 \sin(\omega_0 t + \Delta\varphi_V - \Delta\varphi_R - \varphi_0). \tag{8.7}$$

The first summand in the sine argument is responsible for the presence of a high-frequency component in receiving signal, the second for its Doppler component. The third and fourth, in the first approximation, do not affect the spectrum of receiving signal.

The structure of the high-frequency filling of receiving oscillation (8.7) shows that for the best, purely coherent receiving, when the value of $\Delta\varphi$ in formula (8.1)

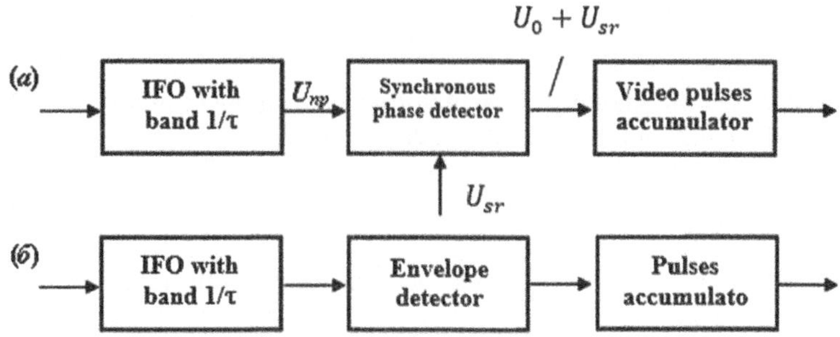

Fig. 8.2 Block diagram of optimal coherent receiver of pulsed signals (IFO with band)

vanishes, the receiver must have a reference voltage that satisfies the following conditions:

1. maintaining of the current phase $\omega_0 t$ of the radiated oscillation, which is ensured by rigid phasing of reference and radiated oscillations;
2. providing Doppler frequency shift $(\pm\Omega_V)$,which is possible with knowledge of target radial velocity;
3. providing additional phase shift $\Delta\varphi_R + \varphi_0$ that is possible with knowledge of distance to a target and phase of reflection.

Compliance with the above conditions would mean delivering of reference voltage to phase detector $U_{sr} = U_0 \sin(\omega_0 t + \Delta\varphi_V - \Delta\varphi_R - \varphi_0)$ obtaining at any time t of the difference phase $\Delta\varphi = 0$; and the maximum voltage at the output of phase detector when exposed by useful signal

$$U_{pd} = U_0 U_{0n}. \tag{8.8}$$

The block diagram of the optimal coherent receiver of pulsed signals against the background of internal noises of receiver or noise interference is shown in Fig. 8.2a. Figure 8.2b illustrates for comparison a block diagram of non-coherent receiver.

The considered coherent signal receiving with the known frequency and phase provides, in comparison with non-coherent reception, the following gains in the signal-to-noise ratio:

1. The operation of multiplying the received signal and the coherent reference voltage (8.1) removes the suppression of the weak signal by noise inherent in the envelope detector. A phase detector, analogue to receiver mixer, is a linear for signal-to-noise ratio and therefore ensures its constancy.
2. Precisely maintaining a zero phase shift between the signal and the reference voltage provides suppression of the out-of-phase (relatively to the signal) noise component. Physically, the phase of noise voltage in receiver continuously changes according to a random law in such a way that all phases within 0 $\div$ 2π turn out to be equally probable. In case of non-coherent receiving, the envelope detector, insensitive to the oscillation phase, outputs all noise voltages.

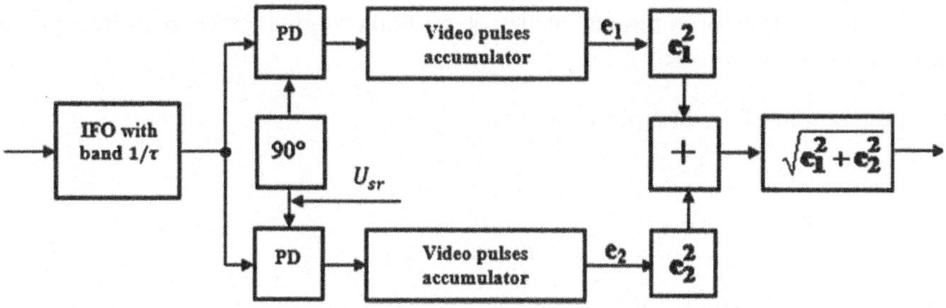

Fig. 8.3 Optimal coherent receiver of pulsed signals (IFO with band; PD-phase detector)

At coherent receiving, at the output of phase detector, those noise components will be weakened, which at the input have a phase shift relatively to the reference voltage $\pi/2 \pm \pi/4$ and $-\pi/2 \pm \pi/4$, and with a uniform distribution, they carry half the noise power. Indeed, at a phase shift $\Delta\varphi = \pm\pi/2$ the voltage in accordance with formula (8.1) is completely suppressed by the phase detector, and near this shift—partially. As a result of the suppression of the out-of-phase noise component, the signal-to-noise power ratio increases 2 times.

With purely coherent receiving, the best use of signal energy is achieved and the maximum receiving sensitivity is realized. However, the phase of received oscillation $\Delta\varphi_R + \varphi_0$, as a rule, is unknown to the observer and in most cases turns out to be random.

With an unknown initial phase and the known target velocity, the reference voltage of the phase detector should provide the following:

1. maintaining of the current phase $\omega_0 t$ of emitted oscillation;
2. create a Doppler frequency shift $\pm\Omega_V$, corresponding to the target velocity.

Compliance with these conditions means delivering of voltage to the phase detector:

$$U_{0tl} = U_0 \sin(\omega_0 t + \Delta\varphi_V) = U_0 \sin\left(\omega_0\left[1 \pm \frac{2V_r}{c}\right]t\right). \tag{8.9}$$

while the voltage of the receiving signal is still described by formula (8.7).

Unlike the receiver examined earlier, at the unknown initial phase, it is necessary to put not one, but two phase detectors at the output of the IF amplifier, the reference voltages of which are phase-shifted by $\pi/2$ (Fig. 8.3). As a result, ignorance of the initial phase does not lead to signal loss, since its phase cannot be shifted by $\pi/2$ simultaneously to the reference voltages of both phase detectors.

So, reference voltages are delivered to phase detectors:

$$\begin{cases} U_{rs} = U_{0rs} \sin(\omega_0 t + \Delta\varphi_V), \\ U_{rs}^* = U_{0rs} \sin\left(\omega_0 t + \Delta\varphi_V - \frac{\pi}{2}\right). \end{cases} \tag{8.10}$$

The interaction of reference voltages with the signal (8.7) gives the phase difference

- at the output of the first phase detector:

$$\Delta\varphi = \varphi_R + \varphi_0 \tag{8.11}$$

- at the output of the second phase detector:

$$\Delta\varphi = \varphi_R + \varphi_0 - \frac{\pi}{2}. \tag{8.12}$$

Approaching of a signal to in-phase (equiphase condition) with one of the reference voltages means simultaneously moving away from in-phase with the other. If, for example, the phase unknown to the observer $\Delta\varphi_R + \varphi_0$ turns out to be equal to zero, then at the output of one of the detectors a signal of the highest amplitude $(\Delta\varphi = 0)$ is obtained, and at the output of the other, there is no signal $\left(\Delta\varphi = -\frac{\pi}{2}\right)$. In other words, in the circuit (Fig. 8.3), the signal is decomposed by phase detectors into two quadrature components:

$$\begin{cases} U_{pd1} = U_0 U_{0rs} \cos \Delta\varphi, \\ U_{pd2} = U_0 U_{0rs} \sin n\Delta\varphi, . \end{cases} \tag{8.13}$$

where $\Delta\varphi$—is unknown to the observer phase shift between the signal and one of the reference voltages.

The video pulses of one target formed at the output of each of the detectors are summed up. The separately accumulated pulses are squared, summed up and passed through an amplifier with a nonlinear amplitude characteristic of the form $\sqrt{X}$.

In accordance with formula (8.8), the voltage at the output of phase detector with the known signal phase

$$U_{pd} = U_0 U_{0rs} = U_1. \tag{8.14}$$

Then, with the ideal accumulation of all n video pulses of one target, a voltage is formed at the output of the processing circuit:

$$U_{\text{out}} = nU_1. \tag{8.15}$$

The voltages at the output of two phase detectors with an unknown signal phase will be:

$$\begin{cases} U_{pd1} = U_1 \cos \Delta\varphi, \\ U_{pd2} = U_1 \sin \Delta\varphi, . \end{cases} \tag{8.16}$$

Assuming that during the reception of a burst of pulses from one target, the phase relationships do not change, we obtain the following results of separate accumulation:

$$\begin{cases} l_1 = nU_1 \cos \Delta\varphi, \\ l_2 = nU_1 \sin \Delta\varphi. \end{cases} \tag{8.17}$$

The voltage at the output of the processing circuit when the phase is not known

$$U_{\text{out}} = \sqrt{l_1^2 + l_2^2} = nU_1. \tag{8.18}$$

coincides with the voltage (8.15) at the output of the processing circuit with the known initial phase.

Note that an important condition for effectiveness of the examined processing method (as well as the method of full coherence) is the maintaining of a constant initial phase of receiving and reference oscillations during the time of receiving the entire target pulse burst (the time the radio beam passes through the target). In particular, a rigid phasing of the reference oscillation with respect to the probing one is required. The latter is a "phase mediator" between the reference and receiving oscillations. Thus, in the examined processing method, the observer, not knowing the initial phase, uses its constancy during the time of receiving target pulses.

If the coherence of reference oscillations in different burst pulses repetition periods is not maintained, the transition from formula (8.16) to formula (8.17) becomes impossible, i.e., the best summation of burst signals in amplitude, carried out in the video pulse accumulator, is violated. Indeed, the violation of the coherence of the reference oscillations will cause different phase shifts $\Delta\varphi$ for different burst pulses; instead of ideal accumulation in accordance with formula (8.17), the impulses will be summed up according to the less favorable law:

$$\begin{cases} l_1 = nU_1 \sum\limits_{k=1}^{n} \cos \Delta\varphi_k, \\ l_2 = nU_1 \sum\limits_{k=1}^{n} \sin \Delta\varphi_k. \end{cases} \tag{8.19}$$

In particular, two impulses, for which $\Delta\varphi_1 = 0$ and $\Delta\varphi_1 = \pi/2$, will give, in accordance with formula (8.19) the following:

$$\begin{cases} l_1 = 1\Delta U_1 + 0\Delta U_1 = U_1 \\ l_2 = 0\Delta U_1 + \Delta U_1 = U_1. \end{cases} \tag{8.20}$$

In this case, the output voltage of the processing circuit will be:

$$U_{\text{out}} = \sqrt{l_1^2 + l_2^2} = \sqrt{2}U_1. \tag{8.21}$$

With the coherent summation of two pulses by formula (8.18), the value $U_{\text{out}} = 2U_1$ would have been obtained.

This particular example clearly shows that refusal from coherence leads to signal accumulation according to the unfavorable law of powers summation: $P_{\text{out}} = nP_1$, inherent to noises.

So, ensuring the coherence of reference voltage of phase detector and generated sounding oscillations is an absolutely necessary condition for the efficiency of coherent-pulse radar station.

In the considered coherent-pulse receiver of a signal with a constant phase and a known frequency, a reference voltage (8.9) is used, rigidly phased with the oscillations of the transmitter, but having a frequency different from it (), taking into account the radial velocity of the target. It has been shown that this method provides efficient separation of the useful signal against the background of internal receiver noise or noise interference.

On the other hand, this processing scheme gives a loss of 2 times in power relatively to signal-to-noise ratio compared to the previously considered receiver of fully known signal. This is explained by the fact that quadrature phase detectors pass both phase (relative to signal) and 90° shifted out-of-phase noise components.

In the considered coherent-pulse receiver of a signal with constant phase and the known frequency, a reference voltage (8.9) is used, rigidly phased with the oscillations of the transmitter, but having a different from it frequency $(\omega_0 \pm \Omega_V)$, considering the radial velocity of target. It has been shown that this method provides efficient extraction of useful signal against the background of internal receiver noise or noise interference.

To detect a useful signal from a moving target against a different background—against the background of reflections from stationary or slowly moving objects—it is advisable to act differently. Keeping the rigid phasing of the reference and sounding oscillations, the original frequency ω_0 is also left in the reference oscillation without giving it a Doppler shift:

$$U = U_{0rs} \sin \omega_0 t. \tag{8.22}$$

In this case, due to multiplying of reference and receiving (8.3) oscillations in-phase detector, we obtain:

$$U_{pd} = U_0 U_{0rs} \cos \Delta\varphi = U_0 U_{0rs} \cos(\Omega_V t - \Delta\varphi_R - \varphi_0). \tag{8.23}$$

Thus, when using a reference voltage that reproduces the phase and frequency of radar transmitter, a voltage is generated at the output of phase detector, which is constant in time for stationary targets ($\Omega_V = 0$) and changes in time with the Doppler frequency for moving targets. Measuring this frequency allows you to determine the

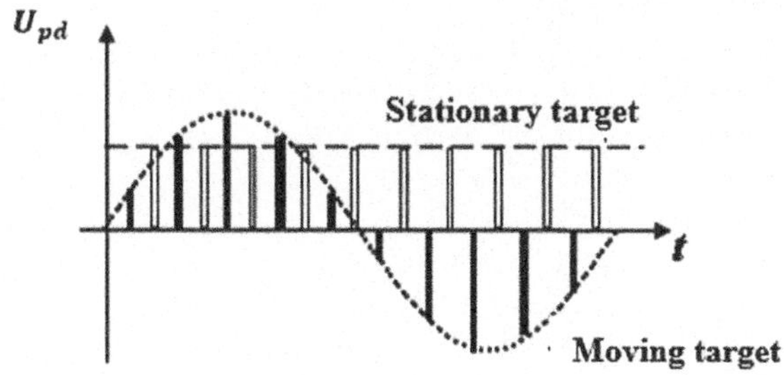

Fig. 8.4 Form of phase detector output voltages of coherent-pulse radar

radial target velocity relatively to the radar, and the target's range is measured by the conventional pulse method.

This principle has been taken as the basis of coherent-pulse radars of moving targets (extraction) discrimination (MTD) against the background of interfering reflections.

The voltage at the output of coherent radar of moving targets discrimination (8.23) was written by us in continuous form, although in fact, in a pulsed radar, voltages in the phase detector are present only during the duration of pulsesτ, following with a repetition period T. Therefore, formula (8.23) makes the envelope of video pulses at the output of phase detector.

The form of the output voltages of the phase detector of a coherent-pulse radar when probing stationary and moving targets is shown in Fig. 8.4. The difference in modulation of pulses sequences, reflected from different targets makes it possible to suppress reflections from stationary and slowly moving objects.

Chapter 9
Compensation of Signals from Stationary Objects

The problem of detecting and discriminating (isolation) of moving targets is solved by signal separation from moving targets and interfering reflections (IRf), and improving the conditions for detecting of a signal, reflected from a moving target. Signal extraction of moving target is a joint task of target resolution and detection. At present, various devices are known for the technical implementation of methods for identifying moving targets based on the use of coherent impulse signals. Such devices are divided into two groups:

(1) devices for suppression (rejection) of interfering reflections;
(2) devices for extraction and accumulation of signals from moving targets.

The signals from IRf represent radio signals in the form of passive interference, reflected when they are irradiated by sounding signals of the radar. Their effect is manifested in the suppression and masking of signals, reflected from the observed target. The intensity of interference can significantly exceed not only the level of the receiver's own noise, but also the useful signal of target, which complicates its radar observation and sometimes makes it completely impossible.

Passive interference (clutters) is created by reflections of radar signals from objects, which are in the radar surveillance zone. Natural passive interference is reflection from land and sea surfaces; terrain feature; moisture targets (meteorological formations) or hydrometeors (rain, snow, fog); atmospheric inhomogeneities (traces of meteorites in the atmosphere, lightning, "angel clutters," etc.). Passive jamming (intentional passive interference) is reflection from clouds of dipole reflectors, aerosols or ionized particles, as well as reflections from decoys.

The features of passive interference include:

- their appearance only when the radar transmitter is operating;
- location of the interfering source either in one resolution element with a useful target or in close vicinity from it;

D. A. Akmaykin et al., *Theoretical Foundations of Radar Location and Radio Navigation*, Springer Aerospace Technology,
https://doi.org/10.1007/978-981-33-6514-8_9

- significant excess of interference power over the receiver's own noise power (dynamic range of passive interference, i.e., the ratio of the passive interference power to the noise power can reach 90 dB);
- the difference of passive interference from signals, reflected from moving targets due to different dynamic characteristics of reflecting objects (radial velocity, acceleration, etc.) or statistical characteristics (correlation function or power spectral density) of the interference itself.

The organization of suppression process of passive interference can be carried out in different functional parts of the radio-engineering system.

Transmitting path. The protection against interference is possible here due to the correct choice of frequency, duration and repetition period of sounding pulse, changing of polarization law, etc. As is known, the contrast between signals, reflected from a target and the underlying surface, increases with decreasing long pulse. Increasing the wavelength is also an effective method for suppression of some types of passive interference (flocks of birds, insects, moisture targets).

Antenna path. Here, the following methods of interference countering are possible: reducing the width of directional pattern and the level of its side lobes; upward deviation of the antenna directional pattern DP (to reduce the reflection level from the underlying surface); use of dual-beam antennas.

Intermediate-frequency path. Temporary automatic gain control, pulse width discriminator (interfering reflections can have a much longer duration than the sounding signal).

Block of primary processing of radar information: notch filters, a set of narrow-band filters, discrimination of moving targets.

Block of secondary processing of radar information: formation of a noise map, rejection of false target blips at the stage of constructing a target trajectory.

At the moment, the main trend in the further development of passive interference countering methods is focused on improving the algorithms for primary and secondary processing.

The existing detection algorithms against the background of interference, depending on the principle of operation, use either static differences between targets and passive interference, or spectral.

Countering passive interference requires, first of all, the weakening of power of interfering reflections received by the radar antenna and narrowing of the dynamic range of interference to prevent overloading the receiving path. The first of mentioned tasks is the most common when ground-based radar needs to detect aerial targets, for example, in air traffic control systems.

To reduce the intensity of signals, reflected from objects located on the ground surface, two main methods are used. The first one consists in deflecting the antenna beam (DP) of the ground radar upward and makes it possible to improve the power ratio of the useful and reflected signals from ground objects by 15–20 dB. However, with wide antenna DP, the lower edge of the pattern still irradiates the ground surface, and the antenna receives signals reflected from it.

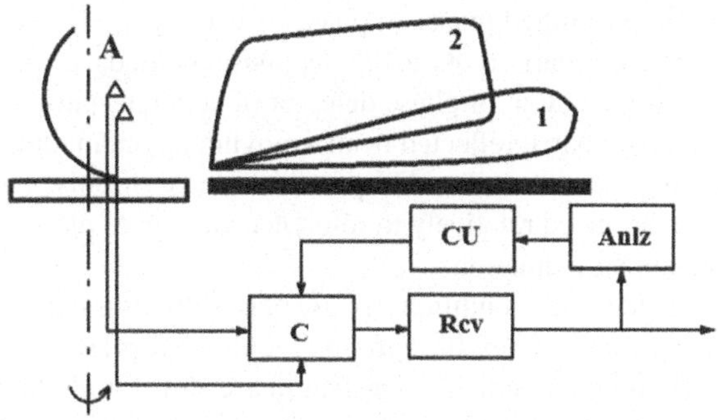

Fig. 9.1 Block diagram of radar receiving path with compensation for passive interference in radio-frequency amplifier (RFA) path (C—compensator; CU—control unit; Rcv—receiver; Anlz—analyzer)

To reduce the intensity of these reflections, a second method is used—a compensation method based, for example, on the use of two-beam antennas (Fig. 9.1.) and a high degree of correlation of signals, received along the lower narrow 1 and upper wide 2 beams of the antenna DP signals, reflected both from the ground surface and from objects, located on it.

The antenna signals received from the same ranges and elevation angles are subtracted in compensator (C). The residual voltage of passive interference after the receiver (Rcv) is fed to the analyzer (Anlz), which generates a signal that is fed to the control unit (CU). Under the influence of this signal, the CU selects such a ratio of the weighting coefficients of the compensator channels, at which passive interference from one source, but received simultaneously by two antenna DP, is maximally compensated.

Using this method, you can reduce the power of passive interference from the ground surface and from objects located on it by about 20–25 dB.

To reduce the dynamic range of passive interference, temporal automatic gain control (TAGC) synchronized in time with sounding pulses and instantaneous automatic or fast automatic gain control (IAGC and FAGC) are usually used. Controlled attenuators are often used, which reduce the level (power) of passive interference in receiving path according to signals from a special control system or according to a specific program.

The attenuation (gain) control is performed in each element of the radar working area resolution in accordance with the continuously updated value of average passive interference power stored in the memory unit with the so-called clutter map (CuMa).

Moving target indication (MTI) devices, which are part of radars, require a sub-interference visibility coefficient of at least 24 dB. The permissible irregularity of velocity characteristic in range from 50 up to 3000 km/h is not more than 10 dB. The presence of reflections from passive interference does not allow solving the problems of detecting and measuring of coordinates of aerial vehicles without the use of MTI systems.

The main effect that can be used to detect moving targets against the background of interfering reflections from stationary objects is the relative bias of reflected pulses during target movement. To extract the indicated pulses shifts, a comparison in phase of receiving reflected signals and reference high-frequency signal, rigidly connected with the emitted pulses in phase, is usually used. The detection of small target shift in range is carried out using the phase methods, as the most sensitive.

At the output of phase detector of coherent-pulsed radar, there is a video voltage of pulses burst, reflected from a moving target, modulated by the Doppler frequency. Using the coherent technique, the voltage of pulse train from a stationary target is synchronized relatively to reference voltage of phase detector, due to which there is no burst modulation.

Under such conditions (phase of reflections is unknown), the optimal signal extraction from the noise background requires the presence of two quadrature phase detectors and maintaining a constant phase shift for all pulses in the burst. However, this can only be done with stationary targets. This means that a circuit based on two phase detectors with a phase shift of its reference voltages by 90° would give the advantages of a stationary target over a moving one. Therefore, in moving target discrimination systems, where the signals of stationary and low-moving targets are interfering, they abandon the quadrature scheme, thereby leveling the possibilities for target discrimination regardless of their velocity. Naturally, this leads to losses in use of signal energy in comparison with noise, to an increase in signal discrimination coefficient against the background of noise. In MTI systems, such losses are incurred, because the main task for them is to suppress not noise, but passive interference. If the usual method of accumulating a burst of pulses against a background of noise is their summation after the detector, performed with a period equal to pulses repetition period, then processing in the interests of eliminating interfering reflections consists in over-periodic subtraction of video pulses.

Consider the particularities of moving targets detection by the phase method. Figure 9.2 shows timing diagrams for several pulse repetition periods at receiving of signals, reflected from a stationary and moving target. The reference signal is used to determine the phase of the incoming signal relative to the emitted one. In Fig. 9.2, we can see that the time delay of pulses from a stationary target does not change from one period to another. The pulses of a moving target are shifted along the time axis (in range). At the initial moment of time, the signal delay from a moving target is equal to:

$$t_{\text{mov}} = \frac{2D_0}{c}, \tag{9.1}$$

where

D_0 is the distance to a target at the initial moment of time
c is the speed of light.

In the next receiving path, the signal will already have a delay:

$$t_{\text{mov}1} = t_{\text{mov}0} + \Delta t = \frac{2D_0}{c} + \frac{2\Delta D}{c}, \tag{9.2}$$

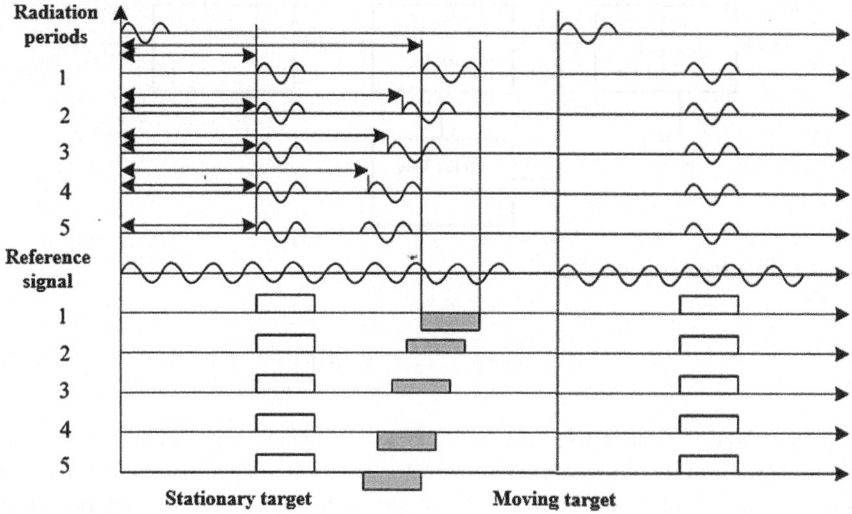

Fig. 9.2 Timing diagrams, explaining the detection principle of moving targets using the phase method

where ΔD is the relative target offset in range during the pulse repetition period $T_{\text{п}}$, the target moves with a radial velocity v, i.e., $\Delta D = vT_n$. Consequently, the delay increment is defined as:

$$\Delta t = \frac{2vT_n}{c}. \tag{9.3}$$

If we assume that the radial velocity does not change $v = \text{const}$, then, at the k-period of radiation, the delay increment will be equal to:

$$\Delta t_k = \frac{2kvT_n}{c}. \tag{9.4}$$

This shift corresponds to a relative change of phase by the value:

$$\Delta\varphi = \frac{4\pi}{\lambda}\Delta \mathrm{D}_k = 2\pi\frac{2vT_n}{\lambda}k = 2\pi f_d T_n k, \tag{9.5}$$

where $f_d = \frac{2v}{\lambda}$ is the Doppler frequency shift.

As it is known, the voltage from the output of phase detector is defined as:

$$U_{pd} = K cos \Delta\varphi, \tag{9.6}$$

Substituting (9.5) in (9.6), we obtain the voltage at the output of phase discriminator of the detector:

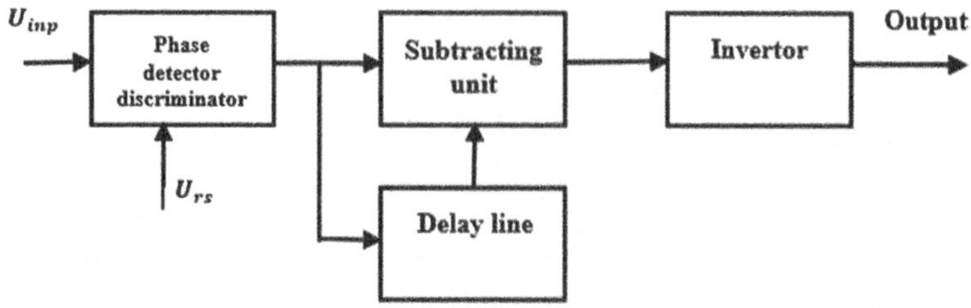

Fig. 9.3 Functional chart of OPC

$$U_{pd} = K\cos(2\pi f_d T_n k) = K\cos\left(2\pi \frac{f_d}{F_n} k\right), \tag{9.7}$$

If the target is stationary, then the Doppler frequency shift equals to zero, so the voltage from the output of phase detector does not change from clock path to observation path (in time). If a target is moving, then the voltage from the output of phase detector will be time dependent. The rate of voltage change U_{fd}, from path to path of sounding, will be determined by the ratio $\frac{f_d}{F_n}$.

To detect moving targets against the background of stationary objects, the phase method is used, as the most sensitive. A moving target will be indicated by a time-varying voltage from the output of phase detector.

If the voltage from cathode-ray tube (CRT) will be lead out from the output of phase detector of phase discriminator to display, then, on display screen, you will see blips in the form of constantly luminous and glittering marks. Glittering blips indicate the presence of moving targets.

However, such a displaying of radar information is inconvenient (there is a lot of unnecessary information on the screen that confuses an operator). To suppress marks from stationary objects, which means suppression of interfering reflections, the schemes of single or multiple inter-periodic subtraction of signals generated at the output of phase detector are used. Let us examine the principle of over-period compensation using an example of the circuit in Fig. 9.3.

The signals from the output of phase detector are fed to the first input of the subtracting unit. The same signals are fed to the second input of subtractor, but delayed by the pulse repetition period T_{rep}. The delay line has no significant effect on the relative changes in signal amplitudes from the output of phase detector. In the subtraction circuit, such signals are suppressed. The pulses of moving targets have a variable amplitude, and when subtracted, a difference signal is formed.

Therefore, the impulses of moving targets are not suppressed and pass further to the indicator after bringing the polarity of the impulses to one value. The timing diagrams of the over-periodical compensation (OPC) circuit operation are presented in Fig. 9.4.

In real conditions of radar operation, interfering reflections cannot be examined as signals of an absolutely stationary target. There are always fluctuations of signal amplitudes of interfering reflections, which create certain residual signals at the

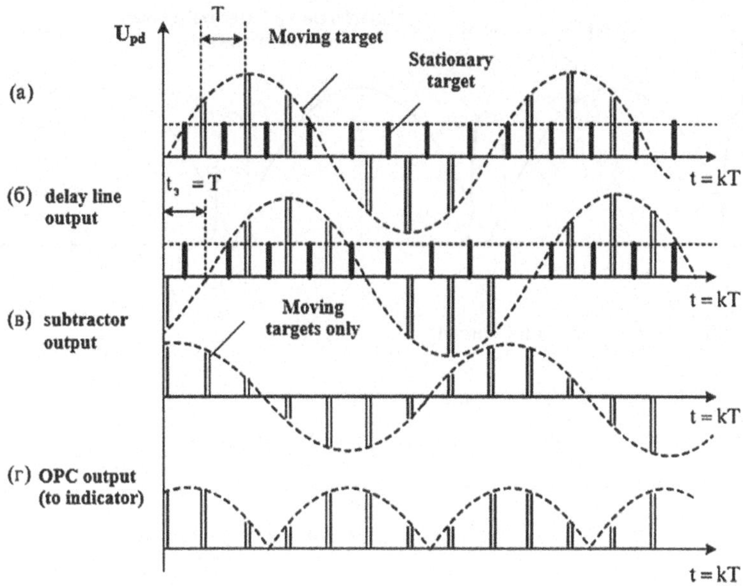

Fig. 9.4 Timing diagrams of OPC circuit operation

output of the OPC unit. Hence, the effectiveness of OPC circuit is limited. Failure to meet these conditions, as well as the presence in the background composition, on which a moving target is detected, of moving elements (e.g., vegetation in the wind) leads to the appearance of uncompensated residues at the output of over-period subtraction (canceling) circuit. Some types of passive interference have a regular velocity component. Among them are rain clouds which are moved by the wind as a single object. If you do not take measures to consider the speed of such formations, the video noise pulses will be modulated by the Doppler beat frequency, and uncompensated noise residues will be present at the unit output. Therefore, in some cases, a wind compensation unit is applied, by using of which the oscillation frequency of coherent local oscillator is changed by a value proportional to the wind speed. The interference phase relative to the oscillation phase of coherent local oscillator becomes unchanged from period to period, the interference compensation is improved.

Let us determine the frequency transfer characteristic of OPC circuit. For this, we use the impulse transfer characteristic of this circuit:

$$h(t) = \delta(t) - \delta(t - T_n), \tag{9.8}$$

The voltage transfer characteristic is defined as the Fourier transform of the function $h(t)$:

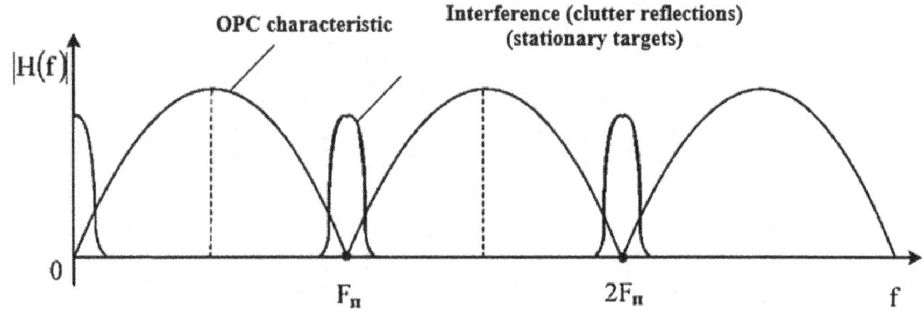

Fig. 9.5 Transfer function of OPC circuit

$$\dot{H}(t) = \int_{-\infty}^{\infty} h(t)\exp\{-j2\pi f t\}\mathrm{d}t$$

$$= 2j\exp\{-j\pi f T_n\}\sin\left(\pi\frac{f}{F_n}\right), \quad (9.9)$$

It is possible to obtain the modulus of voltage transfer characteristic:

$$\left|\dot{H}(t)\right| = \left|2\sin\left[\pi\frac{f}{F_n}\right]\right|, \quad (9.10)$$

In Fig. 9.5, the modulus of voltage transfer characteristic is shown.

It can be seen from the figure that the OPC unit in the frequency domain is a notch comb filter with suppression sections at frequencies equal to repetition rate, which is required to suppress interfering reflections (clutters).

Accordingly:

(1) The OPC circuit is a notch filter in the frequency domain that cuts out interfering reflection signals located at frequencies that are multiple of the repetition rate. Indeed, if a pulsed signal is reflected from a moving target, then each component of the reflected signal spectrum will shift relative to its previous position by the Doppler frequency. If a target is stationary, then the position of the spectral components of reflected signal completely coincides with the position of spectral components of emitted signal. The OPC circuit is tuned to rejection of these spectral components.
(2) The OPC circuit has low efficiency in the case when signal amplitude of interfering reflection fluctuates from period to period of sounding.

The types of plan position indicator (PPI) display with turned-off and turned-on MTI system are shown in Fig. 9.6.

Moving objects in the figure are displayed as a sequence of five blips.

In this paper, the most general type of MTI systems is considered—coherent-pulsed systems. These systems use an interpulse comparison of the Doppler phase

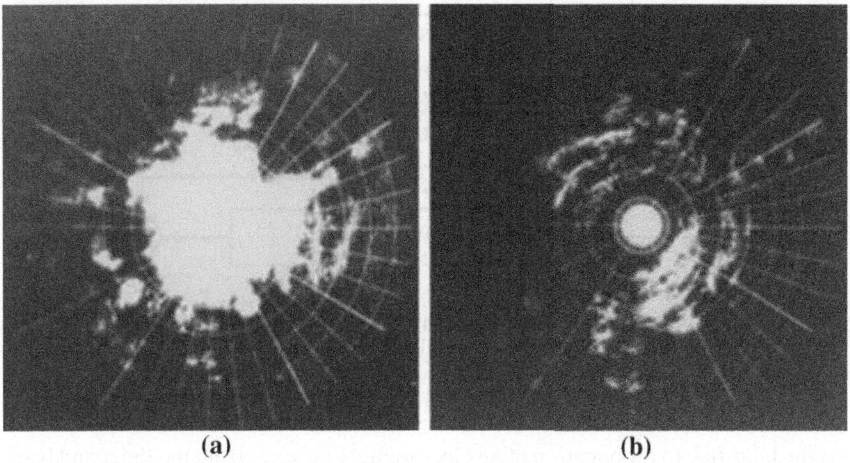

Fig. 9.6 Illustration of screen of radar display at **a** switched-off and **b** switched-on MTI circuit

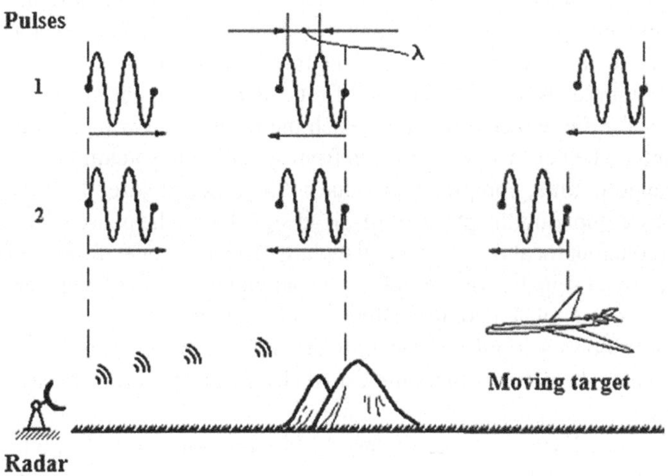

Fig. 9.7 Doppler radar operation principle

shift caused by the movement of targets to distinguish moving targets from stationary ones. In coherent-pulsed radar, the Doppler shift is defined as the change in signal phase observed in two adjacent soundings. Obviously, for stationary or slowly moving objects, the signal phase does not change, and therefore, they can be compensated by the method of over-periodic subtraction.

The operation principle of the simplest MTI unit, refer to Fig. 9.7, explains how a Doppler radar works.

The radar emits a pulse of high-frequency energy that is reflected off, for example, a hill or an airplane. The reflected echo signal arrives at the receiver input with a

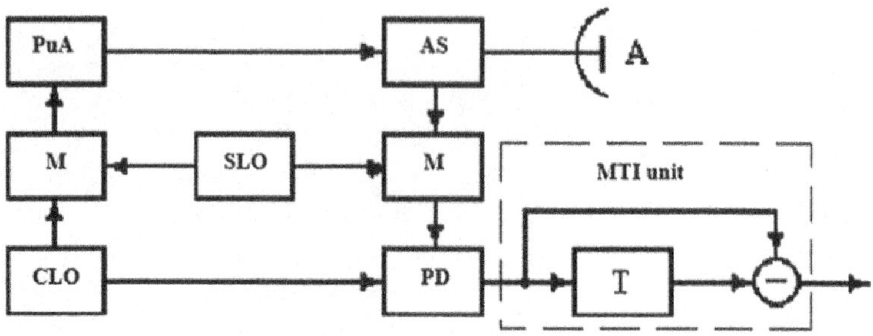

Fig. 9.8 Simplified block diagram of coherent-pulsed radar. Pulse amplifier (PuA), mixer (M), coherent local oscillator (CLO), stable local oscillator (SLO), antenna switch (AS), phase detector (PD), delay line for the radar sounding (sweep) period (T), A—antenna

certain delay due to propagation of an electromagnetic wave from the radar and back. The radar then emits a second pulse. The signal reflected from a hill returns after exactly the same time as in the first sounding, since the range to it does not change, and the signal reflected from an aircraft returns earlier, since the aircraft moves to the radar, and during the time between two soundings, it manages to get a bit closer. Although the distance, at which an airplane travels, is small, moving even a quarter of λ wavelength leads to phase change of reflected echo signal by 180°. The exact time it takes for the reflected signal to reach the radar is not critical. Another thing is important—whether this time changes from sounding to sounding.

The change in time, comprising a value of the order of several nanoseconds, is found out by comparing the phase of received signal with the phase of the reference oscillator (coherent local oscillator) in the phase detector. For stationary objects, the phase of received signals from sounding to sounding does not change, and this fact is taken into account when compensating for interference.

Figure 9.8 shows a simplified block diagram of a coherent-pulsed intermediate-frequency radar. It includes the simplest MTI unit in the form of an over-period compensator.

The echo signal from a stationary target has a constant phase shift relative to the transmitted pulse, so there is no phase change from pulse to pulse. For a moving target, the phase shift will change from one sounding pulse to another, because of which the output signal envelope forms "beats" in the form of a Doppler harmonic signal.

The bipolar signal at the output of phase detector carries information about the phase and amplitude of receiving signal. The bipolar signal, generated by receiving a single transmitted pulse, is shown in Fig. 9.9a. If we observe a point moving target against the background of strong reflections from stationary objects, then when receiving several transmitted pulses, the video signal may have a shape like shown on Fig. 9.9b.

The video signal at the output of MTI unit is shown in Fig. 9.9c. The timing diagram was obtained under the assumption that passive interference signals are

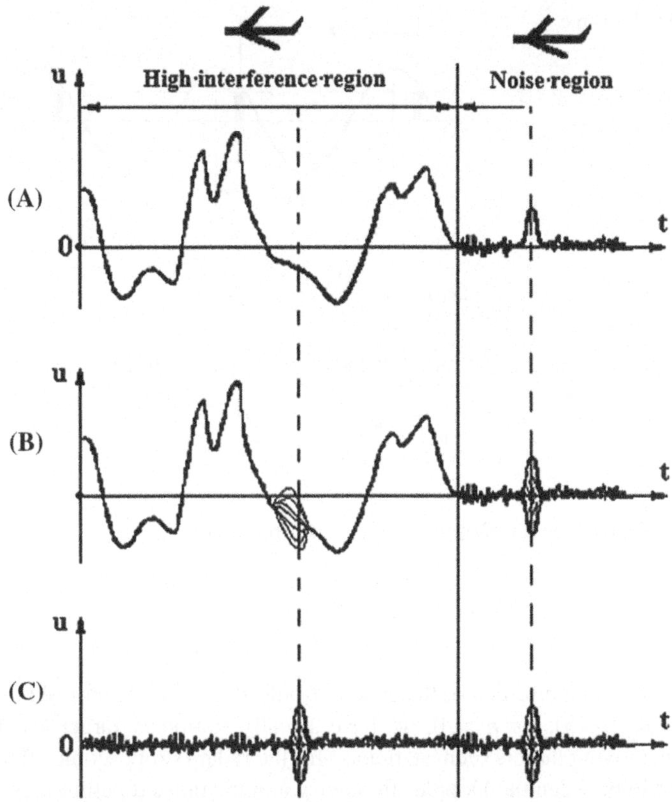

Fig. 9.9 Timing diagrams of MTI circuit

non-fluctuating; otherwise, uncompensated residues may appear at the output of MTI unit in the interference zones.

Modulation of video pulses of a moving target with the Doppler frequency is the basis for the operation of pulsed MTI systems. However, at certain target velocities, called blind speeds, such modulation is absent, resulting in target loss (lorry of lock-on). The condition of modulation absence means the equality in amplitude of target video pulses at the output of phase detector, spaced from each other by the repetition period:

$$\cos\Omega_V t = \cos(\Omega_V(t - T)), \tag{9.11}$$

or

$$\cos\Omega_V t = \cos(\Omega_V(t - T)) + 2\pi n, n = 0, 1, 2, \ldots \tag{9.12}$$

Blind speeds will occur when

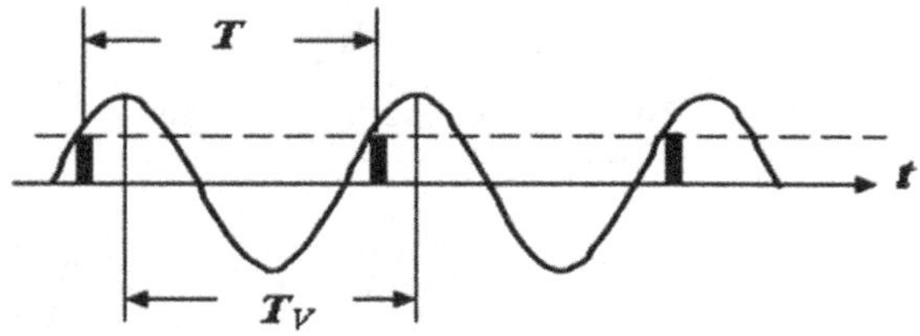

Fig. 9.10 Stroboscopic effect

$$\Omega_V t \equiv \omega_0 \frac{2V_r}{c} T = 2\pi n, \tag{9.13}$$

Any target that has a blind radial (relative to the radar) speed

$$V_r = \frac{c}{\omega_0}\frac{\pi n}{T} = \frac{\lambda}{2T} n, \tag{9.14}$$

and, therefore, creates the Doppler effect with frequency

$$\Omega_V = \frac{2\pi}{T} n, n = 0, 1, 2, \ldots \tag{9.15}$$

is not observed on the screen of the coherent-pulsed radar system of MTI system.

In formula (9.15), for $n = 0$, we have a really stationary target and for other values of *n*, moving targets seem stationary for the radar (stroboscopic effect) due to a certain ratio between the Doppler frequency and the pulse repetition rate. It is for the stroboscopic effect that the equality or multiplicity of the named frequencies is required (Fig. 9.10):

$$F_V = nF, n = 0, 1, 2, \ldots \tag{9.16}$$

The compensating unit will suppress moving targets that satisfy condition (9.16). If the radar with continuous radiation suppresses only those targets whose radial velocity is really zero, then the coherent-pulsed radar has a number of blind speeds.

Similar to the apparent zero velocities, the encounter modulation frequencies of F_m pulses appear in a pulsed radar, which are significantly lower than the true Doppler frequency. This phenomenon occurs when there is less than one pulse per Doppler "half-wave" of modulation (this occurs at high target speeds). A plot of the apparent Doppler frequency with increasing of true Doppler frequency (increasing of target speed) is shown in Fig. 9.11. It indicates that in the coherent-pulsed radar, there are not only blind speeds (points 1 and 2), but also the ambiguity of speed read-out (points 3, 4, 5, 6), when different target speeds are perceived as the same.

In connection with the above, the condition for the unambiguity of velocity measuring in coherent-pulsed radars is essential:

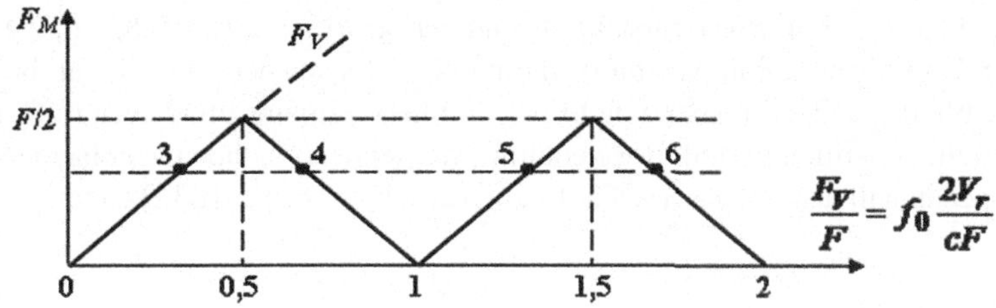

Fig. 9.11 To illustrate the stroboscopic effect

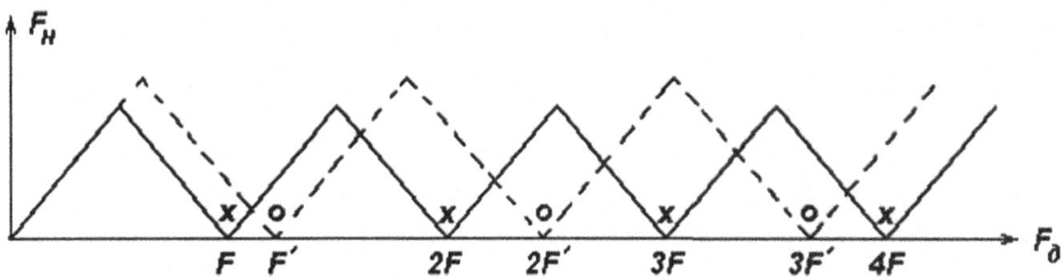

Fig. 9.12 Using two repetition rates

$$F_V \cong f_0 \frac{2V_r}{c} \leq \frac{F}{2}, \tag{9.17}$$

Thus, in order to counter "blind" speeds for excluding the miss of a moving target, it is necessary:

(1) In order to provide that signal frequency from the output of phase detector will coincide with the Doppler frequency, it is necessary to make the repetition rate of sounding pulses at least 2 times greater than the observed Doppler frequency.
(2) To avoid the manifestation of the stroboscopic effect, it is necessary to select a ratio $\frac{f_d}{F}$ not equal to an integer.

In modern radars, in order to eliminate the stroboscopic effect, the repetition period of the sounding pulses is changed from path to path of the radar operation; thus, by changing the pulse repetition period, we move the blind speeds from one position to another.

This can be seen more clearly from Fig. 9.12, which determines the dependence of the ripple frequencies at the output of phase detector for two values of repetition period. Doppler frequencies, corresponding to blind velocities, are indicated as crosses for one repetition period and as dots for another. Doppler frequencies corresponding to blind velocities are different for two values of the period, i.e., the position of the crosses and dots on the abscissa axis does not coincide.

But there may be cases, when some of them (dots and crosses) coincide. Let us consider a numerical example. Let the repetition period of the radar, varying from cycle to cycle, takes two values $T_1 = 1000$ μs and $T_2 = 1250$ μs. Then, for the first period, the Doppler frequencies, corresponding to the blind velocities $F_{Di} = i/T_1$, are equal to 1, 2, 3, 4, 5 kHz, for the second period $F_{Di} = i/T_2$—0.8, 1.6, 2.4, 3.2, 4.0, 4.8 kHz. From a comparison of the numerical sequences F_{Di}, it can be seen that the blind speed, corresponding to $F_{Di} = 4$ kHz, remains blind even if the radar changes the repetition period. If we continue the sequences, then the coincidence of blind speeds will also be observed at frequencies F_{Di}—8, 12, 16 kHz, etc.

Chapter 10
Multi-channel Radar Systems

The radar, as the main all-weather means of information acquisition on targets position in airspace, has high requirements related to the amount of information received, maximum operational range, high noise immunity and accuracy of coordinate measurement. A number of such problems posed for the radar location are solved by multi-channel radars, in which information on a target is received not through one, but through several frequency or spatial channels simultaneously.

Multi-channel stations are receiving simultaneously on several frequencies, but at one point (one antenna). All channels of the multi-channel radar are used simultaneously in the interests of joint processing of information on a target. This method significantly increases the radar efficiency or, in other words, significantly increases the amount of information retrieved by the station from airspace.

In accordance with the abovementioned, multi-channel radars can be divided into spatial-multi-channel and frequency-multi-channel radars. The development of the latter was facilitated by the success of antenna technology and automatic signal processing technology.

In a *frequency-multi-channel radar*, elements that reproduce target signals are spaced in frequency. The target is irradiated simultaneously at several (according to number of channels) frequencies.

In a *spatial-multi-channel radar*, elements that percept target signals are located at different points in airspace. In such radars, stationary elements of the system form many beams (two or more). Radar directivity characteristics are obtained as a result of joint processing of signals, received by separate antenna elements.

A special case of spatial-multi-channel radars are widely used monopulse radars. Especially often monopulse radars are used for automatic high-precision target tracking in angular coordinates.

Monopulse (single-pulse) method of radar operation has got its name as a result of its comparison with the method of conical scanning of radio beam in determining the angular coordinates of target.

D. A. Akmaykin et al., *Theoretical Foundations of Radar Location and Radio Navigation*, Springer Aerospace Technology,
https://doi.org/10.1007/978-981-33-6514-8_10

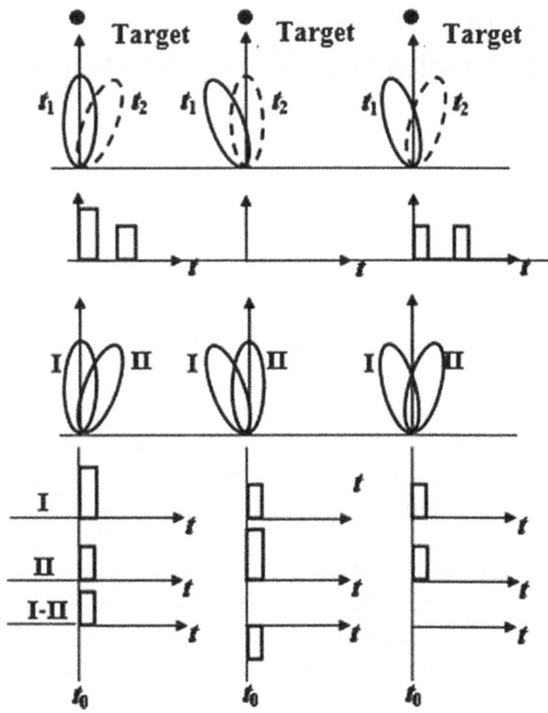

Fig. 10.1 Monopulse radar operation principle

The conical scanning method for determining the angular coordinate in one plane requires the receiving at least two pulses from a target, which are compared between each other in amplitude. The equality of amplitudes of these pulses (equisignal direction), received at two extreme positions of the antenna directional pattern, corresponds to the target angular coordinate (Fig. 10.1). To determine two angular coordinates in two mutually perpendicular planes, the conical scanning method requires receiving from the target at least four reflected pulses with subsequent pairwise comparing them in amplitude.

The monopulse method, in principle, provides the measurement of one or two angular coordinates of target using a single reflected pulse, but received simultaneously through two or four channels, which is ensured by the formation of two or four stationary in space beams. To determine the equal-signal direction, which gives the target angular coordinates, the results of simultaneous receiving on different channels are compared with each other (Fig. 10.1).

Thus, the method of conical scanning of a radio beam is characterized by the reception of target reflections through one channel at different moments of time, and for the monopulse method—simultaneous reception through several channels.

Of course, as a rule, monopulse radars do not operate on one reflected pulse, but use a large number of pulses, but this is not due to a fundamental need, but to energy considerations. In other words, in principle, monopulse radar can measure coordinates using a single reflected pulse, but increasing the range and accuracy of coordinate readings is possible only by using the energy of many pulses.

Information processing in monopulse radars consists in comparing the amplitudes and phases of the reflected signal during its simultaneous reception in spaced-apart channels. In particular, four receiving channels are required for automatic angle track on target in two perpendicular planes.

In onboard devices, usually four emitters (e.g., horns) are used, placed with a slight shift near the focus of the common reflector. The dimensions of the antenna system are changed insignificantly, although instead of a single-channel one, a four-channel reception and radar information processing is provided.

Low-frequency amplitude fluctuations of reflected signal, occurring due to fluctuations in the effective scattering area of a target or due to arising of interference, do not influence on the measurement of angular coordinates by monopulse radars. Therefore, they are distinguished by significantly higher noise immunity and accuracy of angular tracking of a target.

There are many types of monopulse radars that differ in the way they process signals, received by several channels. The simplest type of monopulse radar is an amplitude-difference radar, in which in order to determine the direction to a target, the amplitudes of signals, received by different channels, are compared (Fig. 10.2).

Suppose that at the moment the misalignment angle (error angle) between the direction to a target and the equal-signal direction of the antenna is γ. After frequency conversion, intermediate-frequency amplification and detecting, we obtain the following voltages at the output of the first and the second channels, that is, at two inputs of the subtraction circuit:

$$U_1 = k_1 F(\varphi_0 + \gamma), \text{ and } U_2 = k_2 F(\varphi_0 - \gamma) \tag{10.1}$$

where $F(\varphi)$—is the antenna directional pattern of one channel.

At the output of the subtraction circuits, a voltage is generated:

$$U_{\text{out}} = k_1 F(\varphi_0 + \gamma) - k_2 F(\varphi_0 - \gamma) \tag{10.2}$$

Assuming a small mismatch, we expand the function $F(\varphi)$ in a Taylor series near the point ($\varphi = \varphi_0$) and restrict ourselves to two terms of series:

$$U_{\text{out}} = (k_1 - k_2) F(\varphi_0) + (k_1 + k_2)\gamma \frac{dF(\varphi)}{d\varphi}|_{\varphi=\varphi_0} \tag{10.3}$$

If gain coefficients of k_1 and k_2 channels are equal to each other, then

$$U_{\text{out}} = 2k\gamma \frac{dF(\varphi)}{d\varphi}|_{\varphi=\varphi_0} \tag{10.4}$$

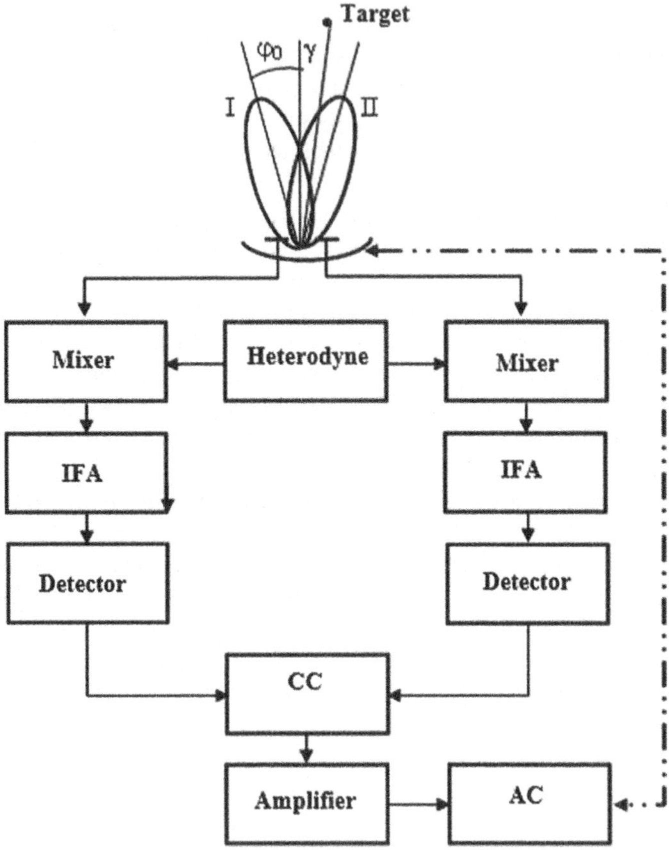

Fig. 10.2 Functional diagram of amplitude-difference radar (IFA—intermediate-frequency amplifier; CC—cancel circuit (subtracter); AC—antenna control)

The error signal, after the necessary amplification, is used for antenna control. The antenna tends to take such a position when the error signal is zero; in this case, the equal-signal direction coincides with direction to a target.

Direct proportionality between the error signal and the actual error γ will be violated if the channels are not identical $(k_1 - k_2)$. According to formula (10.3), in this case, at $\gamma = 0$, the mismatch signal will be supplied to the antenna, although in reality there is no mismatch. This is the main disadvantage of the amplitude-difference monopulse radar, since it is very difficult to maintain the identity of channels during operation.

The direction-finding characteristics of the amplitude-difference monopulse radar with the identity and non-identity of the channels are shown in Fig. 10.3.

To eliminate the instability of the zero boresight direction in monopulse radars, more sophisticated methods of received signal processing are used (Fig. 10.4).

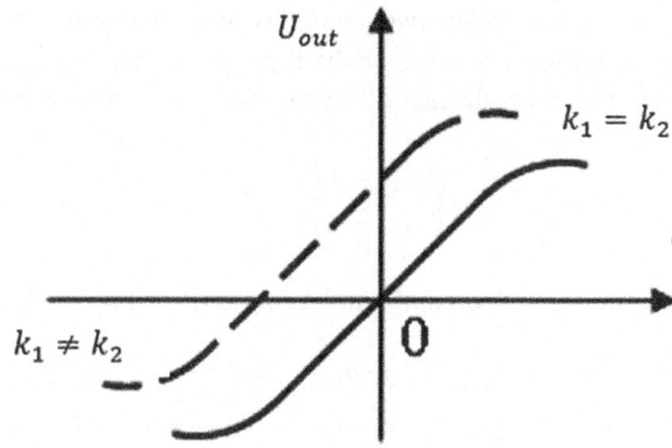

Fig. 10.3 Direction-finding characteristics of amplitude-difference monopulse radar

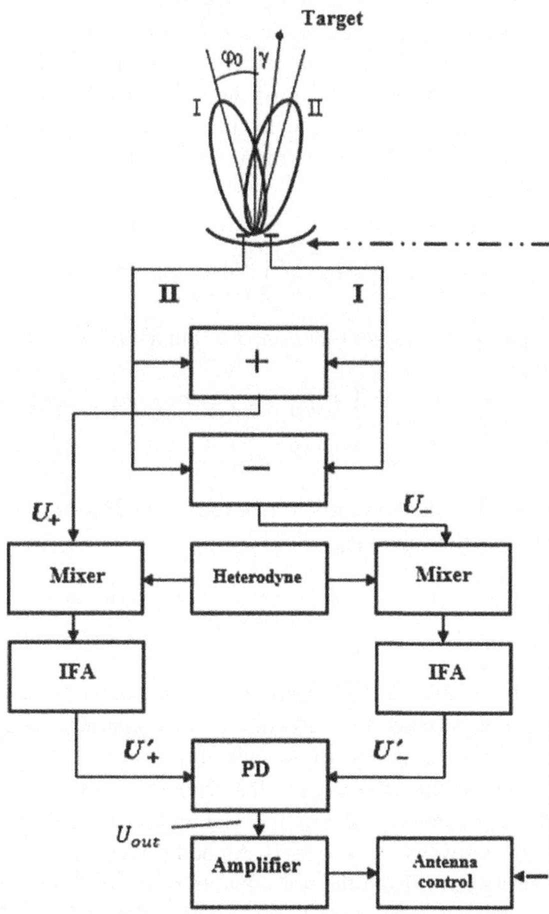

Fig. 10.4 Functional diagram of amplitude sum-and-difference monopulse radar

The signals from two antennas are fed simultaneously to high-frequency sum and equilibrium devices, which represent the waveguide bridges and performing integration and subtract of signals directly at a high frequency:

$$\begin{cases} U_+ = k\big[F(\varphi_0+\gamma) + F(\varphi_0-\gamma)\big]\sin\omega t, \\ U_- = k\big[F(\varphi_0+\gamma) - F(\varphi_0-\gamma)\big]\sin\omega t. \end{cases} \tag{10.5}$$

After conversion and amplification of signals in the IF amplifier

$$\begin{cases} \acute{U}_+ = k\big[F(\varphi_0+\gamma) + F(\varphi_0-\gamma)\big]\sin(\omega_0 t + \psi_1), \\ \acute{U}_- = k\big[F(\varphi_0+\gamma) - F(\varphi_0-\gamma)\big]\sin(\omega_0 t + \psi_2). \end{cases} \tag{10.6}$$

where ψ_1 and ψ_2—are phase shifts in channels.

The phase detector, to which the voltage of the sum channel is supplied as a reference, and the voltage of the difference channel as a signal, performs voltage multiplication. In this case, the phase detector filter removes high-frequency combinational components, and the voltage of the difference frequency is released at the output:

$$U_{\text{out}} = k_{\text{pd}}\frac{k_1k_2}{2}\big[F^2(\varphi_0+\gamma) - F^2(\varphi_0-\gamma)\big]\cos(\psi_1-\psi_2) \tag{10.7}$$

Expansion of the functions $F(\varphi)$ in a Taylor series near the point $\varphi = \varphi_0$ gives:

$$\begin{cases} F(\varphi_0+\gamma) = F(\varphi_0) + \gamma\frac{dF(\varphi)}{d\varphi}|_{\varphi=\varphi_0} + \ldots, \\ F(\varphi_0-\gamma) = F(\varphi_0) - \gamma\frac{dF(\varphi)}{d\varphi}|_{\varphi=\varphi_0} + \ldots. \end{cases} \tag{10.8}$$

Restricting ourselves to two terms of the series due to the small mismatch of γ, we have the following:

$$U_{\text{out}} = 2k_{\text{pd}}k_1k_2F(\varphi_0)\gamma\frac{\mathrm{d}F(\varphi)}{\mathrm{d}\varphi}|_{\varphi=\varphi_0}\cos(\varphi_1-\varphi_2) \tag{10.9}$$

In contrast to the direction-finding (boresight) characteristic of the amplitude-difference radar, the direction-finding characteristic of the amplitude summed-difference radar has a boresight direction that does not depend on the amplitude and phase characteristics of the channels. This means that, despite the non-identity of the channels ($\varkappa_1 \neq \varkappa_2$; $\varphi_1 \neq \varphi_2$), in the absence of mismatch ($\gamma = 0$), the output voltage always turns to zero. An additional summary channel eliminates the target bearing error. Both the non-identity of the channels and the amplitude fluctuations of reflected signal lead to a change in the slope of the resulting direction-finding characteristic (10.9), but do not influence on stability of the zero boresight.

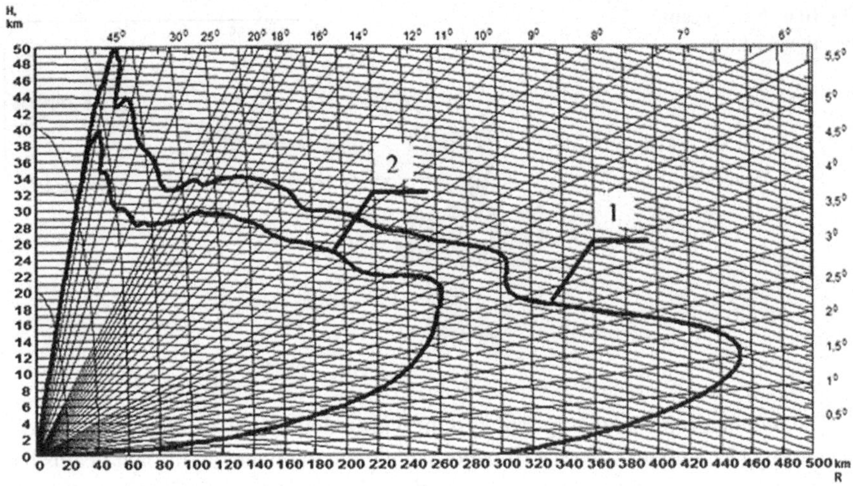

Fig. 10.5 Dual-frequency radar DP

The implementation of frequency-multi-channel radars pursues two main goals: improving target detection performance and expanding functionality by measuring the target elevation angle.

The main features of the construction of a primary pulsed surveillance radar are the use of two transmit–receive channels with a frequency spacing and the use of two-beam antenna DP in the vertical plane (Fig. 10.5).

The first feature of the radar is associated with the use of one of the methods of increasing its energy potential—the frequency separation (offset) method. Two transmitters A and B operate simultaneously for a common antenna in the pulse modulation mode with different carrier frequencies f_A and f_B of sounding pulses, the time shift between which is usually 4 … 6 μs, the frequency separation is 40 … 60 MHz.

Signals reflected from a target with different frequencies are separated using microwave filters and amplified by two receiving channels A and B, tuned to the corresponding frequencies. After detection, the video signals of channels A and B are combined and further are jointly processed.

The advantages of two-frequency radar design over single-frequency radar are: an increase in total radiation power of the radar at the presence of power limitations of a separate transmitter; increasing the detection range and accuracy of coordinates measurement; increasing the reliability of the radar and its noise immunity in relation to interference of artificial and natural origin.

The increase in the detection range and the accuracy of coordinates measurement is explained by the fact that the re-reflection diagram of complex targets at different frequencies has gaps at different viewing angles (angles of sight). Therefore, the sum of output voltages in two-channel radar has significantly less amplitude fluctuations than in case of signals receiving from targets at the same frequency.

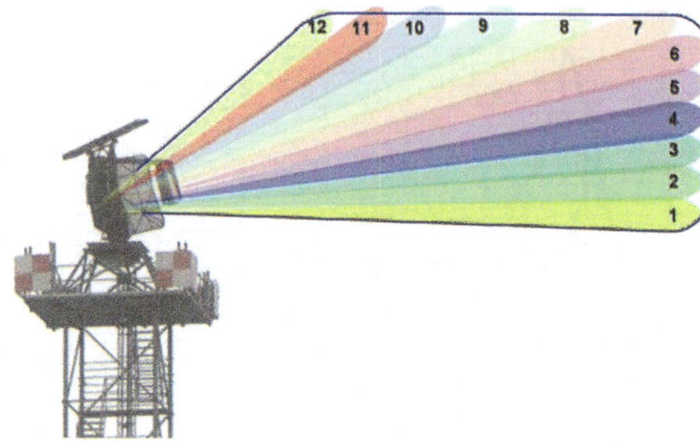

Fig. 10.6 Multi-beam directional pattern

The antenna represents a mirror system with a double curvature reflector and a feed horn. The antenna directional pattern along the main and additional beams has a shape close to that of the cosec^2 type. In the azimuthal plane, the ADP is narrowly directed. To form a two-beam ADP in antenna device and in waveguide-coaxial path, the identical horn feeds of the main and additional channels are used.

When using the multi-frequency method of partial antenna DP, the antenna system forms in the elevation plane a set of narrow antenna DP having different tilt angles ε_n, thereby overlapping the entire viewing sector in the elevation plane. Each DP corresponds to its own frequency channel. Due to the rotation of the antenna system, a spatial navigation problem is solved in the coordinate system: range—azimuth—elevation angle. The presence of reflected signal in a specific frequency channel corresponds to the current angular position of a target and can be considered as a height (altitude) value (Fig. 10.6).

The construction particularity of such radar is the formation of a multi-frequency signal by emitting several signals with different carrier frequencies simultaneously or with a time shift of a carrier frequency of sounding signal according to a certain law. The formation of such a complex directional pattern is achieved by constructing a reflective antenna system with several feeds in the form of horns, offset relative to each other in the vertical plane (Fig. 10.7). The number of horns is determined by the requirement to radar resolution in the elevation plane.

The structure of a modern radar, which implements the partial DP method, as well as the use of different types of sounding signals in the form of a monochrome sequence of short pulses intended for operation in the near field and chirp signals that ensure the detection of targets at large distances is shown in Fig. 10.9. In the presented version, 32 frequency channels are implemented that ensure the measurement of elevation angle of an aerial target in coverage area from 0° to 45° by 32 DP beams, with a width of each in the vertical plane $\theta_{v0.5} = 2°$ and a frequency offset between adjacent channels of 4 MHz.

The main structural elements of the circuit are: a transmitter, a receiver and an antenna-feeder system (AFS). A feature of the construction is the possibility of forming the sounding signals of different structure and parameters, and therefore, the transmitting unit consists of two channels. The monochrome pulse (MP)

Fig. 10.7 Antenna system of multi-frequency radar

channel is intended for AV coordinates detection and finding in the near zone of radar control. This channel forms 32 reference frequencies by multiplying the highly stable harmonic signal of the first coherent local oscillator by 32 coefficients, where each coefficient corresponds to its own DP beam. Subsequently, these signals are transferred to the frequency of the second coherent local oscillator, which is a high-frequency generator, due to the mixer unit. Pulse modulation is applied to each of the resulting oscillations, the parameters of which, namely the duration and time origin of probing pulse, are set by the digital control unit. The formed 32 high-frequency radio pulses are fed to inputs of power amplifiers.

To detect and determine the AV coordinates in the far zone, the chirp channel is used. A feature of this channel is the presence of a block for digital synthesis of chirp signal. This unit generates a chirp signal with the ability to change its duration, frequency and time of radiation start using control signals from the digital control unit. The subsequent processes of transfer and amplification of signal are similar to MP channel; however, the digital synthesis unit performs the function of the modulator in the chirp signal channel.

A circulator is used in circuit as an adder for 32 MP channels and 32 chirp signal channels. To clarify the operation principle of the circuit, the timing diagram, corresponding to signal at the output of the circulator for one frequency channel is shown in Fig. 10.8.

The signals reflected from the objects, received by the antenna system, come through the feeder path to the input of low-noise amplifier, where they are initially amplified and filtered. The timing diagram of these signals is shown in Fig. 10.10.

To decompose the received signal into 32 frequency channels, a third mixer unit is used, to the outputs of which 32 intermediate-frequency amplifiers are connected, which perform the role of the main process of selectivity, amplification and filtering. For further signals processing in digital form, a multi-channel analog-to-digital converter is used. The received digital signal comes to the digital processing unit,

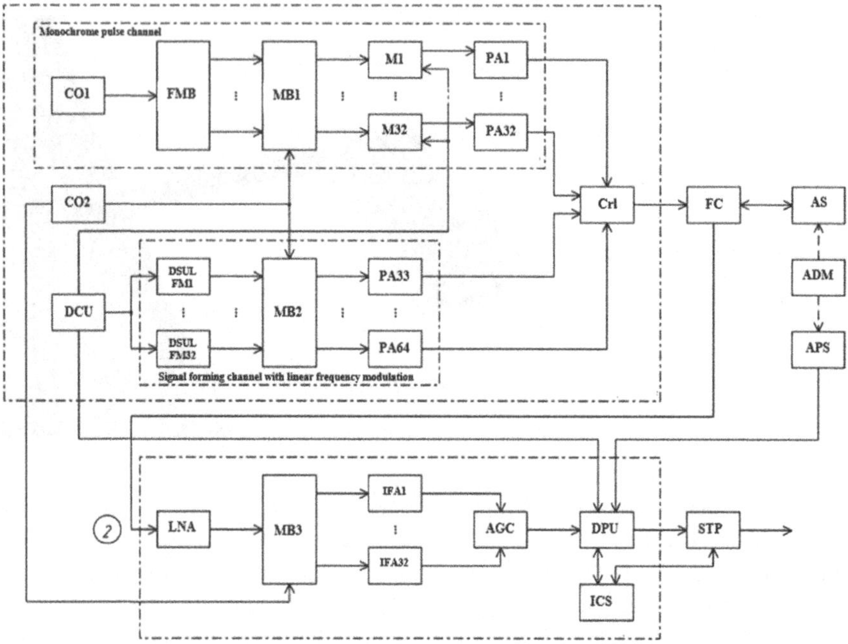

Fig. 10.8 Block diagram of a multi-frequency surveillance radar. CO—coherent oscillator; FMB—frequency multiplier block; MB—mixers block; M—modulator; PA—power amplifier; DCU—digital control unit; DSULFM—digital synthesis unit of LFM signal; Crl—circulator; FC—feeder circuit (path); AS—antenna system; ADM—antenna drive mechanism; APS—angular position sensor; LNA—low-noise amplifier; IFA—intermediate-frequency amplifier; ADC—analog-to-digital converter; DPU—digital processing unit; STP—signal transmission path; ICS—integrated control system

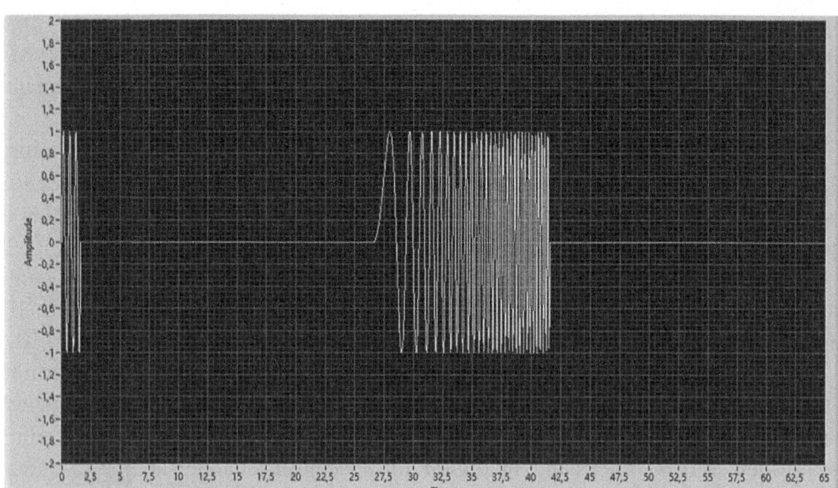

Fig. 10.9 Timing diagram of signal from circulator output

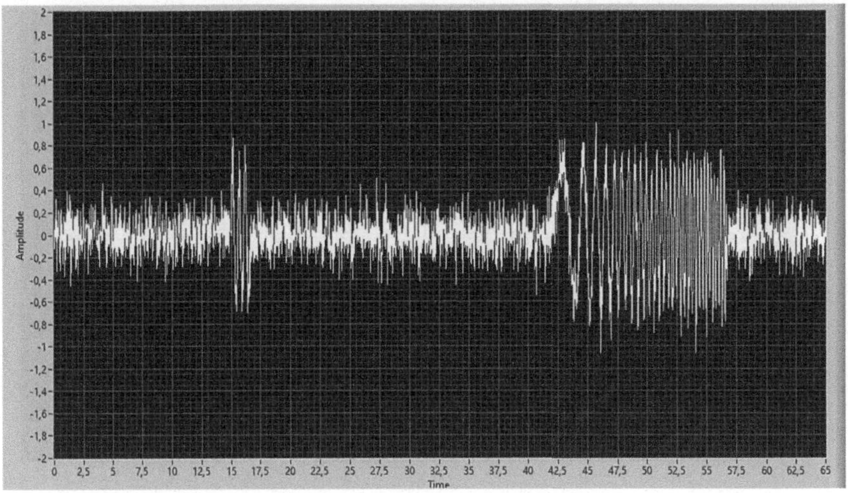

Fig. 10.10 Signal timing diagram at LNA input

where the main signal processing is carried out, namely decision making on detection, solving the spatial problem of determining coordinates, forming an adaptive detection threshold, discrimination of moving targets, tracking the target. To solve the spatial problem of determining the aerial target coordinates at the DPU: To calculate the slant range to the target, a time method is used, to calculate the azimuth, the DPU receives a signal from the angular position sensor of the antenna system, which carries information on rotation angle of the antenna system, and determining the elevation angle is reduced to fixing the channel, on which the target was detected. The control of parameters and state of the radar operation is carried out by the built-in control system. The timing diagram, corresponding to the signal after processing at DPU, is shown in Fig. 10.11.

Figure 10.12 shows the timing diagrams of a real sounding signal of an en route radar, received at a remote point, and they represent a sequence of 8 monochrome, short-duration pulses intended for targets detection in the near field and a sequence of 4 pulses with LFM to detect targets in the far zone with great energy potential.

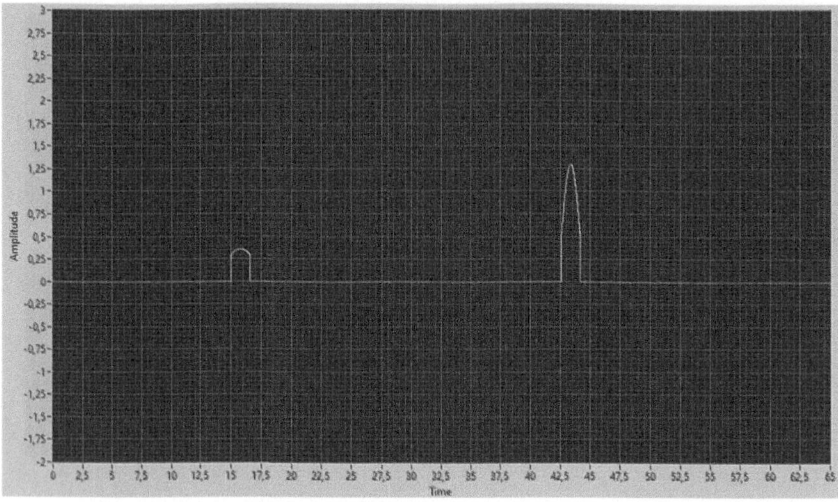

Fig. 10.11 Signal timing diagram with DPU

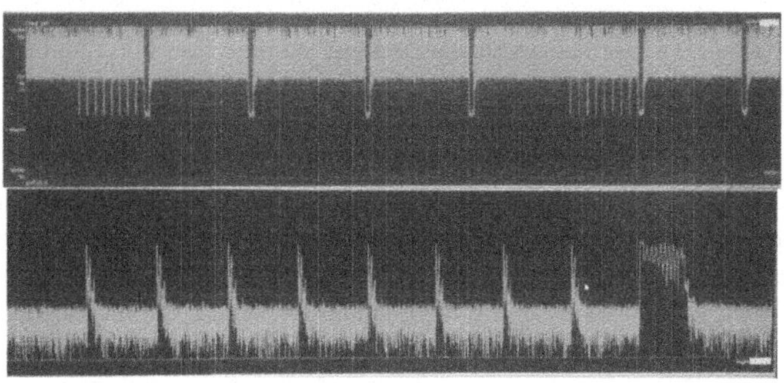

Fig. 10.12 Timing diagrams of radar sounding signal

Chapter 11
Radar Systems of Air Transport

Nowadays, the radar systems of almost all types are widely used in air transport.

None-coherent radar systems have found its application as onboard surveillance systems of aerial vehicles. Let us examine the none-coherent radar systems on example of onboard weather-navigational radar stations (WNRS).

WNRS is intended for the following tasks solution:

- radar surveillance of air space (in horizontal and vertical plane) in order to detect moisture targets and areas in it, dangerous for flights (Fig. 11.1);
- radar surveillance of ground and water surfaces for aircraft navigation by specific ground and water references (Fig. 11.2);
- detection of slant distance and course angles of observed radar reference points (RRP) and moisture targets.

WNRS provides: detection of convective moisture targets (thunders, thick cumulus/clouds) with possibility to detect its dangerous degree for AV flight and dangerous turbulence in moisture targets; detection of distinctive ground reference points like large towns, coast line of large water basins, large sea vessels; built-in operational check for failure detection (trouble shooting) of each block including communications lines.

Such systems have the following main operation modes: “Meteo” and “Ground.”

“Meteo” mode is a basic mode of operation. In this mode, the radar provides on-screen displaying of meteorological situation radar image (Fig. 2.6) in polar coordinates “azimuth-range” in air space, limited by azimuthal angles $\pm 60°$ (or $\pm 45°$) relatively to construction line of aerial vehicle, and elevation angle, defined relatively to horizontal plane by tilt of antenna system. Moisture targets depending on its danger level for aerial vehicle flight are displayed in variously colors.

“Ground” mode is intended for navigational orientation by distinctive ground objects. In this mode, the WNRS provides radar image of ground surface acquisition

D. A. Akmaykin et al., *Theoretical Foundations of Radar Location and Radio Navigation*, Springer Aerospace Technology,
https://doi.org/10.1007/978-981-33-6514-8_11

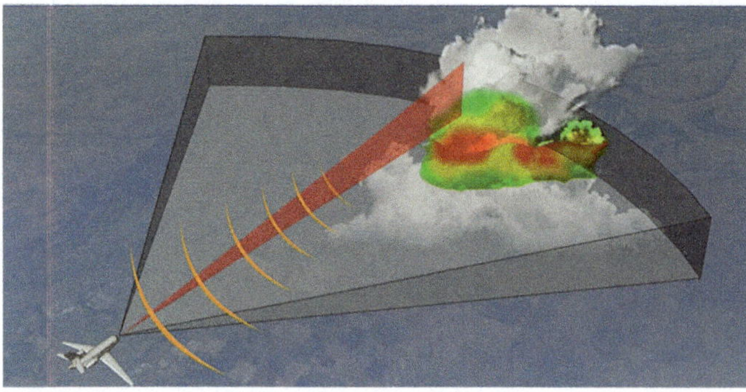

Fig. 11.1. Moisture targets detection

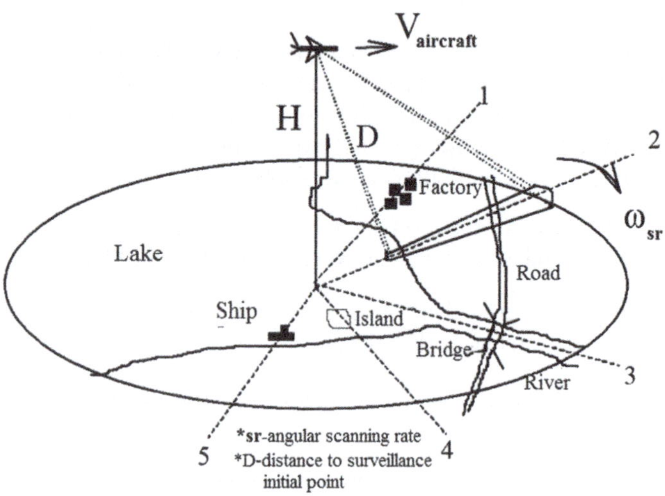

Fig. 11.2. Surveillance of ground and water surface

on multi-function display (MFD). Reflected signals, coming from different zones of ground surface and above ground installations (AGI), are displayed on screen in variously colors (Fig. 11.4):

- green—ground surface background;
- red—above ground installations (radar-contrasting images);
- black—water basins against background of ground surface or shadow area.

As an example, the Russian "Buran" WNRS of An-148 aircraft is represented in Fig. 11.5. WNRS consists of three structurally packaged blocks: antenna (1), transmitter–receiver (2), multi-function display (3). (Fig 11.3)

Waveguide antenna is mounted under radio transparent radome in fuselage nose section (Fig. 11.6) or in special container. Transceiver is interfaced with antenna via

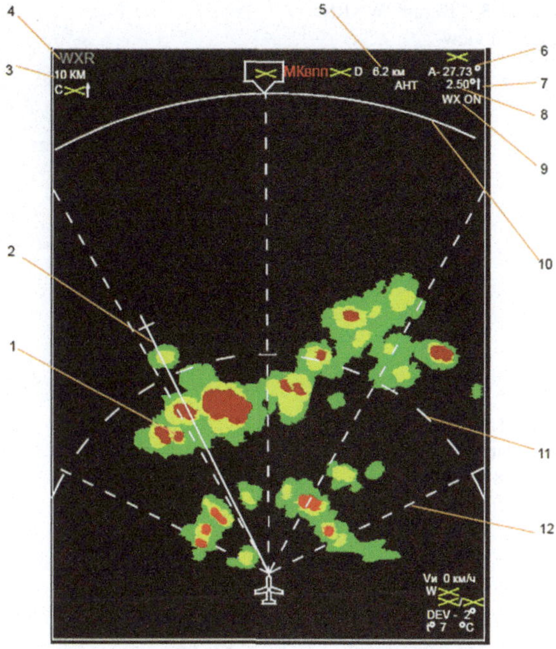

Fig. 11.3 Situation representation on MFD in "Meteo" mode (An-148 aircraft). 1—radar image of moisture targets, 2—sight line (cursor) image, 3—image scale, 4—WNRS operation mode, 5—cursor coordinate in range (in km), 6—cursor coordinate in azimuth (in deg.), 7—antenna inclination direction, 8—antenna tilt (in deg.), 9—UHF is on, 10—scale distance mark, 11—half-scale distance mark, 12—azimuth calibration marker

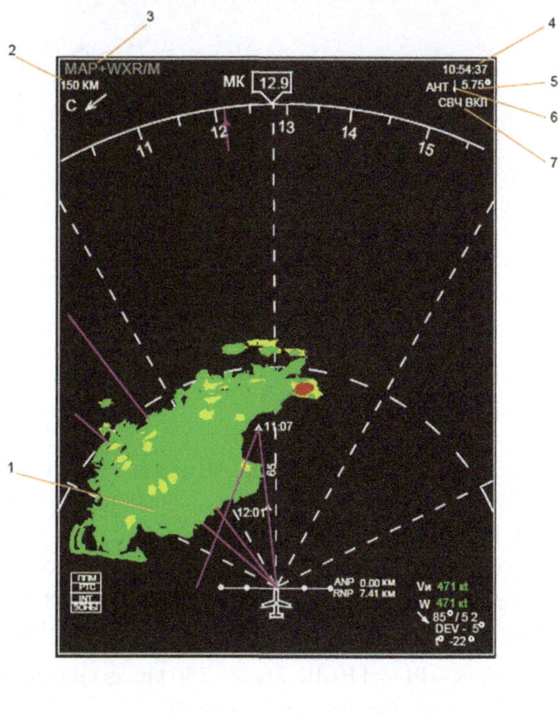

Fig. 11.4 Situation representation on MFD in "Ground" mode (An-148 aircraft). 1—moisture targets, 2—image scale, 3—WNRS operation mode, 4—current time, 5—antenna tilt (in deg.), 6—antenna inclination direction, 7—UHF is on

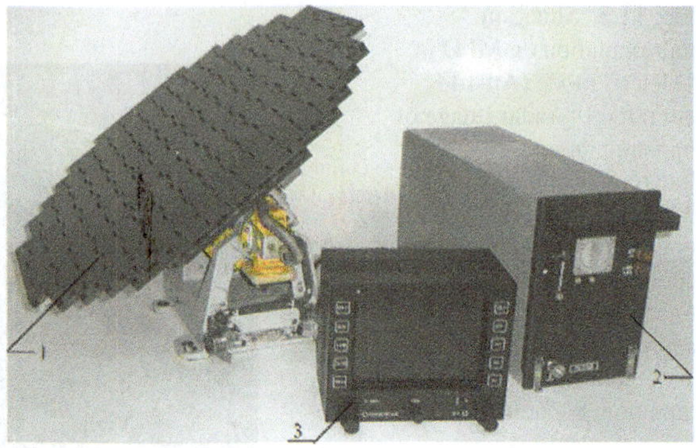

Fig. 11.5 "Buran" WNRS external view

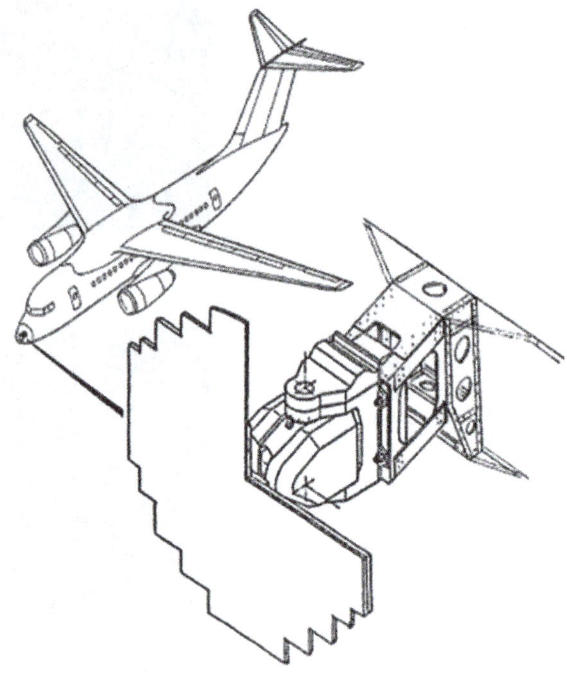

Fig. 11.6 WNRS antenna installation location onboard An-148 aircraft

waveguide communication line and can be located in fuselage in pressurized area or outside it in close proximity from an antenna. MFD is installed in specially allotted place in crew cabin.

Classification of WNRS is based on differences of maximum range (D_{max}) of energy (power) potential indicator (PI) of radar locator in detection mode of hydro-meteorological targets. WNRS is divided into the following classes:

I class—PI — 119 dB, D_{max}—550 km, is intended for installation on supersonic, long- and medium-range aircraft;

Fig. 11.7 WNRS mirror antenna

II class—PI — 111 dB, $D_{\max}$—350 km—small-range and heavy aircraft of local airlines (LAL);

III class—PI — 101 dB, $D_{\max}$—100 km—LAL light airplanes and helicopters;

IV class—PI—90 dB, $D_{\max}$—100 km—helicopters.

WNRS can be designed in a form of mirror antenna and passive phased antenna array (PPAA).

Mirror antenna represents a paraboloid of revolution with 200–1160 mm diameter (depending on AV type), in focal plane of which an exciter (feed element) is positioned (Fig. 11.7).

Application of mirror antenna provides an acquisition of two type directional patterns, depending on selected WNRS operation mode—"Meteo" or "Ground."

In "Ground" mode for correctness of situation (environment) representation on display, it is necessary that similar objects give out the same intensifier brightness, independently on distance difference to each of them. Herewith, in accordance with fundamental equation of radar operational range, the following condition should be fulfilled:

$$P_{\mathrm{rcv}} \leq \frac{G^2(\theta)}{R^4} = \mathrm{const} \tag{11.1}$$

where P_{rcv}—power of reflected signals at receiver input; $G(\theta)$ – gain coefficient (in power) of radar antenna in vertical plane; R—range to objects.

For fulfillment of this condition, it is necessary to select a special law of directional pattern change from current angle θ_i. At constant height of flight H

$$R = \frac{H}{\sin\theta} = H\operatorname{cosec}\theta. \tag{11.2}$$

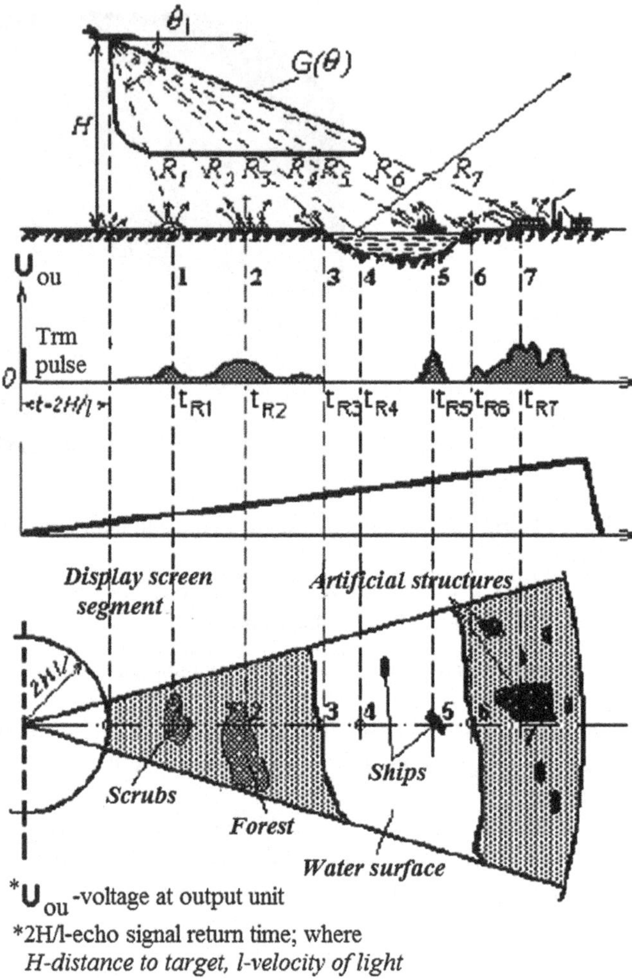

Fig. 11.8 Principle of radar image acquisition on display in "Ground" mode

As it follows from condition $G^2(\theta) \leq R^4 \leq \text{cosec}^4\theta$, whence $G(\theta) \leq \text{cosec}^2\theta$, is antenna gain coefficient in vertical plane should be changed according to a law $\text{cosec}^2\theta$. In Fig. 11.8, an approximate shape of such pattern, which is called "cosecant," is depicted.

Twin antenna reflector consists of symmetrical parabolic reflector and reflector of special form (phase-shaped) in a shape of "canopy" in the top part of paraboloid. At radiation by electromagnetic energy, parabolic reflector shapes a directional pattern in a form of narrow beam of "pencil-type." It is completely manufactured from metalized glass-fiber fabric. Reflecting surface profile of special form reflector is designed with reference to acquire a cosecant directional pattern in vertical plane. This reflector represents a doubly curved surface made of metalized glass-fiber fabric,

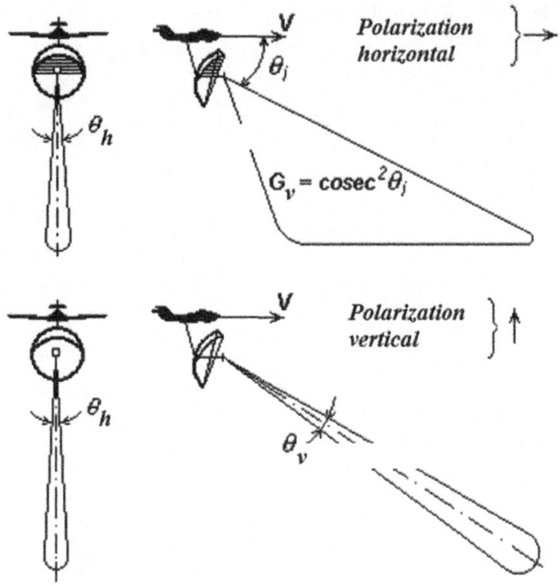

Fig. 11.9 Fan-shaped and pencil-beam directional patterns

the metalized filaments of which are positioned strictly horizontally and have a step of 3-mm. Both reflectors are rigidly connected between each other and mounted on metallic pressed bracket of hollow-shaped type. Special form reflector is located in front of symmetrical parabolic reflector and closes only its upper part by forming a kind of "canopy" of antenna reflector.

Shaping of corresponding directional pattern is carried out via change of polarization plane of radiated HF oscillations. At vertical polarization, a radiating electromagnetic energy easily comes through the special form reflector with horizontal positioning of metalized filaments and is reflected from symmetrical paraboloid. Herewith, a directional pattern is formed in a shape of narrow beam. At polarization change of radiating oscillations into horizontal polarization, a reflection, being non-transparent for this polarization, from special form reflector happens, and fan-shaped directional pattern of cosecant type is formed.

In Fig. 11.9, directional patterns in two planes for both antenna operation modes are depicted.

However, substantial disadvantage of WNRS mirror antennas is weight-size parameters, considerably reducing its application in commercial aviation.

The planar passive phased slotted antennas (Fig. 11.10) have gained a widespread, which besides weight-size parameters have a set of advantages: larger (at 1–1.5 dB) directivity factor (directional gain) and considerably lower side-lobes level than parabolic antennas.

Flight safety of aerial vehicles problem solution in a large degree depends on correct registration of external factors influence on flight outcome, to which, above all things, hazardous weather conditions (turbulence, storm precipitations/showers,

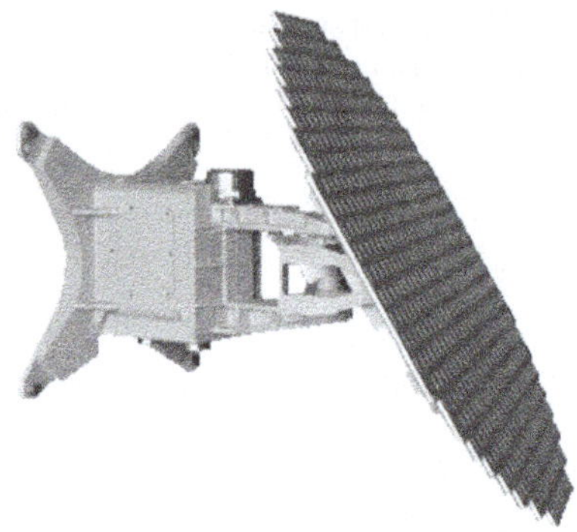

Fig. 11.10 WNRS phased antenna array

Fig. 11.11 AV entry into wind shear at landing

wind shear, meteorological conditions, deteriorating visibility, etc.) should be referred. Of prime importance is ensuring flights safety in conditions of dynamic environmental influence on moving trajectory of aerial vehicle, such influences are "wind shear" and turbulence.

Wind shear is an unexpected and rapid change of wind velocity and/or direction at a small distance in horizontal and vertical plane. Due to wind shear, an aircraft can unexpectedly get into upward (updraft) or downward air current, indicated airspeed can drastically increase or decrease; as a result, lift breakdown or abrupt change of vertical velocity and flight altitude can take place. The abovementioned can lead to severe deviation from assigned trajectory (flight-path error) and require an active interference of pilot to overcome the encountered problem.

Encounter a wind shear is a very dynamic event, arising so unexpectedly and rapidly that even experienced pilots on thrust-to-weight aircrafts can fail to get out of this difficulty smoothly. Entry into wind shear can lead to hazardous ground proximity, bungled (rough) landing or complete damage of AV (Fig. 11.11).

To detect a "wind shear" and turbulence, the special algorithms based on Doppler processing of received signal are used. In some modern WNRS, a "wind shear" and turbulence detection function is realized.

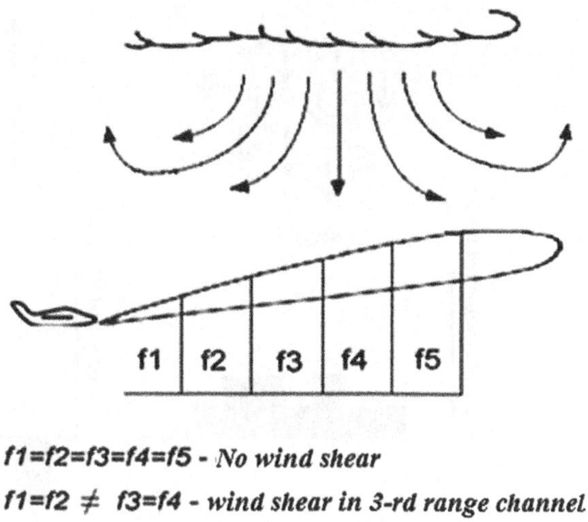

Fig. 11.12 Detection method of "wind shear"

The "wind shear" (WS) detection method is based on principle of Doppler frequencies allocation reflected from a signal of moving moisture targets; this phenomenon is directly connected with moving of air volumes in air space, containing microparticles. The purpose consists in measuring in each element of resolution in range (in range channel) of Doppler frequency and comparing of measured values in neighboring elements. If values of Doppler frequencies are differed, that corresponds to different moving speed of air masses in air space, this testifies on WS at distance, corresponding to a channel number, and difference magnitude of Doppler frequencies—on WS value (Fig. 11.12).

The turbulence detection method is based on Doppler frequencies allocation from a reflected (from moving moisture targets) signal. This phenomenon is directly connected with moving of air volumes in air space, containing microparticles. The purpose consists in measuring in each element of Doppler frequency resolution and determination of a distribution law nature of obtained values. If distribution character is a random process, then a decision on turbulence presence in this resolution element is taken, where a distribution character has a normal law of distribution. Normal law of distribution is characterized by zero mathematical expectation, and maximum spread in values is in confidence interval 3σ.

If in resolution element a phenomenon of turbulence is absent, then accumulation function will have an increasing character with further exceeding of threshold value (Fig. 11.13). At presence of turbulence, the threshold will not be exceeded. Threshold value is defined according to a rule 3σ, where a threshold can have several values, which characterize a value of turbulence current (flow).

WNRS with wind shear and turbulence detection functions is referred to pseudo-coherent radars. An example of such WNRS is a system installed on medium-range aircraft. In Fig. 11.14, an example of information indication on wind shear presence in the direction of an aircraft flight is presented.

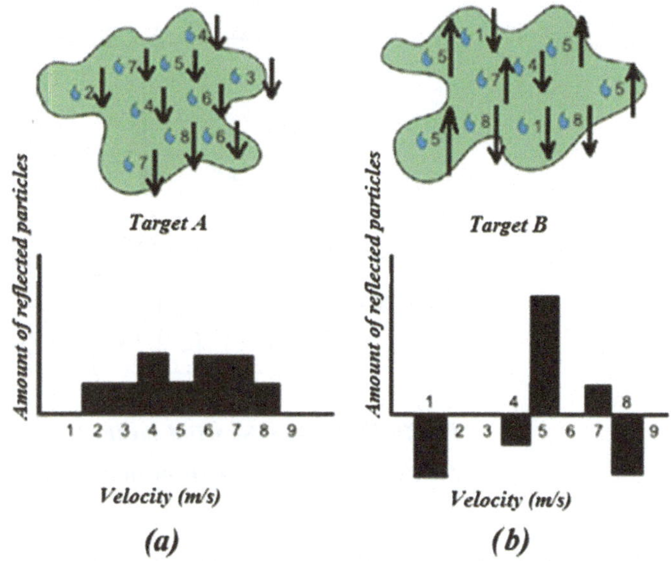

Fig. 11.13 Turbulence absence (**a**) and its presence (**b**)

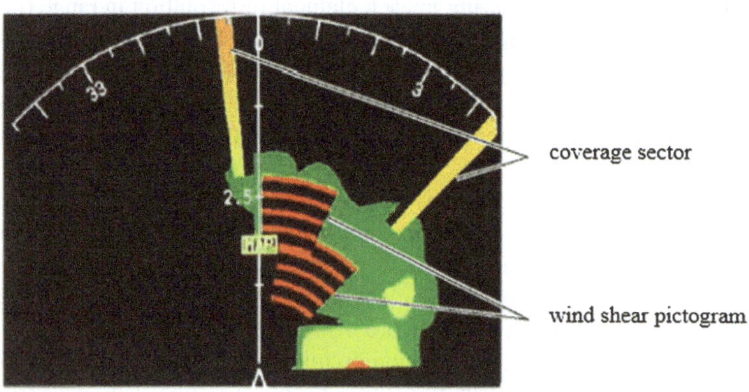

Fig. 11.14 Example of pictogram indication of wind shear on MFD

The mapping mode, implemented in modern WNRS, provides the crew with information on terrain features. Distant coastlines, islands, lakes, small towns, industrial complexes, large rivers and bodies of water, urban areas, open land have varying effective reflective surfaces up to 590 km (320 nm) at altitudes up to 18,300 m (60,000 ft). When the WNRS operates in the mapping mode on the MFD, a continuous radar map of the earth's surface is obtained in the "azimuth-range" format at a distance of up to 590 km (320 nm) and at altitudes up to 18,300 m (60,000 ft) Ahead of an Aircraft. The WNRS, Which Implements the Mapping Mode, is a Truly Coherent Radar.

The process of WNRS operating in flight requires the aircrew to have specific skills in analyzing radar data displayed on indicator units, since radar information is a complex representation of data that require special training, which is different from the perceptions of the world that are familiar to humans.

Among the flight support radio-technical aids, radars occupy a special place, since they are the main sources of dynamic information about the air situation for air traffic controllers. The most commonly the radars are classified by the purpose [42]:

- route surveillance radar locator (SRL-R);
- aerodrome surveillance radar locator (SRL-A);
- secondary radar locator (SRL);
- landing (approach) radar locator (LRL);
- airfield control radar (ACR);
- meteorological radar locator (MRL).

Route surveillance radar locator (SRL-R) is designed to detect and determine the coordinates (azimuth-range) of aerial vehicles in the off-aerodrome area (on air routes and off-routes) with the subsequent transmission of information on air situation to air traffic management (ATM) centers (points) for control and support air traffic control.

The SRL-R permits to perform the following tasks:

- AV detection and positioning;
- AV crews control in specified corridors-keeping and the time of passing the checkpoints on the route;
- prevent dangerous approaches of AV;
- detect the location of moisture targets, dangerous for flights;
- AV identification and receiving of additional data on AV by using the built-in secondary channels.

The SRL-R antenna system is adjusted relatively to the true meridian. The information update period is not more than ten seconds.

It is recommended that the SRL-R be positioned in such a way as to ensure the overlap of the air routes of the given area by the radar coverage area at a height from the lower to the upper echelons of the controlled airspace.

The main requirement for route SRL is to provide a long operational range with a sufficiently good accuracy and resolution. If the route SRL is designed to provide information to automated ATC centers, a secondary channel is often built into it. Route surveillance radar is usually an S-band radar. Figure 11.15 shows the SRL-R exterior view at the position and the antenna in the radar dome.

Aerodrome surveillance radar locator (SRL-A) is designed to detect and determine the coordinates (azimuth-range) of AV in the aerodrome area with subsequent transmission of information on the air situation to ATM centers (points) for the purpose of monitoring and ensuring an air traffic control. The information update period is no more than six seconds. The information obtained by the SRL-A is used by controllers of aerodrome centers of automatic flight control system AFCS, controllers of approach points (DAPP), the main dispatching approach points (MDAP), the

Fig. 11.15 SRL-R on-site and its antenna in radar dome

dispatching points of the circle (DPC), the dispatching points of landing system (DPLS) and the local dispatching points (LDP).

The technical characteristics of aerodrome SRL must ensure the resolution and accuracy of AV position finding in accordance with ICAO and ATC standards. In addition, they must have suppression systems of signals, reflected from ground features and hydrometeors. Aerodrome SRL must detect and determine the position of targets at low altitudes and at a close distance from the radar. Requirements for the maximum range of aerodrome SRL are differentiated depending on the specific purpose and class of the airport.

Modern SRL-A operate in the S-band and are highly stable radar systems built on the principle of internal signal coherence.

It is recommended that the SRL-A be positioned in such a way as to provide a continuous radar scan of the controlled airspace in the terminal area. The absence of radar information from SRL-A is allowed in 3–5 consecutive surveys from an aircraft making a turning maneuver or located in a section with a tangential direction of speed. The SRL-A antenna system is adjusted relative to the magnetic meridian.

Radar information of SRL-A can be used for monitoring and controlling of an air traffic in the off-airfield area (on air route and off-routes) in regional air traffic control centers. In this case, the coordinate information (azimuth) is intended for the district center is recalculated relative to the true meridian by SRL-A processing equipment or by air traffic control equipment of the district center or other special equipment. The SRL-A on-site is represented in Fig. 11.16.

Secondary radar locator (SCRL) is designed to detect, position finding (azimuth-range), request and receive additional information from AV, equipped with transponders, with the subsequent information output to ATM centers (points). SCRL, intended for support of AV flights on air routes and off-routes, should have an information update period of no more than ten seconds, and in the aerodrome zone—no more than six seconds.

SCRL is divided into autonomous and built-in ones according to the construction principle. By interaction performance with onboard transponders, SCRL is divided into radars with a general and discrete-address request; according to the coding

Fig. 11.16 SRL-A on-site

Fig. 11.17 SCRL on-site

system—to meet Russian standards (ATC mode) and ICAO standards (RBS mode). Modern SCRL operates in combined with primary radars mode.

Figure 11.17 shows an example of SCRL on-site.

Landing radar locator (LRL) is designed to control the AV keeping to a given course line and glide path on the last final approach leg, as well as to control the landing by giving commands to aircrew to correct their descent trajectory. Landing radars can be used either as an autonomous means of landing, or as monitoring system of AV landing at airports, equipped with radio beacon landing systems. In the first case, the flight dispatcher completely controls the approach; in the second case, he only controls the approach and, if necessary, informs the crew on AV position relative to the course line and glide path.

The need to use a landing radar at airports is stipulated by a number of its advantages over radio beacon landing systems:

- its functioning does not depend on onboard equipment;
- allows you to continuously observe evolutions of AV trajectory from the ground up to its landing;
- provides observation from the ground for all AV in the landing zone, and the risk of AV collision in this case is minimized;
- provides satisfactory control of in-series AV landings with small intervals in distance in cases where weather deteriorating, fuel shortage or damage to AV necessitate a safe and fast landing;
- the radar accuracy depends little on changes in the weather, changes in the snow cover of the earth's surface, terrain and other factors;
- LRL equipped with pan-and-tilt system can provide landing from any direction, including those not equipped with a radio beacon system.

The LRL is positioned at the aerodrome and is adjusted to provide a sector scan, which begins at a point located at a distance of 150 m from the touchdown point in the direction of landing. The azimuth angle of this sector should be $\pm 5^\circ$ relative to the centerline of the runway (hereinafter referred to as RNY), and the elevation angle should be from -1° up to $+6^\circ$.

In the presence of the LRL and the AV radio beacon system of instrument landing approach, the course line and glide path of these systems must coincide in segment from the entry point into the glide path to the inner locator with radio marker beacon or 1000 m from the runway threshold.

Airfield control radar (ACR) is designed to monitor and control of AV traffic, special-purpose vehicles, technical equipment and other objects located on aerodrome movement area (maneuvering area and apron, runways, taxiways and aircraft parking areas). Information from ACR radar is used by the taxiing dispatchers of the taxiing dispatch points (TDP) and the runway controller of the runway supervisory points (RSP).

For aerodromes with precision-approach runways of II, III ICAO categories, the ACR radar is a mandatory equipment. The main requirement for ACR radar is to obtain the highest possible resolution of the radar image of the airfield and the objects on it in any weather conditions. Radar locators, operating in the millimeter wavelength range, meet these requirements in the best way (Fig 11.18).

Meteorological radar locator (MRL) is designed to detect and determine the location of seat of origin of thunderstorms and heavy rainfall centers, as well as their speed and direction of movement. MRL assists flight controllers in ensuring flight safety in complex meteorological conditions. MRL is also used to measure the parameters of wind shear in wind anomalies by the Doppler effect. The requirements

Fig. 11.18 ACR on-site

for MRL are determined by the specifics of objects, with which these radars operate. The radar locator must indicate the position and determine the main parameters of moisture targets that are dangerous for aircraft flights.

Surveillance and landing radar locators (SLRL) simultaneously perform the functions of airfield and landing radars at small airports of local airlines.

The need for and the possibility of this type of radars is defined by the specific operating conditions of small local airline airports.

The particularities of radar support of the most of them are as follows:

- small airports of local airlines are served by AV with relatively low speeds and low altitudes;
- sometimes there is no stationary runway at local airline airports, and the aircraft landing course can be promptly changed within 360° depending on the wind direction and the surface condition of certain sections of the airfield;
- the staff of dispatchers of local dispatch centers is limited;
- to increase the profitability of local airline airports, the cost of radio-technical flight support equipment, as well as the cost of its operation, should be relatively small.

Chapter 12
Radar Systems of Maritime Transport

The widespread use of ship navigation radars ensures the navigation safety in low or limited visibility, allows you to obtain a fix by the known coastal and floating orientation points or using specially installed radar transponder beacons and provides an automation of maritime navigation processes.

By using the passive system radar, it is possible, for example, to distinguish the boundary between water and land. It is possible to determine the route of passing ships, since the wake current temperature could be higher than the water temperature.

Depending on the structure of emitted (sounding) radar signals, there are continuous-wave radars and pulse radars. Pulse radars are used to the vast majority on the civil fleet vessels.

Pulse radar contains the following main elements that are shown in the block diagram (Fig. 12.1).

The transmitter consists of two main units, a modulator and a microwave oscillator, which is used as a magnetron.

The modulator, under the influence of sync pulses, depending on the range scales, forms powerful triggering (actuating) pulses of a certain duration (in modern radars $\tau_{tp} = 0.07\ldots1.0\ \mu s$), under the influence of which the microwave oscillator generates powerful short radio pulses.

The antenna switch (polyplexer) provides switching of one antenna alternately to the transmitter and receiver, protects (blocks) the input circuits of the receiver from powerful sounding microwave pulses, both of its own transmitter and neighboring radars, and closes the output circuits of transmitter when receiving signals, reflected from targets.

Weak reflected pulses, received by the antenna, passing through the antenna switch, come to receiver, where they are converted in frequency, amplified and detected. The high sensitivity of receiver, capable of receiving short-term pulses is realized on the basis of a superheterodyne receiver with an intermediate frequency (usually 60 MHz). At this frequency, when using transistors or microcircuits, high gain and wide bandwidth can be obtained.

D. A. Akmaykin et al., *Theoretical Foundations of Radar Location and Radio Navigation*, Springer Aerospace Technology,
https://doi.org/10.1007/978-981-33-6514-8_12

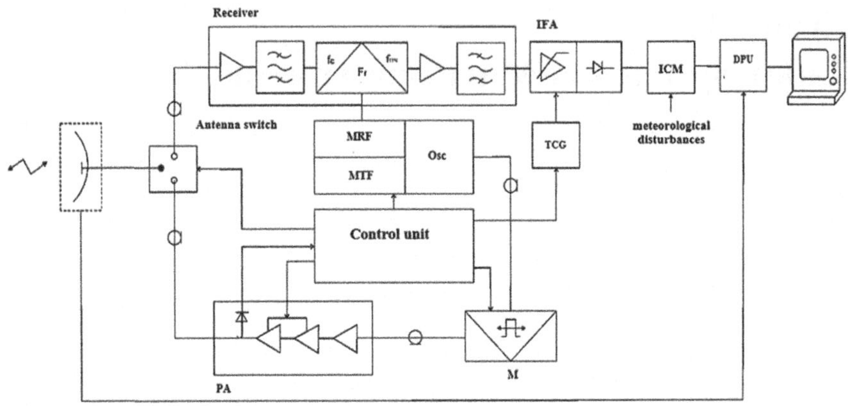

Fig. 12.1 Structural diagram of shipboard pulse radar. MRF—microwave receiver former; MTF—microwave transmitter former; Osc—oscillator; PA—power amplifier; IFA; TCG—time corrected gain; M—modulator; ICM—interference compensation module; meteorological disturbances; DPU—digital processing unit

Therefore, the reflected microwave pulses without preliminary amplification directly at the receiver input are converted into intermediate frequency pulses. The required in this case frequency converter, capable of operating at microwave frequencies, uses a klystron or a Gunn diode local oscillator and a diode crystal mixer, which can operate not only at centimeter, but also at millimeter waves. The mixer receives continuously generated oscillations of the local oscillator with a frequency f_H and reflected pulses from the antenna with an oscillation frequency equal to the frequency of the magnetron f_{osc}. As a result of two frequencies mixing, pulses of the intermediate difference frequency $f_{\text{IF}} = f_H - f_{\text{osc}}$ are generated, which receive the necessary amplification in the intermediate frequency amplifier (IFA), and then are fed to the detector, where they are converted into video pulses.

At the output of the receiver, due to detecting, video pulses are formed, mixed with noise (interference), which are fed to the control electrode of the cathode-ray tube (CRT) of the indicator, creating an amplitude or brightness mark on the screen, depending on the modulation method of the CRT's electron beam.

The main technical characteristics of the shipboard radar are: wavelength (carrier frequency); pulse repetition or repetition rate; transmitter power; the sensitivity and bandwidth of the receiver, the shape of the antenna directional pattern in the vertical and horizontal planes; the suppression degree of side lobes; behavior of field polarization; method and rate of airspace survey; type of terminal unit (indicator); overall dimensions and weight of the station; type of power supply and power consumption. Technical characteristics, or parameters, are selected based on the requirements of the radar operational characteristics.

The wavelength of the shipboard navigation radar is selected based on the task: to ensure the detection of both large and small surface and ground objects within a given radius of radar range; ensure the radar operation with short pulses; to obtain high directivity of antenna in horizontal plane without excessively increasing the antenna size.

Effective reflection of energy from objects is possible only when objects dimensions and the radii of curvature of individual sections are at many times greater than the wavelength. In this case, the reflection intensity reaches a noticeable value, depends little on the wavelength and is mainly determined by the reflecting properties and object dimensions. Taking into account the size of surface objects (buoys, milestones, boats, etc.), only the ultrashort-wave (VHF) range, more precisely, the short-wave section of the VHF range, is suitable for their successful detection using the shipboard radar.

The radar wavelength affects the detection range of low-lying (low-set) surface objects and the maximum range of radar detection. With a decrease in the wavelength, the detection range of low-lying surface objects increases, but at the same time the maximum operational range of the radar decreases due to attenuation associated with an increase in the absorption of electromagnetic energy in the troposphere.

Based on the abovementioned, the shipboard navigation radar uses the centimeter range of radio waves (S-band). Moreover, the standard wavelengths in this range are 3.2 and 10 cm.

To ensure a high potential azimuth resolution and increase the potential accuracy of determining directions, the beam width of shipboard navigation radars should be 1–0.25°.

The pulse-recurrence rate or pulse repetition rate is selected based on the problem of unambiguously determining of range and object detection efficiency in conditions of circular scan. Shipboard navigation radars usually operate on different range scales, the operating frequency $\mathcal{F}_p$ pulse repetition may vary. For example, on short-range scales, frequencies of $1000 \div 3200$ pulses/s are used, on long-range scales, $400 \div 800$ pulses/s.

The transmitter power has an effect on the radar operational range. The distinction is made between pulsed and average powers. Pulse is the average value of power over the time τ_p of pulse duration. The average is the power over the period T_p of pulse repetition.

Since an increase in the average power is connected with an increase in the repetition rate or pulse duration, then in order to maintain the constancy of thermal regime of magnetron generator, it is recommended to leave the average radar power as unchanged as possible at different pulse-recurrence rate or different pulse durations. This is achieved by the fact that when the radar switches a mode to operate with shorter pulses, the repetition rate or pulse sending (pulsing) increases and vice versa.

To ensure a normal operation of the shipboard radar, the side-lobe level should be at 20–30 dB below the main lobe level.

The behavior of electromagnetic field polarization is usually the shipboard radar antenna used the linear horizontal or vertical polarization of the field, at which the direction of E vector is vertical or horizontal.

Table 12.1 Typical characteristics of shipboard navigation radars

Parameter	Parameter value	
Wavelength, cm	10	3.2
Antenna aperture width, m	3.3–3.6	0.9–3.3
Antenna beam width, deg.		
In vertical plane	18–22	17–24
In horizontal plane	1.8–2.3	0.6–1.8
Sidelobe attenuation, dB	23–26	21–28
Antenna rotation frequency, rpm	12–20	15–25
Pulse power, kW	30–80	6–80
Pulse duration, μs	0.1–1	0.05–1
Pulse sending frequency, imp/s	850–4000	625–4000
Receiver bandwidth, MHz	6–12	6–25
Maximum range on the indicator scale, miles	50–64	12–64
Minimum range scale, miles	0.5–1	0.25–1
Screen diameter, cm	42–45	12–45
Receiver sensitivity, dB/W	110–120	110–120

Modern shipboard navigation radars operate in pulsed mode, the most important advantage of which is the separation simplicity of direct (main) and reflected pulses.

Ship navigation radar stations, depending on the configuration, can be dual-band, i.e., operate at 3.2 and 10 cm band, or single-band—3.2 or 10 cm. (Table 12.1)

Active radar station with passive response.

Let us consider an example of a shipboard navigation radar station of the "Liman" type. It is intended for installation on ships of up to 1000 gross tonnage (designated 1000 gt) and serves to improve the safety of ships navigation on the high seas, near the coast, in narrow areas and along limited fairways in difficult meteorological conditions (Fig. 12.2).

The station with the use of microprocessor and computer technology elements at high quality of color television images (no flickering, sufficient brightness and contrast for daytime operation) provides the following:

- circular radar coverage in the 3-cm radio wave range with display of the surface situation (coast, navigation signs, ships and other surface objects) in relative and true motion modes with orientation and image stabilization on the indicator screen according to data from the gyrocompass, log, satellite navigation receiver (SNR) or auto-tracking of a stationary target;
- determination of the coordinates of the observed surface objects (range, heading angle, bearing) in the polar (relative to own ship) reference system and the position of own ship relative to coastal and surface landmarks, as well as measuring the distance between two targets and bearing from one target to another using electronic bearing lines (EBL), range and cursor (electronic ruler).

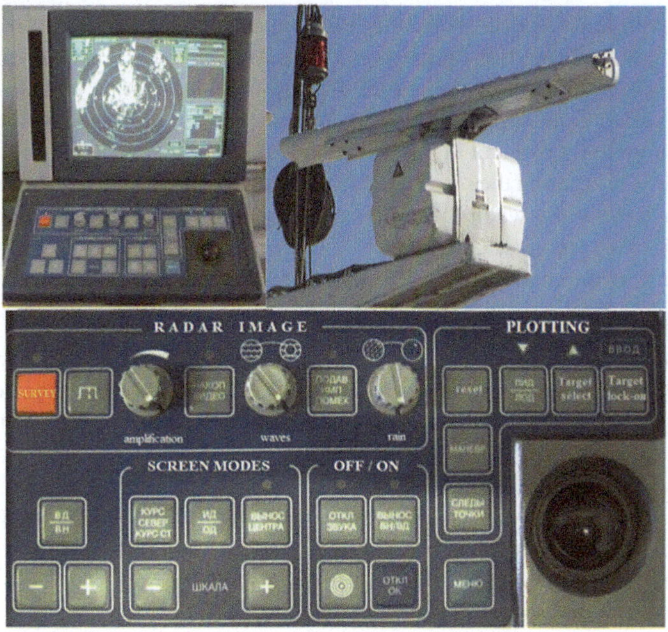

Fig. 12.2 Indicator, controls and antenna of "Liman" shipboard radar

Technical and navigation characteristics.

The maximum target detection range on the indicator screen in conditions of standard atmospheric refraction, in the absence of precipitations, with an antenna installation height of 15 m, with a probability of 0.8—not less:

15 nautical miles—shores 60 m high;
7 nautical miles—6 m high coast;
11 nautical miles—ships of 5000 gross tonnage;
3 nautical miles—ships of 20 gross tonnage;
2.5 nautical miles—medium sea buoy (without corner radar reflector).

The minimum detection range with a probability of 0.8 of an average sea buoy (without a corner radar reflector) or a corner radar reflector with an effective scattering surface of 10,000 m^2, installed at a height of 2.5 m, with an antenna installation height of 15 m, is no more than 20 nautical miles.

Typical antenna installation height above the waterline is 15 m. Increasing the antenna installation height compared to the typical one, increases the maximum and minimum detection ranges and vice versa.

The circular scan rate of airspace (antenna rotation) is 24 + l/−3 rpm. The station has the modes of image orientation in the north, in course or in course, stabilized at the presence of interfacing with the gyrocompass (GC) and can display the radar image on the indicator screen in relative and true motion modes.

The direction resolution probability of the station at 0.8 probability is not more than 2.2°, and the limit of the permissible measuring error of directions is 0.8°.

Guard zone—the auto-lock-on zone for auto-tracking can be quickly switched on and installed at an angle from 0.7 to 360°, in width up to 4 nautical miles, at a distance of up to 24 nautical miles. The guard zone is effective if the boundaries of the zone are observed on the screen within the included range scale. The sound signal is given automatically when the target enters the zone and during its auto-lock-on for auto-tracking (target acquisition).

The built-in automatic radar plotting system (ARPS) provides:

- automatic or manual lock-on/target acquisition of up to 30 targets for auto-tracking;
- display of vectors of its extrapolated relative or true motion on the indicator screen;
- display of trajectories of the past movement of auto-tracked targets in the form of dots;
- display of the form/logbook (bearing, range, course, speed, distance and time of the shortest approach, distance and time of course crossing) of any selected auto-tracked target;
- warning on target maneuver and on danger of collision;
- maneuver imitation with the heading and (or) speed of own ship to deviate from an adjustable delay time of the maneuver beginning from 0 to 60 min.

The station provides the target detection stability when the ship rolls up to 10°, which is ensured by the width of the antenna directional pattern in the vertical plane. The station provides for a constantly switched on wobbling (chaotic change) of sounding pulses repetition rate to protect against "false targets"-type interference caused by the phenomenon of radio wave overrefraction. The station has an operatively switched on asynchronous interference filter that suppresses interference from neighboring radars of the 3-cm range. The station has electronic counter countermeasures aids against reflections from the agitated sea surface and precipitation—operatively adjustable time corrected gain (TCG) and a short time constant (differentiation).

Orientation and stabilization of radar image. Most modern shipboard radar stations provide three modes of orientation of the radar image on the indicator screen: heading (COURSE), stabilized heading (COURSE ST), or north (NORTH), and the second and third modes are provided only when the station is coupled with the ship's heading indicator (gyrocompass, satellite compass, etc.).

In the absence of a heading indicator, the station operates only in the COURSE mode. Examples of the radar image in the indicated modes are shown in Fig. 12.3. The selection of the orientation mode NORTH, COURSE ST, or COURSE occurs by pressing the COURSE NORTH COURSE ST button on the control panel, and the selected mode is indicated on the panel MODES.

In COURSE orientation mode, the heading mark (OK) is always at the top at the zero mark on the azimuth scale. At change of ship course (maneuver, yaw), the OK position remains unchanged, while the blips from targets move in the direction on the screen with each antenna rotation; i.e., the radar image is unstable in direction.

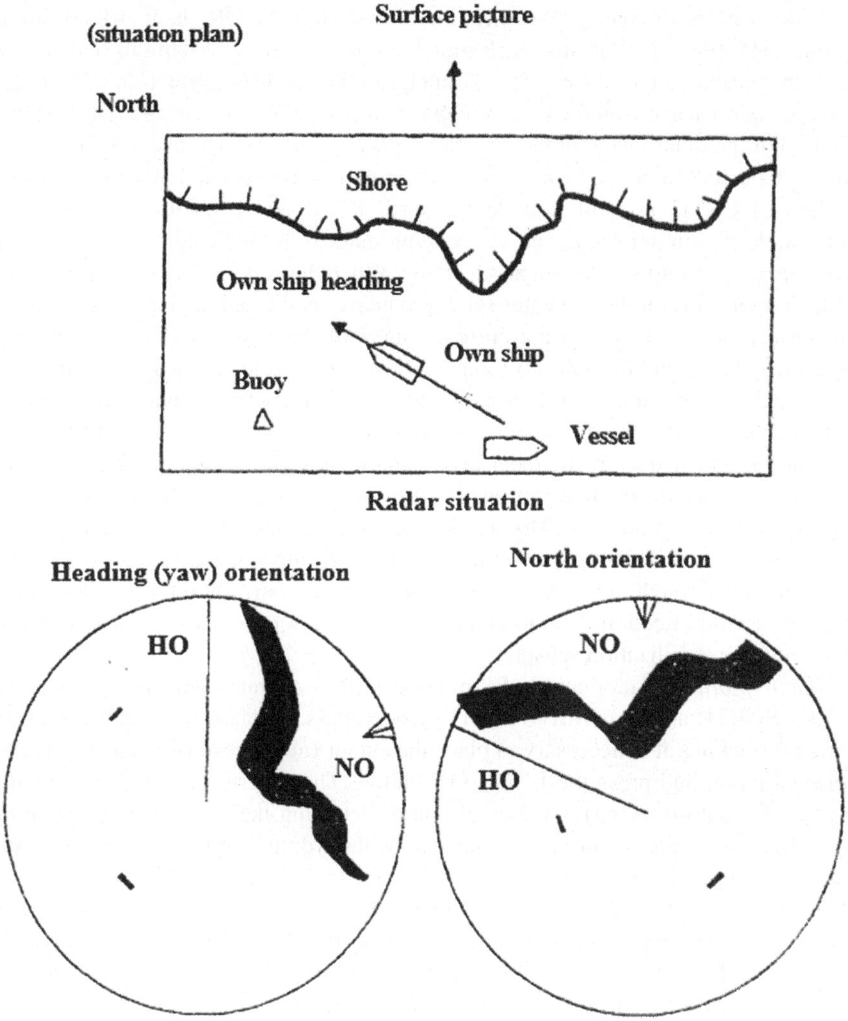

Fig. 12.3 Radar image in "Course" and "North" modes HO—heading orientation; NO—North orientation

The advantage of this mode is the orientation of the radar image in accordance with the situation observed along the course of the ship, and the station ability to operate in the absence of data from the gyrocompass, and the disadvantage is the lack of stabilization of the radar image in the direction (yaw).

In the COURSE ST orientation mode, the radar image is stabilized in the direction, and the OK, as in the COURSE mode, is also always directed upwards, but moves on the screen in the direction in accordance with the change in the ship's course within $\pm 15°$ from the zero of azimuth scale.

When yawing or changing the heading of own ship, the OK changes its position within (345–0–15°), returning, each time by a jump to zero of azimuth scale when reaching the extreme positions. The advantage of this mode is orientation of the radar image in accordance with the situation observed along the course of the vessel, stabilization of the radar image in the direction. It is recommended to use the orientation mode “Course STabilized” when following in narrow corners and along coastlines.

In the NORTH orientation mode, the north direction (OC) is always upwards at zero mark of azimuth scale, and the heading mark (OK) indicates the direction of the ship, i.e. the ship’s gyrocompass heading. When the course changes, the position of the target blips on the indicator screen remains unchanged, and the OK position changes in accordance with the current course of the vessel. BH shows the true bearing to the target blip. The advantage of this mode is orientation and stabilization of the radar image in the north direction in accordance with the navigation charts, but comparison with the real picture observed along the ship’s course is difficult.

The station can display the radar image on the indicator screen in relative or true motion modes. In the relative motion (ОД/RM) mode, the sweep center is stationary, and in the true motion (ИД/TM) mode, the sweep center moves according to the data from the stabilization sensors measuring own ship’s speed (as a vector), and after passing through 2/3 of the screen radius, it automatically returns to the point opposite to the direction of motion and continues to move in accordance with the data from the stabilization sensor.

Radar plotting. Automatic radar plotting (ARP) operates only in stabilization modes NORTH and COURSE ST from gyrocompass. To manually lock-on a target for auto-tracking, it is necessary to place the cursor (not necessarily exactly) on the selected target and press the LOCK-ON button. The criterion for lock-on (taking a target for auto-tracking) is the condition of detecting the target at least twice in three surveys, while the target will have a symbol (dotted square) starting the AC (auto-tracking) operation.

If the lock-on has occurred, then no later than 30 s, and the target must have a target movement and approaching vector: range, bearing, course (heading), speed, distance of shortest approach (DSA), time of shortest approach (TSA), distance to course intersection point (DCI), time to course intersection point (TCI).

Guard zone—the zone of targets auto-lock-on for auto-tracking (target acquisition) is activated through the AUTO-PLOTTING/ZONE SET menu: with the cursor set the desired dimensions of the auto-lock-on zone in angle, in width (1 nautical mile), in range (up to 20 nautical miles) and press the ENTER button. The guard zone is effective if the zone boundaries are observed on the screen. When a target appears in the guard zone, it is locked-on for auto-tracking, and the message TARGET AUTO-LOCK-ON appears on the message board, accompanied by a sound signal. Targets acquisition and display of ARP information on the screen are provided on scales from 3 to 24 miles.

The criterion for target acquisition in the auto-lock-on zone is a target detection in a row on three sweeps (3 out of 3) or—four detection facts on eight sweeps (4 out of 8). Thus, the delay time from the moment the target crosses the border of auto-lock-on zone to the symbol appearance of tracking start (dotted square) will constitute from 8 up to 20 s, depending on the probability of target observation. The guard zone is disabled via the AUTO-PLOTTING/ZONE RESET menu.

When using ARPS, one should consider that over time from the moment the target is locked-on for auto-tracking, the errors in parameters determining of target movement in conditions of uniform and rectilinear movement of one's own ship and the target gradually decrease (the smoothing process is in progress) and after 3 min the errors do not exceed the following values: in direction of relative movement (RM) of target—2 ÷ 5°, in speed—0.3 ÷ 1 knot, in distance of shortest approach—0.5 ÷ 0.7 miles (depending on approach direction and gyrocompass error), in true heading of target—2.5 ÷ 7.5°, in true speed of target—0.8 ÷ 1.2 knots (depending on sonar log error).

The abovementioned time delay in obtaining of the most accurate data on tracked real targets is manifested in the fact that immediately after maneuvering own ship's course or speed, the parameters and vectors of targets movement can have 2–5 times large errors, which with time of steady motion decrease to a stable low values.

The main indicator of the ARP mode is the stability of target tracking under conditions of interference from hydrometeors and rough sea surface, as well as from conditions of target position relative to other surface objects and the coast. By means of adaptive processing of video signals, stable targets auto-tracking is provided under various conditions of interference, even at small increase of target signal level over the interference, i.e., provided that the target blip is barely distinguishable against the background noise on the indicator screen.

ARPS provides targets classification based on motion parameters:

- moving targets (speed from 1.6 knots or more) have an auto-tracking symbol (circle);
- maneuvering targets (or all targets—when maneuvering of own ship on a course with a speed change of not more than 5%) have blinking AC symbols;
- stationary targets (speed of 1.5 knots or less) have an auto-tracking symbol—a circle with a cross.

When the rendezvous parameters (DSA and TSA) of any target of dangerous values are reached, the symbol of a threat (dangerous) target (triangle) appears, the vector of its movement becomes flickering, and the message THREAT TARGET is periodically generated, accompanied by a sound signal. The display of the trajectory dots of the past movement of auto-tracked targets is carried out by the TRACE-DOTS button on the control panel, and the selected mode is indicated on the screen's panel as DOT, next to the target logbook (label).

Trial maneuver. This mode is used to replay own ship's maneuver for collision avoidance with threat target. In case of decision making on maneuver, it should be considered that maneuver calculations do not take into account the maneuvering and dynamic characteristics of own ship. No AC functions are interrupted while playing own ship's avoidance maneuver by changing the simulated course and/or speed.

Active response mode in marine radar location. In maritime radar location, a system designated as radar beacon transponder (RBT/Search And Rescue Radar Transponder, SART) is used for position finding of ships in distress. RBT provides

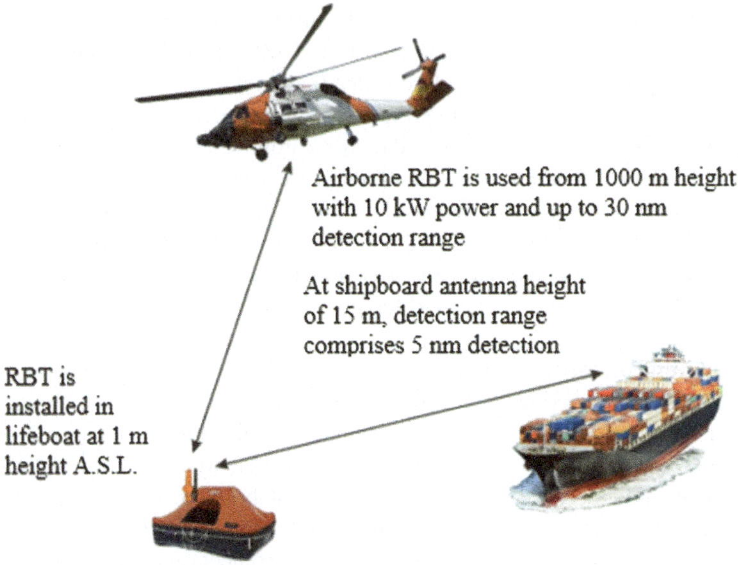

Fig. 12.4 Active response mode in marine radar location

transmission of a series of signals, which are displayed on the radar screen as a series of dots/arcs of a circle/circles, located at an equal distance from each other in the radial direction. RBT operates in the 9.2–9.5 GHz range.

Onboard any ship of 500 gross tonnage per. tons and more must be at least two RBT. On ships of gross tonnage from 300 to 500 per tons must be at least one RBT.

RBT should be installed in locations from where they can be quickly transferred to a lifeboat or raft. The height of the installed transponder antenna must be at least 1 m above sea level. At the same time, it provides normal operation at a distance of at least 5 nautical miles when requested by a ship's radar, the antenna of which is installed at an altitude of 15 m and at least 30 nautical miles when requesting an aviation radar with a pulse power of at least 10 kW, installed on board AV located at an altitude of 1000 m (Fig. 12.4).

Obtaining of 12 characteristic blips on the indicator of the ship's radar is achieved by forming a response signal by the transponder with a saw-toothed frequency change 12 times (Fig. 12.5).

Whenever a signal enters the bandwidth of radar receiver, a blip appears on the screen. Strictly speaking, there should be 24 blips in total—12 blips corresponding to a slow frequency change and 12 blips corresponding to a fast frequency change.

When ship approaches the RBT, the blips blur into arcs and at a very close distance merge into concentric circles, as shown in Fig. 12.6. This is explained by the fact that the response signal of the transponder at close range is received by the ship's radar in any direction of the antenna. The directional pattern of radar antenna has, in addition to the main lobe, side and return lobes. At close range, the reflected signal is received by all lobes in any direction of the antenna.

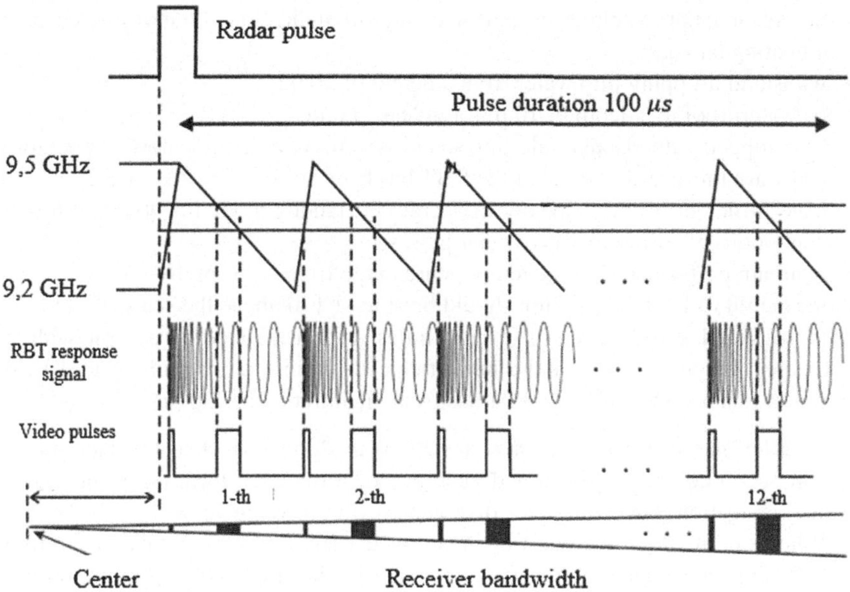

Fig. 12.5 Forming of RBT response signal

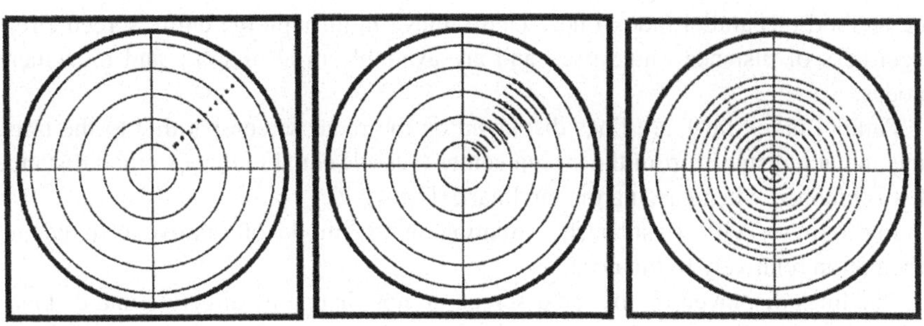

Fig. 12.6 Examples of radar screen when RBT position at more than 2 miles from radar, about 1 mile, close to radar (less than 0.2 miles). Radar scale is 10 miles

The actual location of the rescue vehicle with switched-on of RBT on radar screen is in the area of the blip closest to the center of the screen. In some cases, the closest to the center SART blips are difficult to discern against the backdrop of near reflections from sea waves. The blips near the center of the indicator become indistinguishable against the backlight. Determining the distance to the responder in such a situation is difficult. The last blips are usually clearly visible. In this case, it is possible to extrapolate the position of the first blip and thus determine the location of transponder. This blip is approximately 8.1 nautical miles away from the most distant one.

The operational requirements for RBT are stated in Resolution A.802 (19). In accordance with this Resolution, the RBT should satisfy the following requirements:

- provide manual switching-on and switching-off, indication in standby mode, have a floating lanyard;
- withstand dropping into water from a height of 20 m;
- be waterproof to a depth of 10 m for at least 5 min;
- be equipped with visual or alerting sound systems to determine normal operation and warn those in distress that the RBT has been activated by the radar;
- have sufficient battery capacity to operate in standby mode for 96 and 8 h with continuous exposure to 1 kHz radar pulses;
- maintain performance in the temperature range from—20° up to +55 °C;
- the height of RBT installation should be at least 1 m above the sea surface;
- be triggered at a distance of up to 5 miles when irradiated by a radar with an antenna height of 15 m and when irradiated by an aircraft radar with 10 kW power at a distance of up to 30 miles from an altitude of 1000 m.

Parallel index method. Modern radar stations are designed to detect surface objects and shores in conditions of limited visibility, determine the location of the vessel, ensure navigation in narrow areas, divergences with meeting vessels.

When sailing along the coast or in confined waters, it is necessary to carefully monitor the movement of the vessel relative to the track line. This monitoring should include periodic position fixes, which would be combined with continuous monitoring of the ship's transfer relative to the track line.

Continuous monitoring of the ship's position relative to track line is possible using the method of parallel index lines. These lines run through the entire screen circle regardless of distance range used and are available in all display and movement modes.

Parallel index lines are lines displayed on the radar screen parallel to the track line and at a distance from the sweep center (vessel) equal to the specified distance that is planned when passing the landmark (Fig. 12.7).

Further, we will consider the situation in the mode of relative display and orientation relatively to the north.

The lines displayed on the radar screen do not change its direction and distance relatively to the center of the sweep when changing a course.

In Fig. 12.8, plot of the map with the marked track direction equals to 220°. A tangent line is drawn to the selected cape, which is parallel to the track line. The distance from track to tangent line equals to 0.3 miles (see above Cape). The vessel must navigate such a course that the distance between the tangent and the track line remains constant.

On the radar screen, it will be as follows: The display of the track line is turned on, the direction is set to 220°, and then it is shifted from the center by a distance of 0.30 miles. Thus, the line that was set will remain in the originally set direction and distance from the center of the sweep. When the vessel is moving, landmarks move in the direction opposite to the movement of the vessel on the radar screen in the mode of relative movement. Consequently, in our case, the cape will move. It is

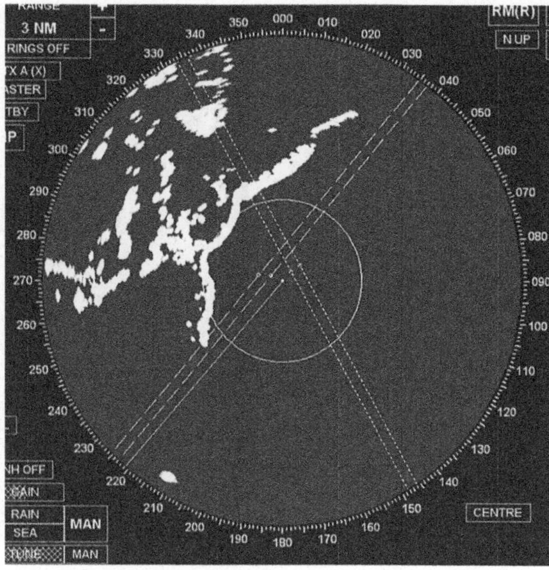

Fig. 12.7 Parallel indices on radar screen

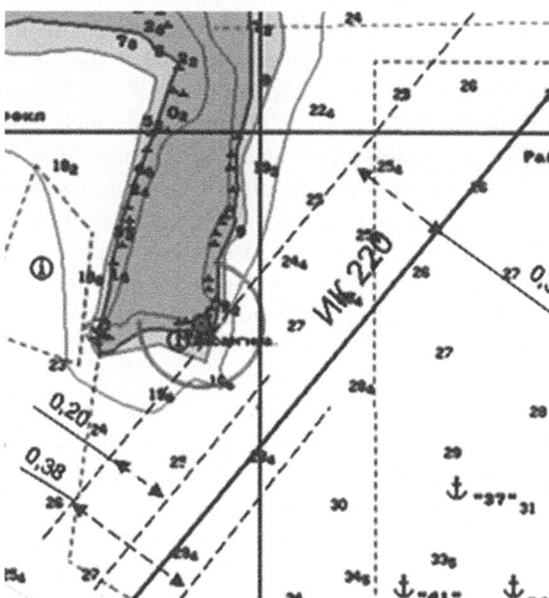

Fig. 12.8 Example of index lines on the map

necessary that during the movement, the line displayed on the radar screen parallel to the track line would be tangent to the selected landmark during the entire navigation on this segment of the track.

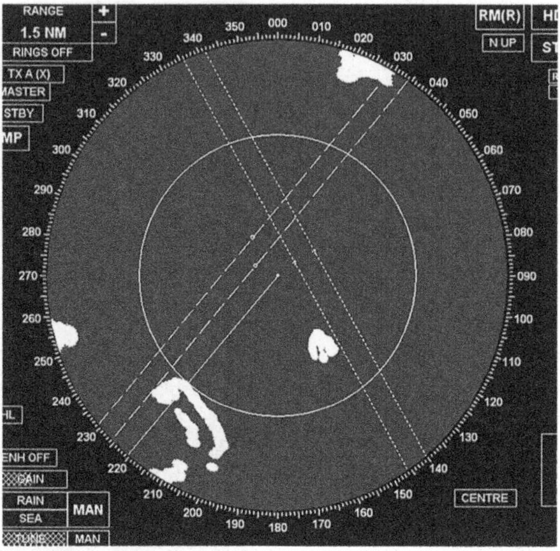

Fig. 12.9 Ship has reached the starting point of the rudder deflection

If the observer discovers a deviation of the landmark from the set line on the screen, this will mean that the vessel is deviating from the track line under the influence of external forces or for other reasons. In this case, the observer's task is to correct the ship's course in such a way that the line, which is parallel to the track line, would remain tangent to the selected landmark throughout the navigation.

In modern radars, you can set up to 4 index lines. This allows you to set two lines to the course the boat is heading and two lines to the next (after turn) course.

When a vessel navigates in confined areas, it becomes necessary to diverge with another vessel, give way, bypass vessels engaged in fishing. In this case, the vessel must deviate from the track; but at the same time, it is necessary for the vessel to remain at a safe depth. Therefore, two index lines can be set to the track line, which would limit the maximum and minimum deviation from the track line (see Fig. 12.9, two index lines are set below the cape).

On the radar screen, the cape will be enclosed between two lines (Fig. 12.7). The vessel can deviate to the right and to the left, and an observer will visually control the deviation of the landmark from the track line, and, at the same time, the landmark must remain between parallel lines.

When navigating in confined areas, parallel index lines may be set for the course the vessel is following and two index lines parallel to the next track line. In addition to this, control distances or radar bearings can be set to determine the moment of the rudder deflection beginning (in Fig. 12.7, in addition to the index lines, a control distance is set).

In Fig. 12.10, the vessel has reached the starting point of the rudder deflection. The steering wheel is shifted to a precalculated angle. The vessel begins to turn, an operator visually controls the accuracy of entering the track. In Fig. 12.10, the operator can see that the boat is turning too quickly.

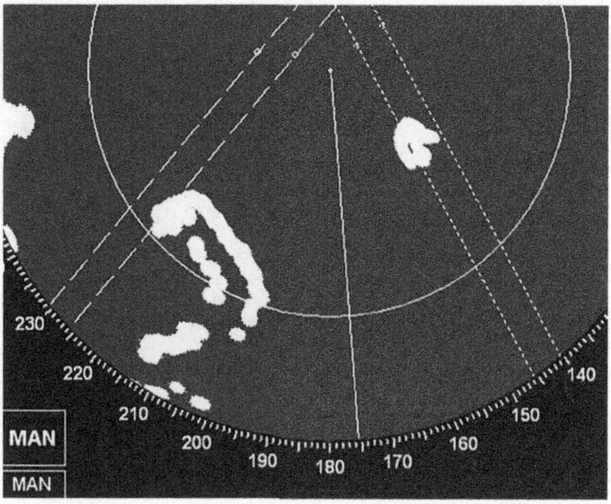

Fig. 12.10 Vessel is turning to heading 151°

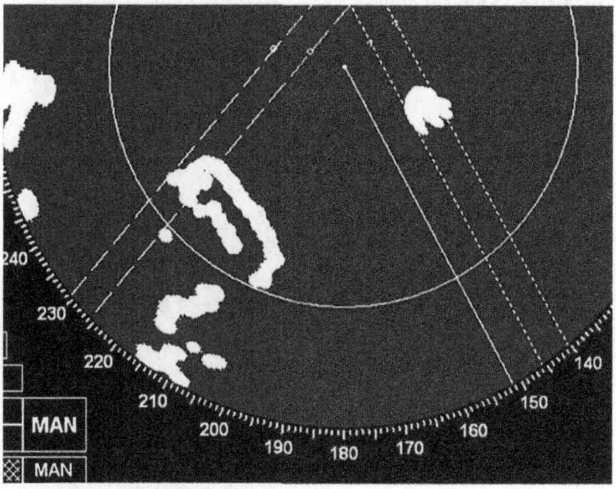

Fig. 12.11 Ship sets course of 151°

In Fig. 12.11, it can be seen that the vessel is too close to the island (during the turn, the rudder angle did not change). The reason may be an incorrectly calculated rudder-deflection angle or the action of unaccounted external forces. In this case, to reach the specified interval, it is necessary at the moment of turning (Fig. 12.10) to reduce the rudder-deflection angle.

Figures 12.12 and 12.13 show a turn when the vessel sets course at a greater distance from the island than was originally set. It can be seen that the vessel goes beyond the limits set by the navigator using two parallel index lines. The moment is shown when it is necessary to plan a turn with a maximum rudder deflection of 20°, because during the turn it becomes necessary to additionally increase the rudder-deflection angle to reach a given track line.

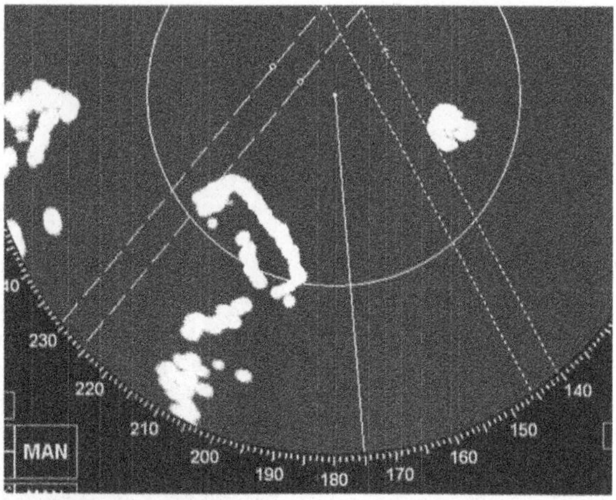

Fig. 12.12 Ship is turning on course 151°

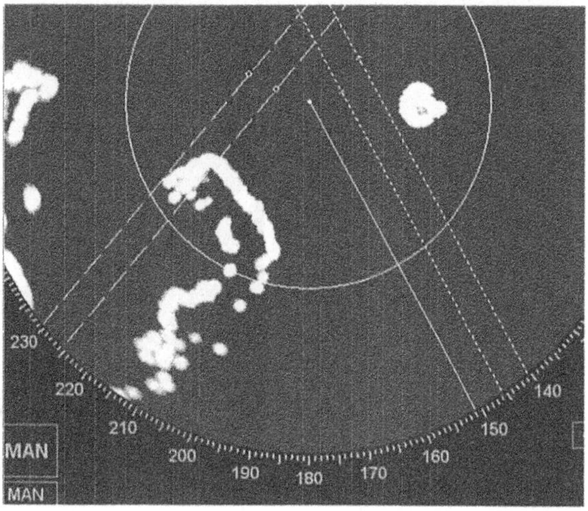

Fig. 12.13 Ship set course of 151°

The reason that the vessel did not turn within the specified limits may be an incorrectly calculated rudder-deflection angle, an incorrectly plotted control distance, or the actions of unaccounted external forces.

Figures 12.14 and 12.15 show a turn when the ship tracks out on course within specified limits.

Using parallel index lines allows the following:

- set two lines parallel to the track lines to the selected landmarks, which will determine the minimum and maximum deviation from the track line;
- set lines in advance parallel to the next course, which will allow you to control the turning process according to the image on the radar screen and more accurately track out the next track;

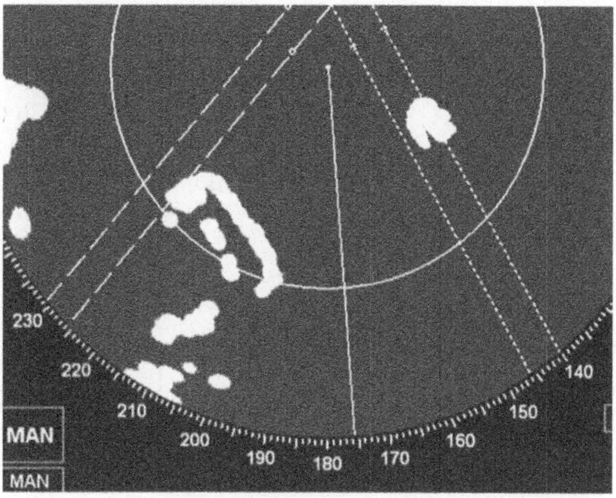

Fig. 12.14 Ship is turning on course 151°

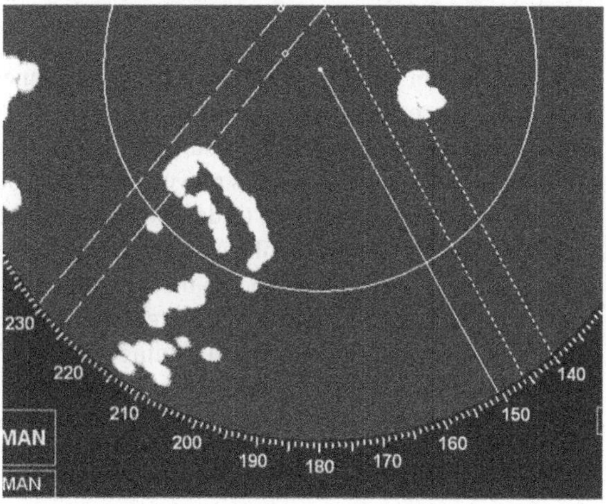

Fig. 12.15 Ship set course of 151°

- use the lines as secants to determine when the rudder begins to deflect to the next course.

This is a simple and quite accurate navigation method that allows continuous control over the movement of the vessel. Parallel indexes are recommended for each part of the coastal passage.

T*he use of radar and ARPS when sailing in confined waters.* Confined waters include navigational areas, the dimensions of which are of the same order of magnitude with the dimensions of the vessel and/or the depth of which is less than 5 of the vessel's draft. Navigation in confined waters is carried out, as a rule, with limited maneuvering capabilities, along recommended routes or fairways, traffic separation schemes (TSS). In this case, it is necessary to almost continuous control the

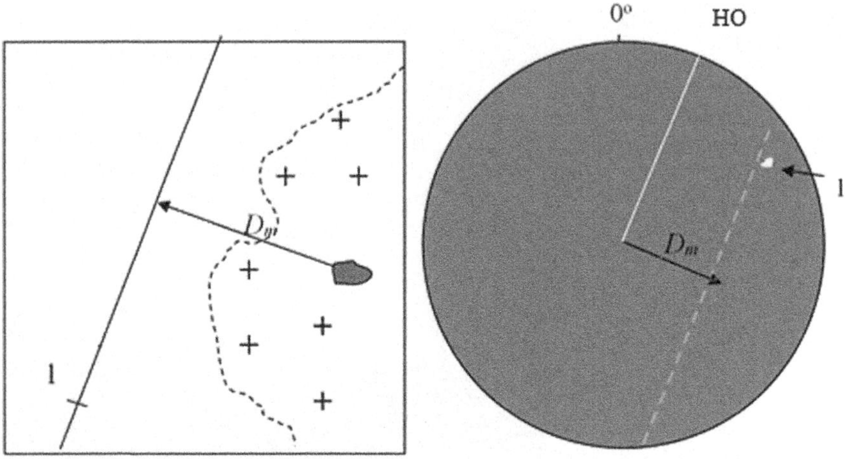

Fig. 12.16 Controlling the vessel movement along the given path using a line of parallel indices

vessel movement along a given path with simultaneous observation of surroundings. Under these conditions, navigation methods in the "from fix by observation to observation" mode cannot fully ensure the safety of navigation and the methods of continuous control of the vessel's movement, based on various methods of visual or instrumental-visual orientation (radar plotting/guidance), come to the fore.

Specific options for the application of certain methods of continuous monitoring of the vessel's movement depend on the operational and technical capabilities of the radar and ARPS; but in any case, they are based on such known concepts in classical navigation as boundary, control and leading contours, as well as parallel index lines. Below are some examples of methods implementation of continuous monitoring of the vessel movement with use of ship's radar or ARPS.

1. *Control of the vessel movement along a given path*

If it is necessary to ensure continuous monitoring of the vessel movement along a given path, for example, near underwater hazards in the presence of radar reference points (Fig. 12.16), then the following method can be applied. At the stage of preliminary plotting, after set the given course (path) on the map, it is necessary to measure the minimum (traverse) distance D_T to the selected radar landmark. Then on the ARPA screen at a distance D_T from the center of the screen, a line of parallel indices is set parallel to the direction of the OK line. During the movement, the vessel, having tracked out the course, will detect the echo signal of the selected landmark on the line of parallel indices (position 1 on Fig. 12.16). Subsequently, by working on the rudder in such a way that this echo remains on the line of parallel indices, the vessel will ensure the passage of this section exactly along the line of the specified path. This eliminates the need for observations, and deviations from the course, caused, for example, by the action of an unaccounted current or wind drift, will be detected in a timely manner from the offset of the echo signal and properly corrected.

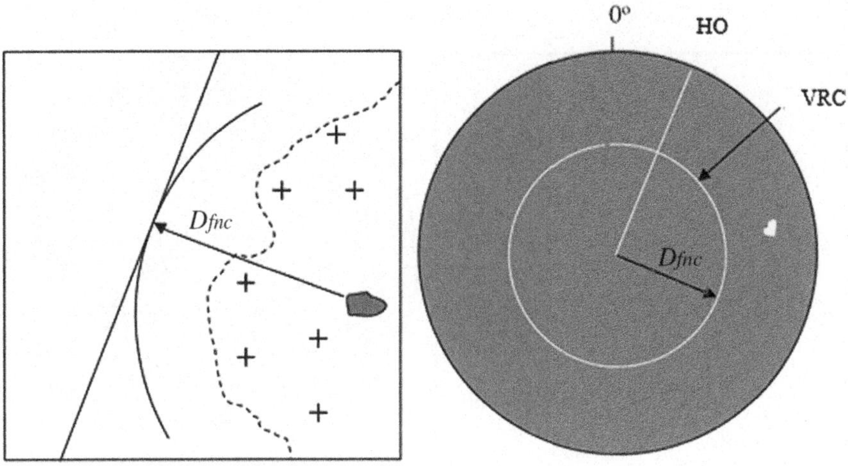

Fig. 12.17 Control of vessel movement in relation to hazards using the fencing distance (VRC)

Experimental studies have shown that this method of visual and instrumental orientation provides guidance of the vessel along the intended path with an accuracy of ±50 m when working on a 4 mile scale and ±115 m on a 16 mile scale.

2. *Movement control using fencing distances*

Fencing distances can be used to continuously monitor the vessel position relatively to navigational hazards, especially underwater hazards. If the ARPS is not stipulated in the radar, then you can use the variable range circle (VRC). To do this, an arc with a radius equal to the distance closer to which the vessel should not approach the selected landmark is plotted on the map from the radar landmark. This arc is called the fencing distance. While movement, it is necessary to set VRC on the radar screen to the value of fencing distance (Fig. 12.17), and continuously monitor the shift of landmark echo signal, steering the vessel in a way that this echo signal remains outside the VRC.

In the presence of ARPS, the same method can be implemented in automatic tracking mode or using parallel index lines. In the first case, the echo signal of the selected landmark should be taken for automatic tracking and perform constant observation of the distance between the landmark and the vessel. If this distance becomes less than the fencing, then appropriate measures should be taken. If ARPS capabilities allow, then it is desirable to set an alarm that would be triggered when the distance between the vessel and the landmark reaches a value equal to the fencing distance. In the second case, it is necessary to draw a straight line on the map parallel to the selected path, beyond which the vessel should not enter when sailing on this section of the route (Fig. 12.18). While movement, you should set the line of parallel indices on the ARPS screen at a distance equal to the fencing one, and steer the vessel so that the echo signal of the selected landmark would be constantly outside the line of parallel indices.

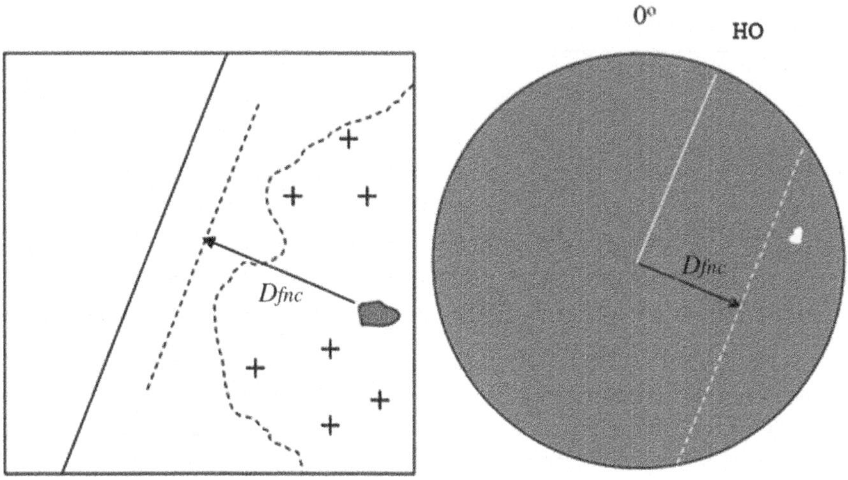

Fig. 12.18 Control of vessel movement in relation to hazards using the fencing distance (line of parallel indices)

3. *Turning control*

When sailing in coastal and especially in confined waters, timely and accurate performing of turns at course changing are essential. To control turns, several methods can be used using radar and ARPS, based on the idea of secant (or control) contours.

In the absence of ARPS, you can use the VRC. If you need to start turning at K point, as shown on Fig. 12.19, then this point can be fixed by the secant distance D_s to the radar landmark located at sharp bow (or stern) corners. At ship motion, it is necessary in advance, before approaching the turning point, to set the VRC to a value equal to the secant distance and, by observing the shift of echo signal of this landmark on the radar screen, give a command to the helmsman at the moment when the echo signal reaches the VRC line. Similarly, this method is performed using an electronic range and bearing line (ERBL). In the presence of ARPS, the selected landmark can be locked-on for automatic tracking and control the approach to the turning point both by the echo signal shift on the screen and by the distance change in the selected landmark on the digital indicator.

Turning control in ARPS can be performed using parallel index lines. One of the possible ways is shown in Fig. 12.19. In this case, to control the movement before the turn, a radar landmark is selected at the bow heading angle (angle on the bow) and a line of parallel indices is set at a distance D_1 from the course line (path), as indicated above in paragraph 1.

To control the movement after the turn at a distance D_2 from the new course line, a second line of parallel indices is set. The turning point is fixed by the D_2 distance, at which one more line of parallel indices is set relative to the second course. When performing such preliminary work, the turning control is performed as follows: Along the first line of parallel indices (D_1), the movement is monitored until the moment of

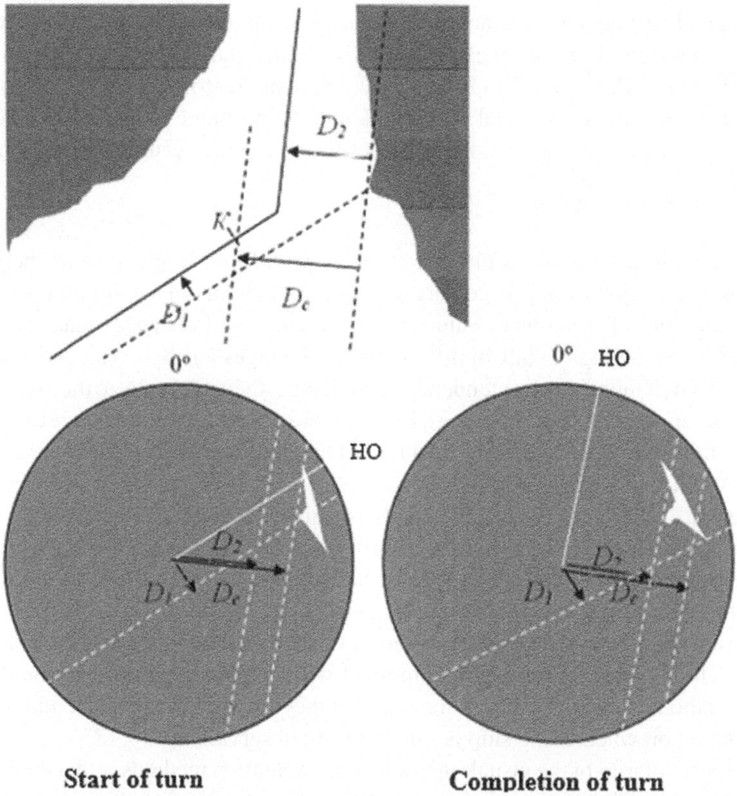

Fig. 12.19 Control of movement and start of turning using a secant distance (line of parallel indices)

turning, while the shifting of echo signal of the selected landmark is observed on the ARPS screen. At the moment when the observed echo signal touches the third line of parallel indices (D_s), the turn should be started. If the point of turn is calculated correctly, then when the ship will track out on a new course, the observed landmark echo will be on the second parallel index line (D_2).

4. *Determination of total drift (sweep) vector*

During the vessel movement with operating ARPS, you need to select a reliable echo signal of a fixed landmark and take it for auto-tracking. If for a fixed reference point the speed and distance values are different from zero on the digital indicator, this means that there is drift in this area from an unaccounted (or incorrectly accounted for) current and/or wind drift. Consequently, according to the indicated values of the "true target heading" ($\mathrm{TH_t}$) and "target speed" (V_t), we can transfer to the drift direction (H_d) and its speed (V_d):

$$H_d = \mathrm{TH_t} \pm 180°,\ V_d = V_t, \quad (12.1)$$

Having determined the elements of total drift, it should be considered when input a graphic plotting. It has been experimentally defined that if the observed landmark is at a distance of up to 6 miles from the vessel, then the drift velocity is determined with a mean-square error equal to ±0,15 knots. At the same time, the specification of drift velocity every half an hour increases the reckoning accuracy by 4 ÷ 5 times.

5. *Control of vessel position at anchor*

ARPS capabilities are useful in other cases as well, for example, during anchorage (anchor hold). For this, in is necessary to take 2–3 radar landmarks for auto-tracking, record the control (reference) values of its bearings and distances and constantly monitor its changing. Small multi-directional changes in these values indicate a reliable hold. If there is a clear tendency in its change, then this is due to the presence of drift at the anchor (Anchor dragging). For automatic control of the distance exceeding the set limits, the sound signaling "distance of dangerous approach" or "guard zones" can be used.

There are other methods of visual and instrumental orientation, providing continuous control of the movement and location of the vessel. The main advantage of visual and instrumental orientation methods is that during navigation they do not require the distraction of the navigator to work on the map, which provides the possibility of continuous visual and radar observation of the surroundings. At the same time, these methods must be rationally combined with conventional observational reckoning methods, their value lies in the fact that they provide fast, prompt and reliable information on whether the ship is safe between observations.

The particularity of the visual–instrumental orientation methods is that their application requires more thorough than usual and deep study of forthcoming passage, selection of characteristic landmarks and corresponding delineation of preliminary plotting. Figure 12.20 shows one of the possible examples of preliminary plotting of a ship passage in confined waters and the radar screen image when the vessel is at point **O**, shown on the map. In the given example, four parallel index lines are used, one of which serves as the fencing distance, and the VRC, which determines the secant distance, which fixes the point of a turn.

6. *Use of ARPS at divergence of vessels*

A complete assessment of the situation is possible only by analyzing both primary (raw/unprocessed target echoes) and secondary (vectors and digital data) information.

The analysis of the primary information for targets selection for acquisition is carried out by an ocular estimate method of targets afterglow traces in the same way as at manual radar plotting. First of all, critical (dangerous) and potentially critical targets are selected for auto-tracking (AT).

Secondary information is used to assess the degree of danger of the situation. During radar surveillance using ARPS, the navigator uses the following data to assess the danger degree of approaching situation:

- location of the RM vector relative to own ship;

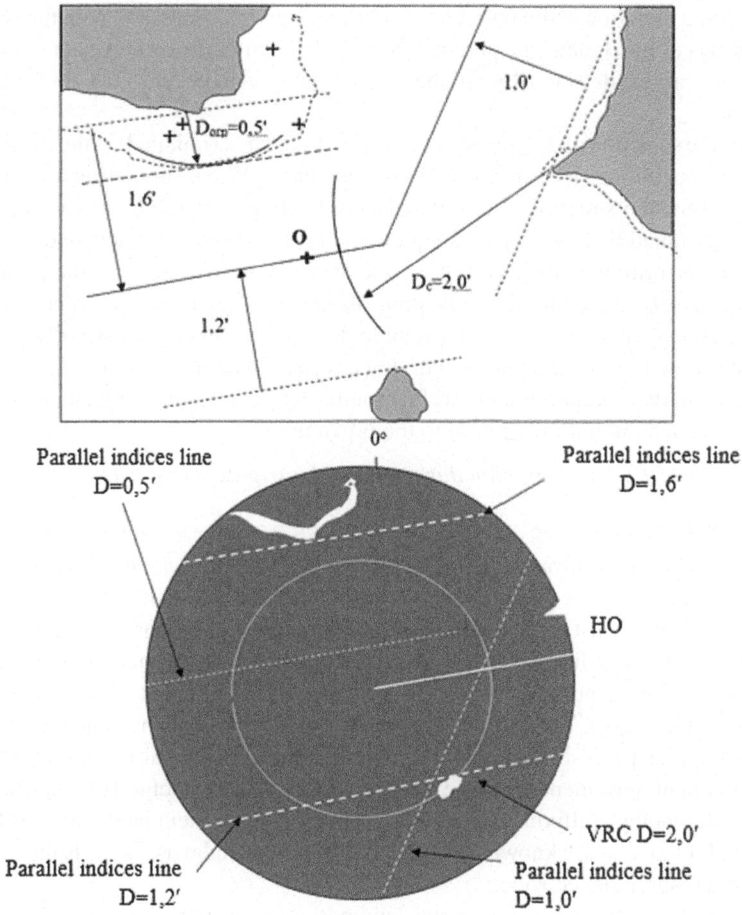

Fig. 12.20 Application of parallel index lines and VRC

- values of D_{KP} and t_{KP};
- heading angle, aspect angle (in true motion mode) and distance to the target;
- behavior of bearing relatively to the target.

Additional useful information for situation assessing and maneuver selection can be provided by predicting the situation development by changing the length of the target vectors.

The maneuver selection for safe divergence should be carried out in advance and decisively in strict accordance with IRPCS-72 (International Regulations for Preventing Collisions at Sea-72), in accordance with the specific circumstances of approaching situation and navigation conditions, and in accordance with the recommendations of good maritime practice. It should be remembered that even a decisive maneuver can be detected by another vessel using ARPS only 3–4 min after its initiation.

After the selection of divergence maneuver, it is trialed (imitated) at the time of the maneuver beginning (lead time) set by the navigator. When simulating a maneuver in all ARPSs, the situation is calculated only for targets that are on auto-tracking, and it is assumed that all of them keep their course and velocity unchanged.

When performing a maneuver, it is necessary to carefully monitor the vectors of meeting vessels, including the indication of their past positions, in order to detect their possible maneuver as early as possible. It is also necessary to carefully monitor the effectiveness of the maneuver and, if necessary, take timely additional measures to ensure safety. Continuous and careful monitoring of the mutual movement of ships must be carried out until the return to the previous course.

7. *Use of ARPS for navigation during coastal navigation in traffic flows*

In coastal navigation, it is effective to use ARPS not only to prevent collisions of ships, but also to control the position and movement of own ship relative to the coast and the line of a given path. Initial identification of radar landmarks when approaching the coast from the sea is recommended to be carried out by series (fan)of bearings and distances. When using 5–6 landmarks, the method is absolutely reliable and allows not only to set the name of landmarks, but also to reveal the "errors of the object," i.e., to establish from which parts of the object (from the water's edge or from some horizontal) the reflected echo signal is coming. During the subsequent movement along the coast, the most effective method of identification is the "referencing" (affixment) method, when each subsequent landmark is reliably identified relative to the known one before this known landmark exceeds the bounds of the radar surveillance zone.

Taking the stationary navigational landmarks for auto-tracking and displaying data on bearing and distance of the tracked object in a form (target label) allow you to effectively control the vessel movement and the performance timeliness of turns.

8. *Navigation management at sailing in ship traffic separation systems*

From a navigational point of view, ship control while navigating in traffic separation systems TSS and in confined waters is reduced to the timely execution of turns and compensation of emerging ship deviations from the line of a given path.

When navigating in TSS, the used navigation methods should meet the following requirements:

- the ability to continuously monitor the vessel position and movement, the ability to instantly assess not only the position, but also the trend of the vessel's movement in order to take timely corrective action (e.g., to clarify the accepted correction for drift);
- possibility of position estimation with errors not exceeding 20% of the traffic lane width (with a probability of 0.95);

- possibility of navigation control without distraction from observing the surroundings and without prejudice to other tasks;
- simplicity, visual aspects, reliability of the applied methods, excluding mistakes. The navigator who manages the vessel (the captain of the vessel or his substitute senior mate) mainly uses accelerated (pilotage) methods to control the position and movement of the vessel (leading, control, fencing bearings and distances), including using radar and ARPS.

The deck officer (officer of the watch, OOW) performs navigational plotting and periodically determines the ship's position using normal pilot methods in order to:

- duplication of control in position and movement of the vessel and the timely detection of possible misses and errors;
- performing the necessary calculations, and obtaining objective data on the ship's deviation from the line of a given track, on the course angle, drift angle and ground speed, on the direction and speed of the total drift, on the distance and bearing to the point of the beginning of the turn, time to turn, etc.;
- timely providing of the navigator, managing the vessel, with all necessary navigational information.

When using radar and ARPS, the most effective and efficient method is visual control of the position and movement of the vessel using an "electronic map" (stabilized relative to the ground in auto-drifting mode), and in its absence—by means of an "electronic fairway" or "electronic lines."

It is important to keep in mind that in most of the existing ARPSs, the switching-on of stabilization mode of the "electronic map" relative to the ground leads to similar stabilization of the true vectors of targets, which can distort the assessment of the situation in case of divergence of ships in conditions of strong lateral drift. In some ARPS models, this limitation is removed: It is possible to stabilize the "electronic map" relative to the ground in combination with displaying the true vectors of targets relative to the water.

In the absence of an "electronic map," an effective control can be ensured by means of an external electronic drift-sight.

Taking a rationally chosen navigation landmark for auto-tracking allows not only to control the leading, fencing, control navigation parameters, but also to systematically determine the ship's position much faster and more accurately than with "manual" methods of taking the bearing and distance from the radar. In this case, it is important to carefully control the location of the auto-tracking strobe on the selected object (especially, if the landmark is not a point one).

9. *Main limitations of ARPS*

Since ARPS provides automatic processing of radar signals, all radar limitations are included as an integral part of ARPS limitations and must be taken into account at divergence. These are, first of all, the limitations imposed by the used range scale, the possibility of not detecting echoes from small vessels, interference with radar detection due to the state of the sea, rain, fog, shadow sectors, etc.

Information processing algorithms, implemented in ARPS, impose additional limitations. The main ones are the following:

- None of the existing ARPS provides guaranteed detection and lock-on for auto-tracking of all targets, including the critical ones. Therefore, the use of ARPS only in automatic acquisition mode cannot be considered as proper radar surveillance.
- If the echo signal is unstable (small vessels, tracking in conditions of strong interference), the target may be reset and information on it will not be displayed. With a close divergence of two targets, the loss of one target is possible. In this case, the other target will have two vectors, one of which will be false;
- Signals from the radar, gyrocompass and log come to the ARPs with errors. With the roll of the vessel, the presence of interference, maneuvering and yaw of own ship, the errors of the sensors increase. Therefore, when calculating the elements of target movement and the parameters of the approach situation, the "smoothing" is used, which leads to a delay in the output of reliable data up to three minutes from the moment the target is taken for tracking;
- Errors of the calculated elements of target movement and situation parameters can reach the following values:
- true course of the target—$\pm 5 \div 7°$;
- true target speed—± 1.2 knots;
- distance of the shortest approach—± 0.7 miles;
- time of the shortest approach—± 1 min.
- The target maneuver is detected with a significant delay, and the data generated by the ARPS on the maneuvering target will be unreliable for $3 \div 4$ min after its completion.
- When maneuvering own ship, the ARPS output information for all tracked targets will be unreliable.

Part III
Radio Navigation Systems

Chapter 13
Range-Finding Radio Navigation Systems

The range-finding radio navigation systems (RRNS), also called the radio-navigational range finders, are intended for determination of the current range (distance) $D(t)$ from an object to radio-navigational guide point, e.g., radio beacon. RRNS position line is a circle, and in spatial case—a sphere. RRNS are widely used autonomously and within composition of other radio navigation systems.

The main tactical characteristics of RRNS are accuracy, operational range, operating (coverage) zones, bandwidth capacity, jamming resistance and reliability. RRNS capacity is evaluated by the maximum amount of objects, which it can simultaneously provide with corresponding navigational information. RRNS capacity with responder is limited by its capability to give out response signals and mutual interference of interrogators. RRNS, used at the present time, are often operated in frequency bands. The prime RRNS technical characteristics are the following: size and stability of a carrier frequency, modulation type and its parameters, directional patterns of receiving and transmitting antennas, transmitter power and receiver sensitivity etc.

The RRNS are divided by design concept into RRNS without responder and RRNS with responder.

In RRNS without responder, a signal propagation time is measured, radiated by ground-based station, to an object, on board of which a receiving unit is mounted. The generalized functional chart of RRNS without responder is shown in Fig. 13.1. In composition of ground-based station transmitter and onboard receiver, there are high-stable reference (calibration) oscillators, which at specified moment are synchronized between each other with special synchronization signal, radiated by ground-based station.

Based on received signal, either time interval $\tau_D = D/c$ between reference pulse and pulse at receiver output (at pulsed method of range measuring), or phase difference $\Delta\varphi$ between oscillations of onboard reference oscillator and oscillations at receiver output (at phase method of range measuring) are measured onboard an object. The measuring range is correspondingly defined as follows:

D. A. Akmaykin et al., *Theoretical Foundations of Radar Location and Radio Navigation*, Springer Aerospace Technology,
https://doi.org/10.1007/978-981-33-6514-8_13

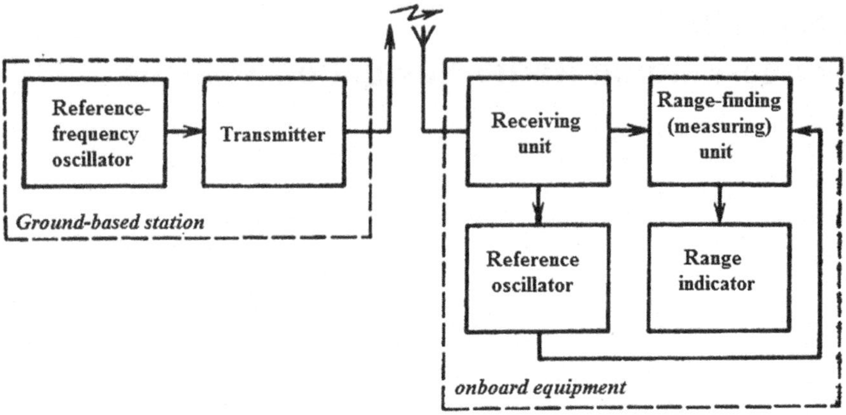

Fig. 13.1 Generalized functional chart of RRNS without responder

$$D = v\tau_D; \tag{13.1}$$

$$D = v\frac{\Delta\varphi}{\omega_{\text{com}}} = \frac{\lambda_{\text{com}}}{2\pi}\Delta\varphi, \tag{13.2}$$

where v — velocity of radio wave propagation; ω_{com}— frequency comparison, at which the measuring of phase difference is provided. As it follows from Eqs. (13.1) and (13.2), range measurement error equals $\delta D = v\delta\tau_D$ or $\delta D = \frac{\lambda_{\text{com}}}{2\pi}\delta\varphi_D$. Values $\delta\tau_D$ and $\delta\varphi_D$ mainly depend on oscillations instability of reference oscillator and time, passed after synchronization moment.

Value $\delta\tau_D$ can be represented in the following form:

$$\delta\tau_D = \frac{\delta f_{\text{car}}}{f_{\text{car}}} t_{flt} + \delta_{\tau_1}.$$

where $\frac{\delta f_{\text{car}}}{f_{\text{car}}}$—relative instability of a carrier frequency of reference oscillator; t_{flt}—time, passed after synchronization moment, i.e., flight; δ_{τ_1}—propagation time measurement error, stipulated by waves propagation velocity instability, timing inaccuracy of onboard oscillator, interference effect and other factors.

To meet up-to-date requirements, specified for range measurement accuracy, reference oscillators should have a relative frequency instability $\frac{\delta f_{\text{car}}}{f_{\text{car}}} = 10^{-9} - 10^{-10}$. Such a high-frequency stability can be achieved in atomic (clock) and molecular oscillators. Necessity to ensure high frequency stability of reference oscillators is the main disadvantage of RRNS without responder. Advantages of RRNS without responder are concluded in the following:

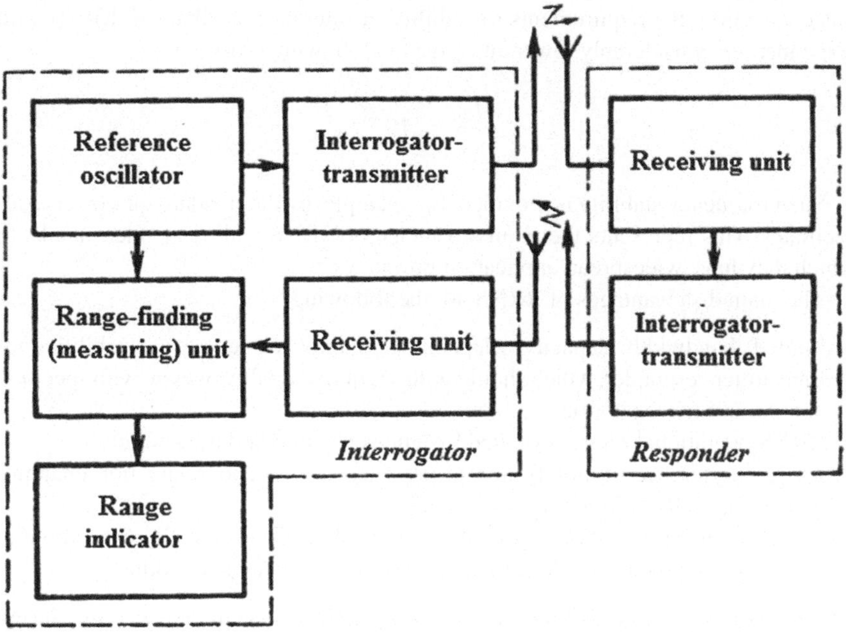

Fig. 13.2 Generalized functional chart of RRNS with responder

- unlimited bandwidth capacity;
- relatively large operational range, since ground station power can be selected as sufficiently great.

Advancements, achieved to the present time in development of quantum oscillators, are making the RRNS quite promising, especially as one ground-based station can operate with unlimited number of objects, located in its coverage zone.

In RRNS with responder, there are two communication channels: interrogation and responding channels. Stable reference oscillator is included only in onboard station and is called an interrogator. The station, which radiates reply pulses, is called a responder. Responder can be installed both on the ground and on object. Generalized RRNS functional chart with responder is depicted in Fig. 13.2. The total propagation time of radiofrequency signals τ_D from responder to interrogator and back with regard to delay time t_{dly} in responder equipment is measured onboard an object:

$$\tau_D = \frac{2D}{v} + t_{\text{dly}}. \tag{13.3}$$

As opposed to RRNS without responder, in RRNS with responder the measuring error in general is determined by instability of reference oscillator and does not depend on time spent by an object en route (e.g., flight time of aerial vehicle). In this connection, the requirements to stability of reference oscillator in RRNS with responder are considerably lower and have the following order.

$$\frac{\delta f_{\text{car}}}{f_{\text{car}}} = 10^{-5}.$$

Such frequency stability is provided, for example, by the standard quartz (crystal) oscillator. This represents the main advantage of RRNS with responder, thanks to which they have widespread application nowadays.

The main disadvantages of RRNS are the following:

- limited bandwidth capacity (depending mainly on energy capabilities of transmitter-responder, which should with a required quality operate with specified amount of interrogators);
- RRNS operational range is limited by a power of onboard transmitter;
- lower interference immunity due to presence of two communication channels comparing to RRNS without responder;
- operation of interrogation and respond channels (links) is usually carried out at different frequencies that lead to expansion of occupied bandwidth.

On air transport, the RRNS are represented by DME system (distance-measuring equipment), which in combination with AV onboard equipment is intended for measuring of slant distance between AV and ground radio beacon. DME radio beacons are installed at airfields and air routes of civil aviation.

The principle of range measurement is based on pulsed (temporal) method. The information on range is within a time interval between a radiation moment of AV onboard equipment of "range request" (RQ) signal and receiving moment of "range response" (RS) signal from ground DME radio beacon. RQ and RS signals are transmitted at different bandwidth using the differing from each other codes. This technique permits to avoid initiation of DME transmitter (and formation of false range responses) by signals of DME radio beacon, reflected from clutters, and to reduce effect of neighboring radio beacons and increase interference immunity of range channel.

Depending on signal format there could be DME/N and DME/P.

DME/N[1]—distance-measuring (ranging) equipment, primarily intended for operational navigational needs servicing en route or in terminal control area, where "**N**" is narrow spectral characteristics.

DME/P[2]—ranging element of MLS landing system, where "**P**" means an precise distance measuring. Spectral characteristics are similar to DME/N. DME/P has two operational modes, IA and FA. **Mode of final approach stage at landing (FA)**—DME/P operation conditions, which provide flights in final approaching zone and in runaway area. **Mode of initial approach stage at landing (IA)**—DME/P operation conditions, which provide flights beyond the final approaching zone and at which

[1] Narrow band distance-measuring equipment.

[2] Precision distance-measuring equipment.

Table. 13.1 Performance characteristics of DME radio beacons

No.	Radio beacon type	DME-90	DME-734	DME 2000	DME 2700
1	Bandwidth, MHz	960…1215	960…1215	962…1213	962…1213
2	Coverage area in HP, in deg	0…360	0…360	0…360	0…360
	in km	260	300	340	360
3	Coverage area in VP, in deg	0…40	0…40	0…40	0…40
4	Range measurement error, in meters - at operation with VOR, DVOR	±75	–	±150	±150
	- at operation with ILS	–	–	±75	±75
5	Operational environment: temperature, °C	−50… +50	−50… +70	−50… +50	−50… +50
	Wind load, m/s	50	55	50	50
	Precipitations, mm/min	–	–	up to 3	up to 3

the interaction with DME/N, is provided. DME operation channels are formed via pairing of interrogating and responding frequencies with pulsed coding on twined frequencies. DME/P channels have two different interrogation pulsed codes. One should be used in initial approach stage mode IA, another one should be used in final approach stage mode FA.

DME operation channels are selected from a table of 352 channels, performed in [Supplement 10 (ICAO)]. For operation of radio beacons of DME system, a bandwidth from 960 up to 1215 MHz is allocated. Tactical and technical characteristics of some modern DME radio beacons are presented in Table 13.1.

RRNS without responder are represented by differential ranging radio navigation systems and satellite navigation systems, which realize a pseudo-ranging position finding method of an object. Sometimes high- and small-altitude radio altimeters (altitude-finding radars), which are intended for measuring of true altitude of AV flight, are referred to special group of RRNS without responder. Radio altimeters, in fact, represent a radar, based on the pulsed (high-altitude altimeters) or frequency (small-altitude altimeters) range measurement method.

Chapter 14
Pseudo-Ranging Radio Navigation Systems

14.1 Pseudo-Ranging Position Finding Method of an Object and Design Concept of Pseudo-Ranging Radio Navigation Systems

In the previous chapter, the range-finding radio navigation systems without a transponder were considered, the principle of which is to measure the distance from a ground station (radio beacon) to an object by the time delay of highly stable radio signals from ground stations received on board the object. The implementation of the ranging method stipulates the presence of highly stable reference generators both at the ground station and onboard an object, which are synchronized with each other at a given time by a special synchronization signal. At the same time, in practice, it is not always possible to provide the required level of synchronization, which leads to errors in the range measurement. For example, a mismatch of 1 ms will result in an error of 300 km.

Attempts to improve the accuracy characteristics and expand the operational area of the range-finding radio navigation systems without a transponder while maintaining their main advantage—the presence of only a receiving device on board an object—led to the development and scientific substantiation of a pseudo-range method for determining the coordinates of an objects. In this case, the pseudo-range method considers the mismatching of the clocks of radio beacons and object equipment.

The core of the method concludes in the following. Highly stable generators of radio beacons, or, more simply, clocks, are synchronized with each other, and the clock of the equipment on the object can be ahead of the clock of radio beacons or lag behind them by some unknown value $\delta\tau$. Taking into account the signal propagation speed v, the measured by the equipment distance $d\,(t)$ to the radio beacon will have an error due to the inaccuracy of the clock of the radio beacon and the equipment of the object $v\delta\tau$

D. A. Akmaykin et al., *Theoretical Foundations of Radar Location and Radio Navigation*, Springer Aerospace Technology,
https://doi.org/10.1007/978-981-33-6514-8_14

$$d(t) = \sqrt{(x_{\mathrm{rb}}(t) - x_{\mathrm{ob}}(t))^2 + (y_{\mathrm{rb}}(t) - y_{\mathrm{ob}}(t))^2} + \nu\delta\tau, \tag{14.1}$$

where $x_{\mathrm{rb}}(t)$, $y_{\mathrm{rb}}(t)$, $x_{\mathrm{ob}}(t)$, $y_{\mathrm{ob}}(t)$—are the coordinates of the radio beacon and the object, respectively.

Value $d(t)$ is called ***pseudo-range***. The measured values of pseudo-ranges from an object to three radio beacons by solving a system of equations permit

$$\begin{aligned}
d_1(t) &= \sqrt{(x_{\mathrm{rb}1}(t) - x_{\mathrm{ob}}(t))^2 + (y_{\mathrm{rb}1}(t) - y_{\mathrm{ob}}(t))^2} + \nu\delta\tau, \\
d_2(t) &= \sqrt{(x_{\mathrm{rb}2}(t) - x_{\mathrm{ob}}(t))^2 + (y_{\mathrm{rb}2}(t) - y_{\mathrm{ob}}(t))^2} + \nu\delta\tau, \\
d_3(t) &= \sqrt{(x_{\mathrm{rb}3}(t) - x_{\mathrm{ob}}(t))^2 + (y_{\mathrm{rb}3}(t) - y_{\mathrm{ob}}(t))^2} + \nu\delta\tau,
\end{aligned} \tag{14.2}$$

to determine an object coordinates on a plane and calculate the value of clock drift of the object's equipment. It is easy to understand that an error $\nu\delta\tau$ will be the same for all measured pseudo-ranges, because it is caused by a common reason—the clock departure of the object's equipment. To solve the problem of determining the spatial coordinates of an object (e.g., an aircraft), it is required to measure at least four pseudo-ranges.

The pseudo-range method has found wide application in satellite radio navigation systems (SRNS) GLONASS, GPS, Galileo, etc. The navigation space vehicles (NSV) are used as radio beacons, the coordinates of which are calculated in the object's equipment at the time of signal emission from ephemeris information transmitted by each NSV in the service message. In addition, the service message contains the value of the clock correction of each NSV relative to a single system time scale. Pseudo-range measurements of up to four satellites are required to assess the spatial coordinates and corrections to time scale of the user equipment

$$d_i(t) = \sqrt{(x_{\mathrm{rb}i}(t) - x_{\mathrm{ob}}(t))^2 + (y_{\mathrm{rb}i}(t) - y_{\mathrm{ob}}(t))^2} + c\delta\tau, \tag{14.3}$$

where $c = 3 \times 10^8$ м/с, $i = 1,\ldots,4$.

System (14.3), strictly speaking, has two solutions, one of which can be discarded by indirect signs, for example, the location of an object below the earth's surface or in space.

SRNS users (ships, aircraft, etc.) need not rectangular, but geodetic coordinates—latitude, longitude and altitude. It is known that the relationship of geocentric coordinates x, y, z with latitude φ, longitude λ and height H is expressed by the following relations:

$$\begin{aligned}
x &= (N + H)\cos\varphi\cos\lambda, \\
y &= (N + H)\cos\varphi\sin\lambda, \\
z &= \left((1 - e^2)N + H\right)\sin\varphi,
\end{aligned} \tag{14.4}$$

where N—is the radius of curvature of the first vertical, determined by the formula $N = a\left(1 - e^2\sin^2\varphi\right)^{-\frac{1}{2}}$; a—semi-major axis of the earth ellipsoid, e—eccentricity of the earth ellipsoid.

The calculator of the object equipment also solves the inverse problem with respect to the above formulas—using already known rectangular coordinates, it calculates the latitude, longitude and height above the surface of the earth's ellipsoid.

Let us now consider how pseudo-range is measured in user equipment. All NCV SRNS are equal in their system. Each satellite transmits a coded signal through the transmitting antenna on two carrier frequencies in the L1 and L2 bands. Signal emission in two frequency ranges is necessary to compensate for ionospheric signal propagation errors.

The signals transmitted by the NSV are continuous modulated oscillations. The information transmitted by the NSV includes:

- pseudo-random ranging code, which is used to measure the range to the satellite,
- a navigation message containing the information necessary for the user (NSV ephemeris, system almanac, data on the condition of the NSV, shifts of NSV time scales relative to UTC, etc.).

A pseudo-random ranging code represents a long, periodically repeating sequence of video pulses. This sequence looks completely random, but in fact it is formed according to a very definite law. This law is the code, without knowledge of which it is impossible to obtain information from the NSV. The codes of all NSV systems are recorded in the computer memory of the user equipment. It also generates a code (sequence of pulses) identical to that received from the NSV.

The process of information obtaining (pseudo-range values) is as follows. When the receiver is turned on, it starts generating a code corresponding to the first satellite in the list and evaluates the coincidence of the generated pseudo-random code with the pseudo-random code in the received radio signal (Fig. 14.1). Of course, the sequences of pulses will not coincide by the reason that they are shifted relative to each other by the value, corresponding to the signal passing time from the NSV to the object's receiver. If they do not match, then the receiver shifts the generated sequence in time by a small amount and tries to find matches again. Such shifts continue until the sequences match. If they did not coincide, then this may simply means that the NSV is out of sight. In this case, the receiver starts to generate the code of the next

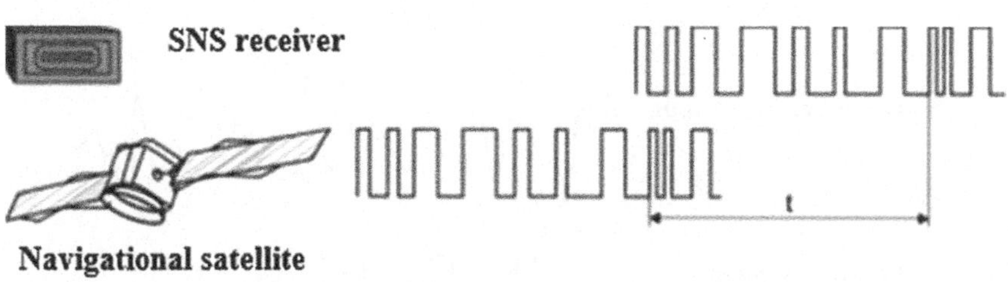

Fig. 14.1 Pseudo-range measurement principle

satellite, performs its shift, and the whole procedure is repeated. This process can take several minutes.

When a signal from at least one satellite is received, the process goes faster. The pseudo-range $d(\mathrm{t})$ is determined by the value by which the sequences had to be shifted so that they coincide. Then a navigation message is received, containing the almanac of the system (orbits parameters of all NSV and other service information). Based on these parameters, the receiver can already estimate which NSVs are now within sight, and begins to purposefully "search" for their signals. After signal receiving from four NSV, it is possible to determine the spatial position of the user and other necessary parameters.

The received almanac is stored in the receiver's memory; therefore, when the receiver is turned on the next time, it will immediately calculate which NSV can be in the field of view and will first of all try to receive signals from them, and in a few seconds the navigation information will be received. If the receiver has not been turned on for a long time and the almanac is outdated, or if the receiver has been moved far to another location in the off-state, then the process will take longer.

It should be noted that the power of the signal received from the NSV is negligible and is on the order of 10^{-14} W. The signals are so weak that they are simply lost against the background of the earth's natural radio emission, atmospheric interference and thermal noise from the receiver itself. However, all these noises are random variations of electronic pulsations, and the received pseudo-random code—is a strictly defined sequence of video pulses. Since the pseudo-random code sequence is periodically repeated, then by using algorithms based on methods of optimal reception, it is possible to perform multiple comparison of the received signals and select the pseudo-random code against the background of the natural radio noise of the Earth.

Figure 14.2 schematically depicts a received radio signal, which possibly also contains a code sequence. The detection of the NSV signal is carried out using correlation reception. The correlation receiver is based on calculating the cross-correlation of the received signal from the NSV $s(t)$ and the signal generated in the receiver $s_0(t)$, which is a copy of the useful signal from the satellite:

$$q = \int_0^T s(t)s_0(t)\mathrm{d}t, \tag{14.5}$$

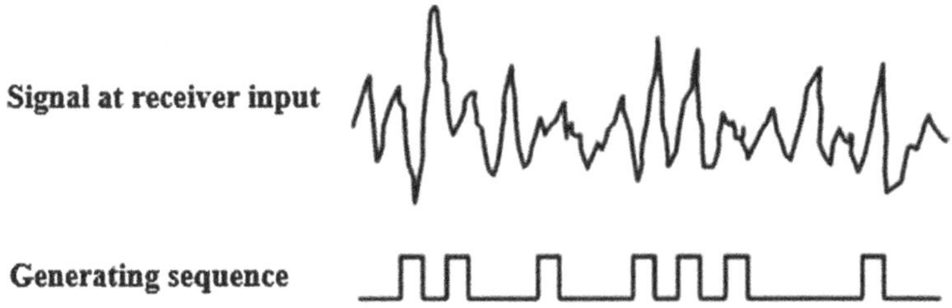

Fig. 14.2 Received signal and reference sequence

The correlation receiver allows to obtain the best value of the signal-to-noise ratio at the output and thereby ensure reliable detection of the NSV signal.

The use of optimal methods of NSV signals processing permits to use antenna of a small size, and in general, the user equipment has a relatively small size, weight and relatively low cost.

One of the most important reasons for using a pseudo-random code in an SRNS—is the using possibility by all satellites the same carrier frequency in their transmitters. But since each NSV transmits a code inherent only to it, the receiver can easily distinguish the signals of a particular satellite. In addition, the use of a pseudo-random code in the SRNS allows to control the mode of access to it.

14.2 Factors Affecting the Accuracy of Satellite Radio Navigation Systems

The user's spatial location in the SRNS is determined relatively to the NSV. It is clear that the more precisely the position of the satellite in orbits is known, the more accurately the coordinates of the object will be determined. The current coordinates of the satellite are calculated by the onboard receivers using known orbital elements (ephemeris). Orbital elements of the NSV are calculated on the ground and periodically transmitted to the NSV. In addition to the orbital elements themselves, their derivatives (rates of change caused by disturbing factors) are also transmitted. All these data as part of the navigation message are sent to the onboard receiver, which calculates the current coordinates of the satellite. The errors in determining the AV coordinates increase over time, because the calculated data are outdated. The error in calculating the coordinates of the NSV is the greater the more time has passed since the time, at which the orbit parameters were determined. At the same time, these parameters are updated quite often, so a significant error, as a rule, does not have time to accumulate.

The error in determining the ephemeris (coordinates) of the NSV is closely related to the error in time measuring. The SRNS operates in its own system time, determined by a high-precision ground central synchronizer (instability 10^{-13}–10^{-15}). However, this system time deviates from UTC and the value of the deviation is periodically transmitted to each satellite. Each NSV uses its own frequency standard (i.e., clock), its own time scale with the worst stability. Therefore, in fact, each satellite uses its own time. It is periodically corrected from the Earth by entering corrections, but does not completely coincide with the system time, for which the orbital elements are calculated. It turns out that the onboard receiver determines the coordinates of the satellite not quite for the moment in time for which it is required, which is equivalent to an additional error. Along the way, we note that the error of the onboard clock of the receiver (on AV) does not make any contribution to the error, since it is continuously considered in the process of measuring the pseudo-range (see Formula 14.1). As a result of the inaccuracy of ephemeris information and time, the error in determining the coordinates of a satellite, as a rule, does not exceed one meter. If, for some reason, the ephemeris information was not in-time updated, then the error increases sharply and after 4 h it will be hundreds of meters. In this case, it was about determining the coordinates of the NSV in real time, which is important for air and sea transport objects. If you process the data received from all tracking stations (it takes about 12 days), then "retroactively" you can determine where the satellite was at any time with an accuracy of 2–3 cm. Such data are widely used in geodetic and other scientific research.

Pseudo-range measurement errors. The navigation parameter in the SRNS is the pseudo-range, calculated from the measured propagation time of radio waves and the propagation velocity, embedded in the onboard receiver. When radio waves pass through any medium, the speed of radio waves changes and depends on the characteristics of this medium. The radio waves, emitted by the NSV, pass through the atmosphere, the characteristics of which change in space and time. There are ***ionospheric*** and ***tropospheric errors***. The ionosphere is the upper part of the atmosphere and contains free electrons. When radio waves pass through the ionosphere, their trajectory is curved, and the speed of radio waves changes. It no longer coincides with the speed that is included in the SRNS receiver for calculating the pseudo-range. This causes the ionospheric error in measuring the pseudo-range, the magnitude of which is different in different places of the planet, changes depending on the time of year and day, and is influenced by solar activity and cosmic radiation. Usually, ionospheric errors are 8–10 m, but they can reach 40–100 m and more. Tropospheric errors occur in the lower atmosphere. They do not depend on the signal frequency, but depend on temperature, pressure, humidity. These quantities are difficult to predict in order to be used in mathematical models. The residual value of the tropospheric error is of the order of 1–4 m.

Another reason for inaccurate pseudo-range measurement is ***radio signal multipath propagation***. If local objects (hills, buildings, airport facilities, etc.) are located near the onboard receiver, then the radio signal arrives at the receiver not only by a direct path, but also after re-reflection from local objects (terrain features). This causes an error when flying near the ground, for example, during an approach for landing, which is especially important, since higher navigation accuracy is required at this stage of the flight. In good conditions, the error due to multipath is a few meters, in urban conditions it can reach several tens of meters. To reduce this error, various methods are used—from the use of special antennas up to the use of complex signal processing algorithms (e.g., optimal filtering). As a result of these measures, the error can be reduced (for the high-precision GLONASS code and the P(Y) GPS code), at best, up to 1–3 m, at worst—up to 8 m.

Geometric factor. The accuracy of determining the coordinates of the SRNS receiver depends not only on errors in measuring the distances to the NSV, but also on the relative position of the object and the satellite. It is known that the error in determining the location of an object is inversely proportional to the sine of the angle between the lines of position, along which this location is determined. All other things being equal, the best accuracy will be at an intersection angle of 90° ($\sin(90°) = 1$). If, for example, this angle is 30°, then the accuracy will be two times worse.

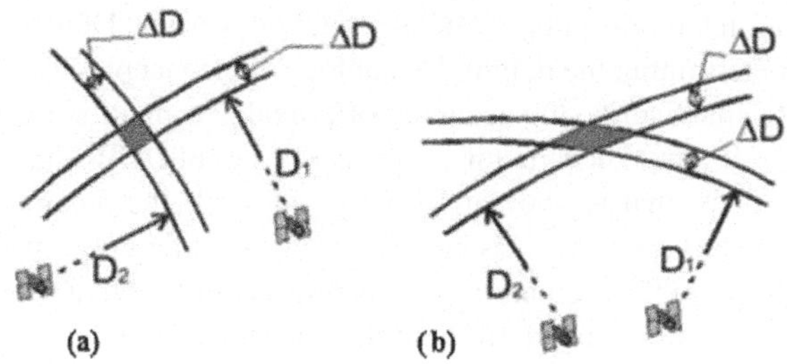

Fig. 14.3 Geometric factor

In Fig. 14.3, it can be seen that with a decrease in the angle between the lines of position, the zone of uncertainty of the object's location increases.

To assess the influence of the relative position of the NSV and the object receiver on the accuracy of determining the coordinates, the DOP value—Dilution of Precision (geometric reduction in accuracy) is used. This is a dimensionless quantity that characterizes the geometric factor and shows how many times the accuracy of determining the coordinates is worse than the accuracy of measuring the pseudo-range. In the abovementioned example for a plane (with an angle of 30° between the position lines), the DOP would be equal to 2. For the SRNS, where the spatial position of the object is determined and four position surfaces are used, things are a little more complicated, but the meaning remains the same. The DOP value can be illustrated by the following geometric considerations. The position points of the four satellites: S1, S2, S3, S4, and the onboard receiver form a polyhedron in space called a tetrahedron. The larger its volume V, the better the relative position of the satellites and the receiver, the higher the accuracy, the lower the DOP. The largest volume of the tetrahedron will be when one satellite is at the zenith and three satellites are located near the horizon and are equally distributed in azimuth. The value of the DOP criterion is taken inversely proportional to the volume of the tetrahedron V, taking into account a some proportionality coefficient k:

$$\mathrm{DOP} = \frac{k}{V}. \tag{14.6}$$

The DOP value can theoretically take values from 1 up to, as a rule, 10. It is believed that at DOP ≤ 4, the high accuracy of coordinate determination is ensured, DOP values > 6 indicate an unsatisfactory accuracy.

In practice, several varieties of DOP are used. After all, using the SRNS, four main parameters are measured: two horizontal coordinates (latitude and longitude), a vertical coordinate (height) and time. For a comprehensive assessment of all four parameters, the GDOP (Geometrical DOP) is used. To assess the accuracy of coordinates determining of an object spatial location (without considering the time accuracy), the PDOP (Position DOP) is used. If it is required to determine not the radial error in determining the object spatial location, but the accuracy of determining individual coordinates, then similar indicators are used: HDOP—(Horizontal DOP)—characterizes the accuracy of determining coordinates in the horizontal plane, that is, in latitude and longitude; VDOP—(Vertical DOP)—characterizes the accuracy of determining the height. By analogy, the concept of TDOP (Time DOP) is introduced to characterize the accuracy of time determination by an onboard receiver.

It should be remembered that all types of DOPs characterize an accuracy in relative values, that is, they are dimensionless. If the mean-square error of measuring the pseudo-range σ_D was known precisely under these specific conditions, then it would be possible to quantitatively determine in absolute values (in meters) the root mean-square error in determining the coordinates σ_R.

$$\sigma_R = \text{DOP}\sigma_d. \tag{14.7}$$

Depending on what kind of error you need to determine (the spatial position of the aircraft, in the horizontal plane or in height), PDOP, HDOP VDOP should be used as DOP in this formula.

Most navigation receivers indicate the current DOP value, which allows the user to judge the accuracy of the navigation definitions. This is especially important for navigation in areas where certain requirements for its accuracy are established.

Interference. The operating experience of the SRNS shows that the signals emitted by the NSV are subject to various types of interference. Unintentional man-made interference includes emissions from radio transmitters that can generate signals with unwanted L-band power levels. Artificial unintentional interference is created by radio lines, harmonics of television channels, request signals of short-range navigation systems, harmonics of existing VHF radio stations, the Globalstar satellite communication system, radar stations of air traffic control systems. Portable electronic devices used by passengers on board an AV (computers, mobile phones) can also interfere with the SRNS, as well as other onboard radio navigation systems. Another factor leading to errors in pseudo-range measurement is the ***noises*** of the object onboard receiver itself.

We have examined only the main factors affecting the SRNS measurements accuracy. Modern equipment of SRNS users is multichannel and uses methods of optimal signal processing. As shown by experimental studies and the practice of using the SRNS, the global navigation of objects of sea and air transport can be provided with maximum errors in determining the spatial coordinates of an object of 30–60 m on a plane and 50–100 m in height.

SRNS are constantly being upgraded in order to provide higher accuracy. In addition, various options for functional updates to the SRNS are being developed, which will be discussed below.

14.3 Satellite Radio Navigation Systems

SRNS are autonomous medium-orbit satellite systems that allow with a high accuracy to determine the spatial coordinates of moving and stationary objects on the Earth's surface and in near-earth space, as well as to carry out accurate time coordination. The operation organization of various systems, primarily GPS, GLONASS and Galileo, is similar. The e SRNS consists of three main segments:

- subsystems of space vehicles, that is, NSV;
- control and management subsystem, which includes a control center and a network of ground stations for measurement, control and monitoring;
- navigation equipment of users, including onboard receivers of the SRNS.

Ground stations SRNS solve the following main tasks:

- determination and forecasting of NSV coordinates (ephemeris) and parameters of its orbits,
- synchronization of the time scales of each satellite with the system time,
- transfer of an array of service information to the NSV,
- transfer of an array of service information to the NSV,

The information transmitted from the ground to each satellite includes the parameters of the orbits of all the NSV and its state (serviceability), corrections to time scales and to the carrier frequency, as well as other data. Since, due to gravitational perturbations, the orbital elements are continuously changing, not only the orbital parameters themselves are transmitted, but also the coefficients of the polynomials, which can be used to calculate the rate of change of these parameters and refine the orbital elements at any time. The NSV includes an onboard navigation transmitter, chronizer (i.e., NSV clock), orientation and stabilization system, control complex, and other systems that ensure the operation of the NSV.

Users' navigation equipment consists of navigation receivers and computing devices intended for processing of navigational signals. This equipment performs non-query measurements of the pseudo-ranges and radial velocities of the satellite, as well as the calculations necessary to obtain navigation information by users.

Below we will briefly examine the main SRNS and perform its characteristics.

Global satellite navigation system GLONASS.

The GLONASS system provides the user with two types of services—standard and high precision. Standard precision services are provided to users by transmitting standard precision signals in the L-band. Each GLONASS-M satellite transmits navigation radio signals with frequency division in two bands: L1 (1.6 GHz) and L2 (1.25 GHz).

A standard accuracy signal with a clock frequency of 0.511 MHz, intended for use by domestic and foreign civilian users, is available to all consumers equipped with the appropriate airborne instrument, in the observation zone of which there are GLONASS satellites.

The regular GLONASS orbital constellation consists of 24 satellites (see Fig. 14.4) located in medium-altitude near-circular orbits with nominal altitudes of 19,100 km, inclination of 64.8° and a period of 11 h 15 min 44 s. The value of the period made it possible to create a stable orbital system that does not require, in contrast to GPS orbits, for its maintaining of correcting impulses, practically during the entire period of active orbital lifetime. The nominal inclination ensures 100% availability

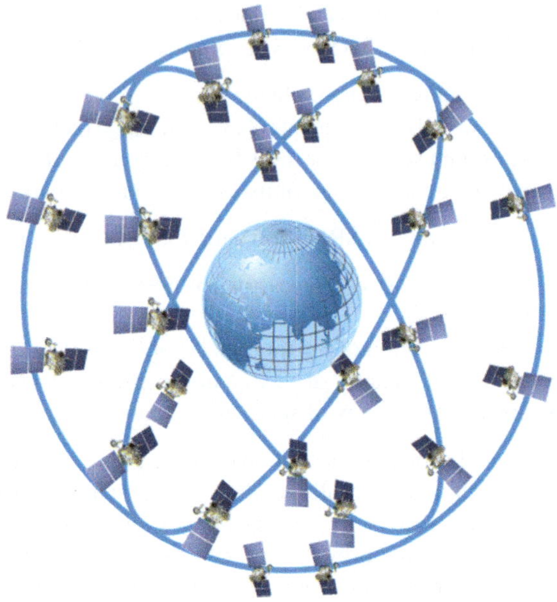

Fig. 14.4 GLONASS orbital constellation

of navigation on the territory of the Russian Federation, even if several NSV satellites leave the orbital group.

The characteristics of the GLONASS orbital constellation are given in Table 14.1.

The current state of the GLONASS system can be found on the website of the Information and Analytical Center for Coordinate Time and Navigation Support. Table 14.2 shows the technical characteristics of the NSV of the GLONASS system.

At the design stage, a frequency method was adopted for the GLONASS system for signals separating from various satellites: Each of them uses its own pair of carrier frequencies, one of which belongs to the L1 band, the other to the L2 band. For NSV that are located at diametrically opposite points of the orbit, the same letter frequencies are used, 12 in each frequency range.

The GLONASS-K satellite of the first stage, launched into orbit in 2011 for flight tests, along with L1 and L2 radio signals with frequency division, completely analogous to the GLONASS-M signals, additionally emits open-access radio signals

Table 14.1 Characteristics of GLONASS orbital constellation

Number of NSV (regular)	24
Orbital altitude, km	19,100
Number of planes	3
Semi-major axis, km	25,420
Period	11 h 15 min 44 s
Inclination	64.8°

Table 14.2 NSV technical characteristics of GLONASS system

Characteristics	NSV «GLONASS»	NSV «GLONASS -M»	NSV «GLONASS -K»	NSV «GLONASS -K2»
Deployment year	1982–2005	2003–2016	2011–2018	After 2017
State	Out of service	In-service	Being developed based on tests carried out	Development phase
Active state period, year	3.5	7	10	10
Weight, kg	1500	1415	935	1600
Daily instability (actual)	$\sim 10^{-13}$	$\sim 5 \times 10^{-14}$	$\sim 5 \times 10^{-14}$	$\sim 5 \times 10^{-15}$
Signal type	FDMA	FDMA (CDMA on NSV No755-761)	FDMA and CDMA	FDMA and CDMA
Open-access signals	L1OF (1602 MHz)	L1OF (1602 MHz) L2OF (1246 MHz) L3OC (1202 MHz) (beginning with NSV No 755)	L1OF (1602 MHz) L2OF (1246 MHz) L3OC (1202 MHz) L2OC (1248 MHz) (beginning with NSV No 17L)	L1OF (1602 MHz) L2OF (1246 MHz) L1OC (1600 MHz) L2OC (1248 MHz) L3OC (1202 MHz)
Signals with authorized access	L1SF (1592 MHz) L2SF (1237 MHz)	L1SF (1592 MHz) L2SF (1237 MHz)	L1SF (1592 MHz) L2SF (1237 MHz) L2SC (1248 MHz) (beginning with NSV No 17L)	L1SF (1592 MHz) L2SF (1237 MHz) L1SC (1600 MHz) L2SC (1248 MHz)
Orbital launch vehicles		«Souz-2.16», «Proton-M» launch-vehicle		

with a code division. The upgraded GLONASS-M satellites also emit a code division navigation radio signal in the L3 band.

Figure 14.5 shows the spectrum of navigation radio signals of the GLONASS system with frequency and code division of channels.

Characteristics of navigation radio signals of the GLONASS system with code division are given in Table 14.3.

Ephemeris transmitted by each GLONASS vehicle as part of operational information describes the position of the phase center of the transmitting antenna of this NSV satellite in the PZ-90 geocentric coordinate system associated with the Earth, defined as follows:

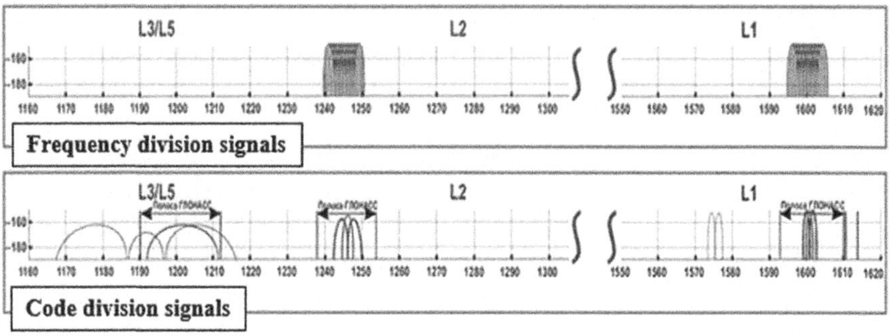

Fig. 14.5 Spectrum of navigation radio signals of GLONASS system

Table 14.3 Characteristics of navigation radio signals of GLONASS system

Band	Carrier frequency, MHz	Signal	Duration of PRS code, symbols	Clock frequency, MHz	Modulation type	Transfer rate, bit/s
L1	1,600,995	L1OCd	1023	1023	BPSK(1)	125
		L1OCp	4092	1023	BOC(1,1)	
L2	124,806	L2 CSI	1023	1023	BPSK(1)	125
		L2OCp	4092	1023	BOC(1,1)	
L3	1202,025	L3OCd	10,230	10,023	BPSK(10)	100
		L3OCp	10,230	10,023	BPSK(10)	

- The origin is located at the center of mass of the Earth;
- *Z*-axis is directed to the reference pole of the Earth, as defined in the recommendation of the International Earth Rotation Service (IERS);
- *X*-axis is directed along the line of intersection of the plane of the Earth's equator and the prime meridian established by the International Time Bureau (ITB);
- *Y*-axis complements the geocentric rectangular coordinate system to the right.

As the GLONASS system time scale, a conditional continuous time scale is adopted, formed on the base of the time scale of the central synchronizer of the system. The central synchronizer is equipped with hydrogen frequency standards.

The reference time scale for the GLONASS system is the national coordinated time scale of Russia UTC (SU). The discrepancy between the system time scale GLONASS and UTC (SU) should not exceed 1 ms.

The GLONASS system time scale is adjusted simultaneously with the planned adjustment by an integer number of seconds of the UTC scale.

Global satellite navigation system GPS.

The GPS provides two types of services:

- Standard Positioning Service (SPS), available to all users;
- Precise Positioning Service (PPS) available to authorized users.

The service of standard positioning SPS and time synchronization is available for all categories of consumers free of charge and globally and is realized through the emission of navigation radio signals by all GPS SV, modulated by the rangefinder code C/A (Coarse / Acquisition). The C/A code is a 1023 symbol Gold pseudo-random sequence (PRS) with a 1.023 MHz clock rate. Thus, the PRS of C/A-code has a repetition period of $T = 1$ ms, which corresponds to a pseudo-range unambiguous measurement interval of about 300 km. The GPS Development Program envisions the provision of SPS services to civilian customers using L2C, L5 and L1C signals.

The PPS is realized by the emission of navigation radio signals in the L1 and L2 bands, modulated by the ranging P(Y) code, by all space vehicles of the GPS orbital constellation. The PPS service is intended for use exclusively by the US military, US federal agencies, and some Allied military forces.

The regular GPS orbital constellation consists of 32 main satellites located in six circular orbits, denoted by Latin letters from A to F. Additionally, in some orbits there may be one or two backup satellites designed to maintain the system parameters in the event of the main satellites failing. The inclination of the orbital planes is 55°, and the longitudes of the ascending nodes differ by 60°. An orbital altitude of 20,200 km corresponds to an orbital period of 11 h 58 min, i.e., the orbits of GPS vehicles are synchronous.

Currently, the replenishment of the orbital constellation is carried out by the launch of Block IIF space vehicles ("F"—follow on). In accordance with the current plans, Block IIF satellites must replace Block IIA satellites in orbit, Block III satellites will replace Block IIR ("R"—replacement). The main task of the Block III NSV is to provide navigation services using the new L1C navigation radio signal and to improve the accuracy of ephemeris-time information, the availability of the navigation radio signal, the radiation power, as well as increase the active lifetime. Table 15.5 shows the satellites technical characteristics of the GPS. The spectrum of GPS navigation radio signals is shown in Fig. 14.7 (Fig. 14.6).

The characteristics of the orbital GPS constellation are given in Tables 14.4 and 14.5.

The characteristics of the GPS navigation radio signals are given in Table 14.6.

The introduction of new GPS navigation signals is accompanied by the improvement of the structure of digital information and the use of new types of modulation, as well as the transition from the structure of the navigation message of the NAV type to the structures of the CNAV and CNAV-2 type. Navigational messages such as CNAV are improved versions of the navigation message NAV, allowing more accurate transmission of real-time and non-real-time information about the state of the GPS. The CNAV navigation message contains information of the same type as the NAV message (current time, signs of the satellite state, ephemeris-time information, system almanac, etc.), but this information is transmitted in a new format. Instead of using a superframe/frame architecture, the message is transmitted as packets of different lengths. The most significant changes in the CNAV structure are the expansion of the number of space vehicles used for its intended purpose from 32 up to 63, as well as the ability to promptly transmit data on the performance of a particular vehicle (integrity) with a delay of less than 6 s.

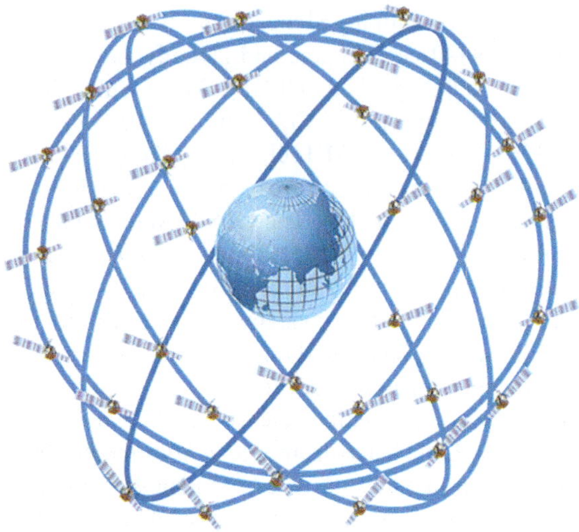

Fig. 14.6 GPS orbital constellation

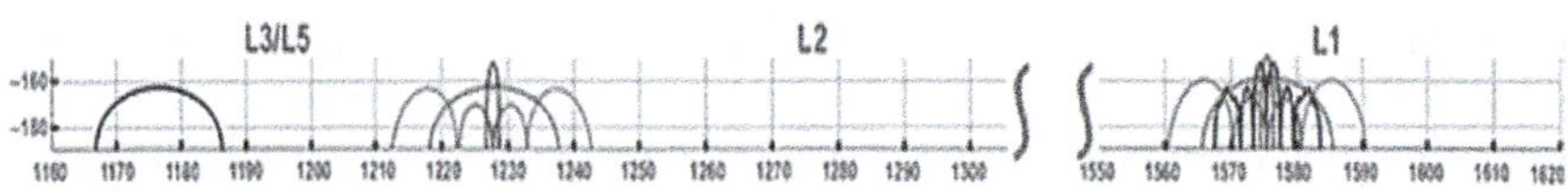

Fig. 14.7 Spectrum of GPS navigation radio signals

Table 14.4 Characteristics of GPS orbital constellation

Number of NSV (regular)	32
Orbit altitude (km)	20,200
Number of planes	6
Semi-major axis (km)	26,560
Period	11 h 58 min
Inclination	55°

The GPS uses the 1984 World Geodetic System WGS-84. Another refinement of the parameters of the WGS-84 (G1678) system took place in 2012. The GPS time is associated with coordinated universal time UTC as observed by the United States Marine Observatory (USNO). Nominally, the GPS time scale has a constant, equaling to 19 s, discrepancy from the international atomic time TAI.

The GPS orbital constellation is controlled by the 2nd Operational Space Squadron of the US Air Force Space Command. Currently, the GPS orbital constellation is controlled by the second-generation ground control complex (Operational Control Segment—OCS), which includes:

- the main control center for the GPS at Schriever Air Force Base;

Table 14.5 Technical characteristics of GPS NSVs

Characteristics	NSV Block IIA	NSV Block IIB	NSV Block IIB-M	NSV Block IIF	NSV Block III
Contractor	Rockwell International	Lockheed Martin	Lockheed Martin	Boeing	Lockheed Martin
Active state period, year	7.5	10	10	12	15
Weight, kg	985	11,267	11,267	14,651	2161
Signals	L1 C/A L1/2 P(Y)	L1 C/A L1/2 P(Y)	L1 C/A L1/2 P(Y) L2C L1/2 M-Code	L1 C/A L1/2 P(Y) L5I L1M L2M L2C	L1 C/A L1/2 P(Y) L1C L2C L2M L5 L1/2 M-Code

Table 14.6 Characteristics of GPS navigation radio signals

Band	Carrier frequency, MHz	Signal	Duration of PRS code, symbols	Clock frequency, MHz	Modulation type	Transfer rate, bit/s
L1	157,542	C/A	1023	1,023	BPSK	50/50
		P	~7 days	1023	BPSK	50/50
		M	–	5115	BOC(10,5)	–
		L1Co	10,230	1023	BOC(1,1)	100/50
		L1Cp	10,230·1800	1,0231,023	TMBOC(6,1/11)	–
L2	12,276	P	~7 days	1023	BPSK	50/50
		L2C	M:10,230 L:767,250	1023	BPSK	50/25
		M	–	5115	BOC(10,5)	–
L5	117,645	L5I	10,230·10 10,230·20	10,023 10,023	BPSK BPSK	100/50
		L5Q				

- backup control center for the GPS;
- monitoring stations of the National Geospatial Intelligence Agency;
- global network of upload and measuring stations;
- US Air Force GPS monitoring stations;
- L-band GPS interrogation stations.

The ground-based GPS control complex implements a request-free technology of ephemeris-time support. The global network of command and measurement stations permits to upload information on board with 4–6 h interval.

Global satellite navigation system GALILEO.

The global navigation satellite system GALILEO is built up by the European Union to ensure the independence of the member states in the field of coordinate time and navigation support. The European GNSS Program was officially approved in 1994, when the European Council asked the European Commission to take steps to develop information technology, including satellite navigation. It was decided to develop two directions. The first of them is systems of functional supplement to existing GPS and GLONASS. This program is called the European Geostationary Navigation Overlay Service (EGNOS). The second direction was to build its own SRNS, intended for civilian use and built on the principles of public–private partnership. In 1999, the European GNSS project was codenamed GALILEO in honor of the Italian astronomer Galileo Galilei. Experimental satellites GIOVE-A and GIOVE-B were launched into orbit on December 28, 2005 and April 27, 2008, respectively. The main task of GIOVE-A was to assess the accuracy characteristics of the GALILEO navigation radio signals in all frequency bands, and GIOVE-B—to test the navigation payload.

The first two NSVs were launched on October 20, 2011, using a Soyuz-STB rocket from the Kourou launch site. The launch technology of the GALILEO satellites assumes group launches of two satellites on the Russian Soyuz carrier rocket and four satellites each on the European Ariane-5 missile.

The fully deployed GALILEO orbital constellation will provide three modes of navigation service and provide the following types of navigation services:

- Open Service—open signals, without subscription and other fees, available to all types of users;
- Commercial Service—encrypted signal, access to two additional signals, higher data transfer rate. The commercial service will provide two functions—global high-precision navigation and navigation signal authentication. For the technical implementation of the commercial service CS signals of an open service plus two encrypted signals in the E6 band (GALILEO signals) will be used;
- service with regulated access by the state (Public Regulated Service)—for the coordinate time provision of regulated users (two PRS signals with encrypted rangefinder codes).

The orbital architecture of GALILEO assumes that there will be 27 NSVs in orbit in three circular orbits with an altitude of 23,229 km, an orbital period of 14 h, and an inclination of 56°. A total of 24 satellites are used for its intended purpose, one satellite in each orbital plane is a backup one. Such a constellation configuration was chosen based on the guaranteed fulfillment of requirements in accuracy and accessibility at the minimum expenditures on orbit correction during the lifetime of the space vehicle.

The characteristics of the orbital GALILEO grouping are given in Table 14.7. Table 14.8 shows the NSV technical characteristics of the GALILEO system.

The spectrum of the navigation radio signals of the GALILEO system is shown in Fig. 14.9. The characteristics of the navigation radio signals of the GALILEO system are given in Table 14.9 (Fig. 14.8).

Table 14.7 Characteristics of the GALILEO orbital constellation

Number of NSV (regular)	27 (+3 backup)
Orbit altitude, km	23,222
Number of planes	3
Semi-major axis, km	29,640
Period	14 h 4 min 45 s
Inclination	56°

Table 14.8 NSV technical characteristics of the GALILEO system

Characteristics	NSV GALILEO IOV	NSV GALILEO FOC
Contractor	EFDS Astrium GmbH	OHB AG
Active lifetime period, year	12	More than 12
Weight, kg	700	730
Signals	E1, E5, E6	E1, E5, E6

Table 14.9 Characteristics of the navigation radio signals of the GALILEO system

Band	Carrier freq., MHz	Signal	Duration of PRS code, symbols	Clock frequency, MHz	Type of modulation	Transfer rate, bit/s
E1	157,542	E1A	–	25,575	BOC(15,2,5)	50/100
		E2B	4096	1023	MBOC(6,1,1/11)	125/250
		E1C (pilot)	4092	1,023	MBOC(6,1,1/11)	–
E6	127,875	E6A	–	5115	BOC(10,5)	50/100
		E6B	5115/1	5115	BPSKBPSK	500/1000
		E6C (pilot)	5115/100	5115	BPSK	-
E5	119,179	E5a-I	10,230/20	1023	AltBOC(15,10)	25/50
		E5a-Q (pilot)	10,230/100	1023	AltBOC(15,10)	–
		E5b-I	10,230/4	1023	AltBOC(15,10)	125/250
		E5b-Q (pilot)	10,230/100	10,23	AltBOC(15,10)	–

The GALILEO system uses the traditional geocentric Cartesian coordinate system called the Galileo Terrestrial Reference Frame (GTRF). This coordinate system is associated with the international terrestrial coordinate system ITRF and is defined in such a way that its discrepancy with the ITRF does not exceed 3 cm with a probability of 0.95. To support the GTRF, a special geodetic service, GALILEO, has

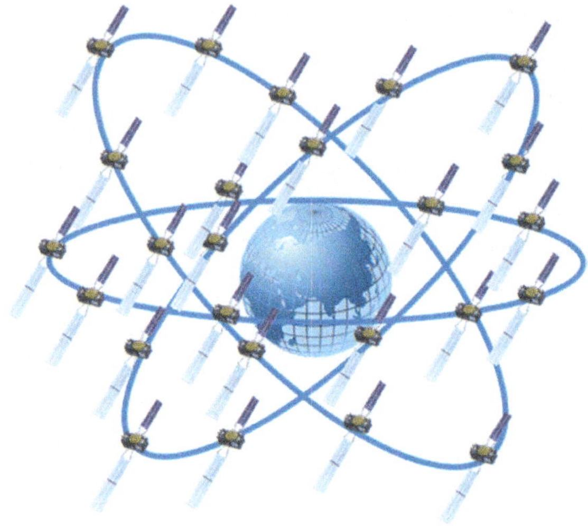

Fig. 14.8 GALILEO orbital constellation

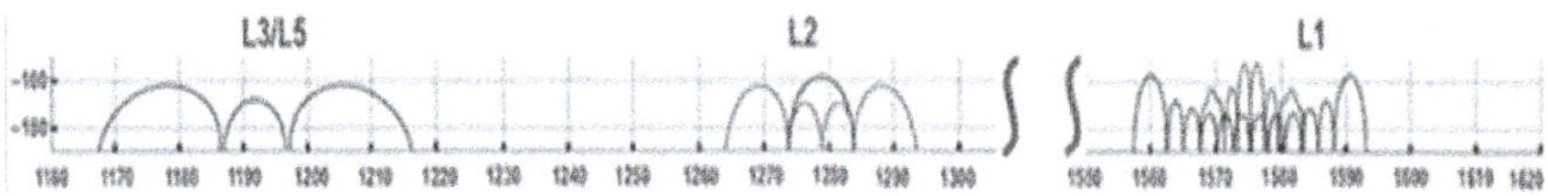

Fig. 14.9 Spectrum of navigation radio signals of GALILEO system

been formed, which also ensures the participation of the international community in defining and maintaining the GTRF coordinate system.

The Galileo System Time (GST) system time scale is a continuous atomic time scale with a constant offset of an integer number of seconds relative to TAI international atomic time. The GST has a variable discrepancy with the UTC time scale by an integer number of seconds.

The GST scale is maintained by a system of atomic frequency standards based on active hydrogen generators. To correct the GST, the GALILEO ground control complex synchronization system receives information on the TAI time scale from the International Bureau of Weights and Measures. According to the technical requirements for the GALILEO system, the discrepancy between GST and TAI should not exceed 50 ns with a probability of 95%.

Information on the discrepancy magnitude of the GST time scale relative to the TAI and UTC scales is included in the navigation message for transmission to users. The time in the navigation message is transmitted in a format similar to GPS, in the form of the week number and the number of seconds within the current week. In the navigation message, in comparison with GPS, the number of digits intended for transmitting information on the week number has been increased. This provides time measurement for 4096 weeks (over 78 years), which is more than the analogous parameter of the GPS, where the interval is 1024 weeks or 19.5 years.

Time synchronization accuracy for GALILEO signals will be 30 ns with a 95% probability at any daily interval. The discrepancy between the GPS and GST time scales is transmitted as a separate parameter, in the same way as it is done in GLONASS.

The GALILEO ground control complex includes two independent loops:

- space vehicle control loop (ground control segment—GCS);
- ephemeris-time loop; ground mission segment (GMS).

The GCS control loop receives and processes telemetry from the GALILEO space vehicle, controls the operation of the satellite subsystems, generates command information and transfers it to the satellite. The interface between the space segment and the GCS control loop is through a network of tracking stations, receiving telemetry and transmitting control commands (Telemetry Tracking and Command—TT&C) in the S-band.

The GMS loop fulfills missions of collecting data from the global non-request network of measuring stations (ground sensor stations—GSS), processing the received information, generating and uploading of ephemeris-time information, as well as information on the integrity on board the satellite through upload stations (Uplink Station—ULS).

At the first stage of the GALILEO system deployment, the control center in Oberpfaffenhofen (Germany) acts as the coordinating center for the GCS control loop, and the control center in Fucino (Italy) acts as the GMS loop center. At the stage of full operational readiness, all tasks of the ground control complex will be coordinated by both centers in the hot standby mode. Thus, at the stage of normal operation, the GALILEO ground control complex will include:

- system control center in Fucino (Italy) and Oberpfaffenhofen (Germany);
- data upload stations;
- global network of request-free measuring stations;
- stations for tracking, receiving telemetry and transmitting control commands;
- stations of the medium-orbit search and rescue system, receiving the distress signal relayed by the satellite GALILEO.

In addition to the considered SRNS, it is necessary to mention the Chinese SRNS BEIDOU, which is at the stage of the third stage of deployment and will soon provide users with navigation information globally. The BEIDOU system will provide two types of global and two types of regional services. Global services are services with open and authorized access. Regional services are wide area differential correction services and short message transmitting services.

The ground control complex BEIDOU is built according to the classical centralized scheme: a network of no-request measuring stations forms the readings of the primary measurements of the navigation parameters of the radio signals of navigation space vehicles and transmits them to the system control center, in which information is generated that is uploaded on board the space vehicles via special-purpose ground stations.

The BEIDOU network of no-request measuring stations is also located in China. The long-term development strategy of the system assumes the creation of a global network of stations to improve the accuracy of the navigation services of the BEIDOU system.

There are two more regional SRNS:

- Japanese quasi-zenith satellite system (QZSS), designed to serve users in the Pacific-Asian region.
- Indian satellite system (IRNSS), designed for autonomous navigation and time support on the Indian Peninsula. The new name of the system is NavIC. The NavIC system provides services with open and authorized access.

14.4 Functional Supplement to Global Satellite Radio Navigation Systems

Airborne augmentation system (ABAS). The airborne augmentation system (ABAS) ensures that the navigation services of the global navigation satellite system (GNSS) comply with aviation requirements through special techniques for processing GNSS data by onboard aircraft systems or integrating GNSS data with data from other navigation systems.

ABAS is based on one of the following technologies:

- Receiver autonomous integrity monitoring (RAIM), which uses redundant GNSS information to ensure the integrity of GNSS data;
- Autonomous onboard integrity monitoring (AAIM), which uses information from additional onboard sensors to ensure the integrity of GNSS data;
- Integration of GNSS equipment with other sensors (e.g., inertial dead reckoning) to provide improved performance of the onboard navigation system.

Satellite-based augmentation system (SBAS). Satellite-based augmentation system (SBAS) is one of the most promising and complex augmentation systems. These systems can be used not only in civil aviation, but also for other transport systems, increasing their efficiency and safety.

The SBAS satellite augmentation system monitors the signals of the main satellite constellation (GPS or GLONASS) using a network of observation stations distributed over a wide geographic area. For each monitored satellite of the main satellite constellation, SBAS estimates the errors of the transmitted ephemeris and satellite clock parameters and then transmits these corrections and other data to users via the geostationary satellite (Fig. 14.10).

Most sources of GPS positioning errors are due to atmospheric (ionospheric and tropospheric) interference and satellite timing delays. Positioning errors can be measured by interrogated ground control points with high accuracy and then transmitting this information to the user at the same frequency as the conventional GPS signal.

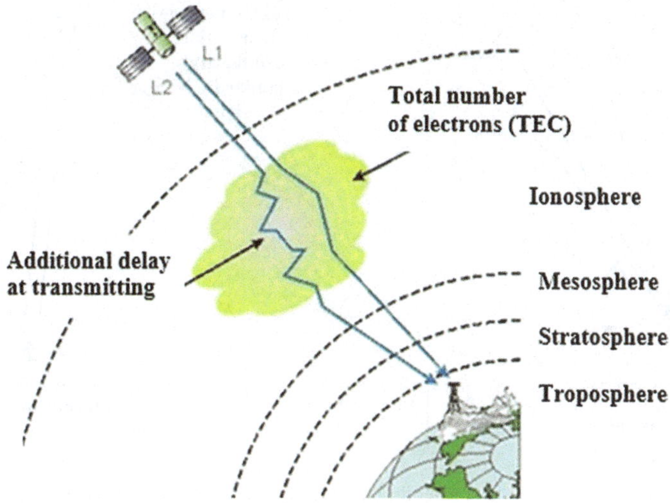

Fig. 14.10 Impact of external factors

The coverage area of SBAS is determined by the coverage of the geostationary satellite, and the servicing area is determined by the SBAS service provider.

SBAS messages ensure integrity, increase GNSS operational availability (readiness) and provide the performance required for AV landing approach procedures with vertical guidance (APV).

An example of SBAS is the wide-area SBAS (WAAS), which represents a network of 35 reference stations (RfS), each of which is interfaced with two master stations (MS) that process information received from the RfS.

Further, this information arrives to the ground earth stations (GES) and through the satellite communication line is transmitted to the geostationary satellites (GEO), which broadcast this information to all users in the WAAS coverage area. At the same time, geostationary satellites emit a range-finding signal in GPS format at L1 frequency, which improves system performance in terms of accessibility, readiness, continuity and service integrity requirements by increasing the number of satellites in the constellation (Fig. 14.11).

The network of reference stations (wide-area reference stations—WRS) is used to receive and process the information from satellites. These data are sent via ground communication lines to control stations, where, based on the information received, the integrity, residual errors, differential corrections and information on the ionosphere for each monitored satellite (GPS), as well as the navigation parameters of geostationary satellites (GEO) are determined. This information is sent through ground stations to the GEO network, through which it is transmitted to users at the GPS L1 frequency with modulation corresponding to the GPS modulation. Additionally, WAAS monitors the integrity of its own information.

The WAAS communication system consists of two independent networks: the WAAS network and the control network. These networks include both ground and

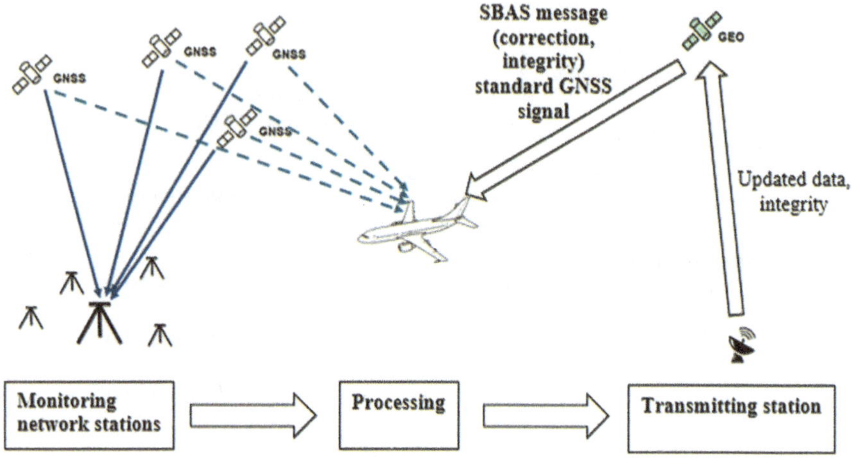

Fig. 14.11 Functional principle of satellite-based augmentation system (SBAS)

space segments. Each reference station is connected to each control station by two communication lines having an availability (readiness) of 0.9985 and a reliability of 2×10^{-6}. Communication lines between control stations must have availability 0.9999 and reliability 2×10^{-6}. The total delay in the terrestrial network for any connection is approximately 150–250 ms, and the maximum switching time from the main to the backup network should not exceed 4 s.

The time division multiplex network must support data rates of 2400/4800/9800 bps, 19.6 kbps, and 56.64 kbps.

WAAS has eight main functions:

- acquisition of data from RS;
- calculation of ionospheric corrections;
- determination of satellite orbits;
- calculation of orbital corrections;
- determination of the integrity of satellites;
- independent confirmation of data;
- transmission of WAAS messages and pseudo-ranges from GEO;
- management and maintaining of the system in serviceable condition.

There are two types of data correction used in the WAAS format—fast and slow. Fast correction is designed to compensate for rapidly changing GPS clock errors. These adjustments are the same for all users.

Slow-varying corrections are designed to compensate for ephemeris errors, slow clock drift and errors, associated with ionospheric parameters. For this purpose, a wide-area model of ionospheric delays and real-time data are transmitted to the user, which allow estimating the delays in the ionosphere for each satellite.

The tropospheric delay depends on the local conditions in which the user is located, and therefore, the WAAS GEO message does not contain explicit tropospheric corrections and only data are transmitted that allow these corrections to be calculated in the onboard receiver.

The WAAS system allows the use of up to 52 satellites and has the ability to transmit corrective information via GLONASS satellites.

Whereas in GBAS systems, the reliability of the information used by the pilot depends on one ground station; then in SBAS, this reliability will be determined by the entire network of ground and space objects, which puts forward very strict requirements for the reliability of SBAS elements.

To ensure communication, received on board the corrections to pseudo-ranges with the aircraft position, these corrections are transmitted as separate corrections:

- time frequency;
- ephemeris;
- corrections related to the selective access error;
- ionospheric corrections.

The latter are transmitted in the form of a rectangular grid containing about 1000 points.

The elimination of the selective access mode from the GPS signal facilitated the implementation of the system, since the amount of transmitted information required to correct the errors caused by this mode is reduced.

The provision of the second frequency to civilian users even more simplifies the implementation of the system, since the amount of information transmitted over terrestrial and space data lines is significantly reduced and the function of monitoring the ionospheric component of the error can be excluded.

One of the most important requirements set by ICAO to the existing and developed satellite systems of functional augmentation is the requirement of "seamlessness" during the flight of AV from the service area of one system to the service area of another, i.e., full compatibility of various systems at the level of aircraft onboard equipment.

Ground-based augmentation system (GBAS). The ground-based augmentation system (GBAS) or local control and corrective system (LCCS) is designed to provide an AV precision landing approach or GLS using GNSS signals. It can also provide aircraft positioning services within the GBAS ground station coverage area limited by line-of-sight range (Fig. 14.12).

The GBAS ground station monitors GPS and/or GLONASS signals in the terminal area and transmits area-specific pseudo-range corrections, integrity parameters, inoperative satellite numbers and final approach segment (FAS) data to AVs in the VHF band.

The GLS approach is similar to the radio beacon landing system (RBLS) approach. The difference is that in GLS, the deviation of the AV from the nominal heading and glide path is determined by comparing the true position of the AV, calculated from the GNSS / GBAS signals, with the data of the onboard navigation database and the FAS block.

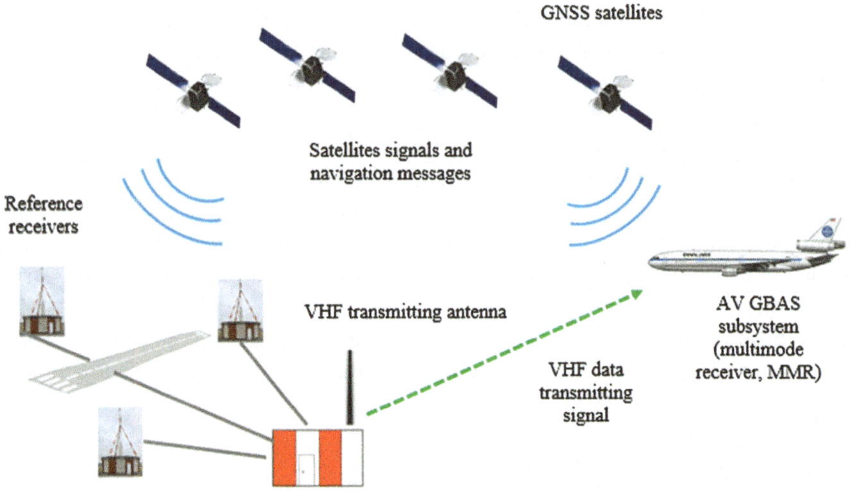

Fig. 14.12 Operating principle of the ground augmentation system (GBAS)

One GBAS ground station can provide a precision approach for AV on all runways of an aerodrome with both landing courses, and under certain conditions—at several nearby aerodromes.

The ground-based augmentation satellite system (GBAS) transmits GNSS differential corrections in real time, providing a high-precision approach and landing of AV. The GBAS ground station (control center) uses data from two to four GNSS reference receivers located around the airport to generate a correction message. Corrections are transmitted via the VHF data channel (VDB, 108.025–117.95 MHz) to the receiver of the approaching AV to correct the GNSS signal it receives. VDB data (D8PSK modulated signal) are transmitted in packets, each of which consists of eight time slots. Each slot contains application data that can belong to one or more message types (MT).

The advantages of using the GBAS system are the following:

- for GBAS operation, one station per airport is sufficient, and not one (or two) per runway (RNY);
- curved or complex approach procedures can be used to avoid obstacles, noise-restricted areas or approach trajectories for nearby airports;
- GBAS requires less frequent flight checks than instrument landing systems (ILS) equipment;
- GBAS provides an alternative (GLS) to the traditional landing system (ILS).

When using the GBAS system, three options for the implementation of the landing maneuver are considered:

- non-precision approach;
- approach with vertical guidance (APV I, II);

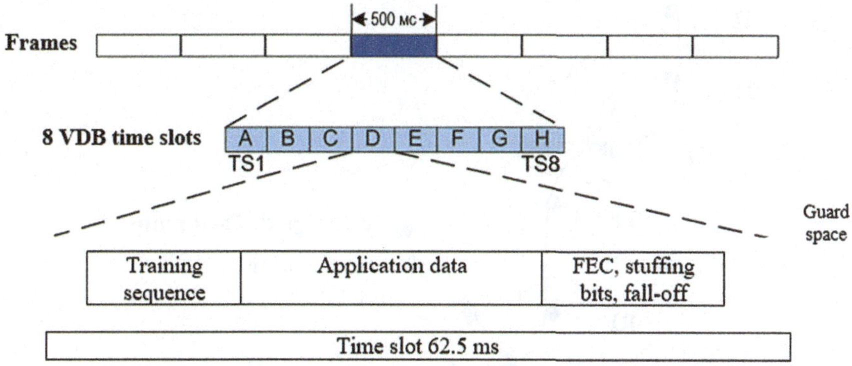

Fig. 14.13 GBAS ground equipment signal structure and content

- precision approach (Cat. I, II, III).

Each of these approach types requires a different level of positioning accuracy. GBAS approach service type (GAST) describes the accuracy provided by GBAS equipment:

- GAST-A –APV I (vertical accuracy 20 m);
- GAST-B –APV II (vertical accuracy 8 m);
- GAST-C –CAT I (vertical accuracy 4.3 m);
- GAST-D –CAT III (vertical accuracy 1.8 m).

Information on airport and GNSS correction data are transmitted from the ground subsystem to the AV subsystem using one-way data broadcast in the VHF band. Each frequency channel is time-multiplexed by frames of 500 ms, each of which is divided into packets (Fig. 14.13). Packages contain application data in the form of messages.

In accordance with existing standards, there are the following types of GBAS messages:

- Type 1: Differential correction data (slow—100 s);
- Type 2: GBAS data;
- Type 3: Null message;
- Type 4: May contain one or both messages;
- Final approach segment formation data (FAS);
- Data of the formation of the path to the terminal area (TAP);
- Type 5: Range source availability (optional);
- Type 6: Reserve for carrier corrections;
- Type 7: Reserve for military use;
- Type 8: Reserve for testing;
- Type 11: Differential correction data (fast—30 s);
- Type 101: GRAS pseudo-range correction data.

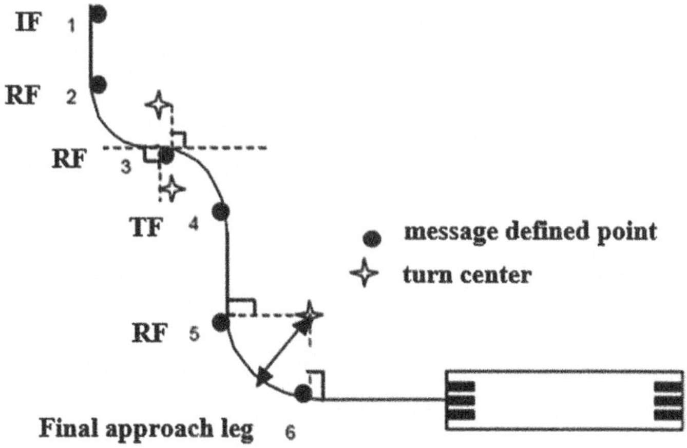

Fig. 14.14 Route points and track segment (TAP) scheme

Messages of 1 and 11 types contain GPS differential corrections data (satellite information and pseudo-range corrections), which are used to improve positioning accuracy and reliability over standard GNSS and SBAS signals. The data of type 1 messages are smoothed with a time constant of 100 s, and the data of type 11 messages are smoothed with a time constant of 30 s. Type 2 message (GBAS data) contains information on the location and configuration of the ground station, as well as parameters that can affect the quality of the received GPS signal, in particular, changes in the magnetic field and ionospheric parameters.

Type 4 message contains data for the path to the terminal area (TAP) and/or information on the final approach segment (FAS). The route to the terminal area is used to transition from the route path to the approach path. The prelanding segment is used during descent/direct landing. The terminal area path (TAP) (Fig. 14.14) consists of a set of legs (defined by route points) that lead to the final approach leg path (FAP). These legs are: approach to the point (intermediate fix, IF); approach to the track line (TF); approach to radius (RF). This is similar to existing RNAV/RNP routes, but TAP can also provide vertical guidance. It should be noted that information on approach procedures can be transmitted to the AV (and not stored locally).

The FAS segment is the line followed by the AV at the final stage of a (precision) approach (Fig. 14.15). The FAS segment consists of the following elements:

- runway threshold landing point/imaginary runway threshold point (LTP/FTP);
- glide path intersection point (GPIP);
- flight path alignment point (FPAP);
- threshold crossing height (TCH);
- glide path angle (GPA).

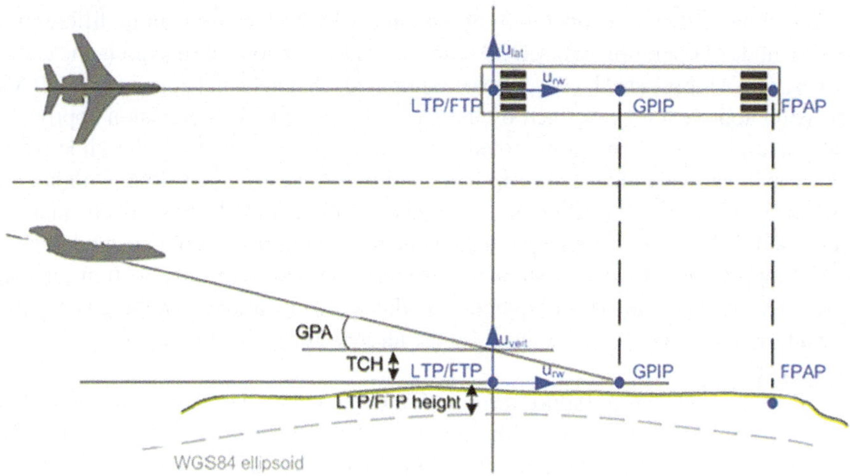

Fig. 14.15 Final approach flight control (FAS) implementation

The ground-based regional augmentation system GRAS is designed to support AV operations using GNSS en route, in the terminal area, non-precision approaches, departures and approaches with vertical guidance in a certain area of airspace (region).

GRAS is the result of a combination of SBAS and GBAS principles to improve the performance and enhancement of GNSS navigation capabilities to users. GRAS, like SBAS, uses a distributed network of reference stations to monitor the GNSS satellite constellation signals and a processing center to calculate GNSS integrity and differential correction information. The difference is that GRAS does not transmit this information via a geostationary satellite, but reformats it and transmits it via a network of ground stations in the VHF band, similar to GBAS.

When performing a precision approach using GBAS, keep in mind the following. The flight trajectory by GBAS is defined differently from the ILS approach trajectory. Trajectory data, including glide path, horizontal sector width, horizontal sensitivity, and other guidance sector characteristics, are transmitted by GBAS ground stations (LCCS) to GNSS (multi-mode receiver, MMR) avionics via a high-integrity digital data link. The digital message defines the FAS trajectory and guidance characteristics. Onboard GNSS equipment, based on geometric relationships, calculates the trajectory parameters and determines the guidance characteristics indicated in the transmitted digital data. The same equipment generates guidance parameters with characteristics similar to other precision approach systems, such as ILS.

The decision to perform a precision approach using GLS is made by the crew after listening to the automatic terminal information service (ATIS) report before passing the starting point of the selected procedure for the landing aerodrome and the lack of information from the air traffic controller about the impossibility of performing this operation due to the presence of any anomaly.

The crew activates the operation of the onboard GNSS equipment in differential mode (mode of operation with LCCS) and selects a precision approach scheme using LCCS, which is provided by the onboard equipment channel number switch. A GBAS precision approach is performed by the method much like ILS precision approach, using lateral guidance in an intermediate segment prior to entering the glide path, after which vertical guidance begins and continues to be provided along with lateral guidance for landing. The minimum required functionality for displaying information from the LCCS in the AV cockpit is equivalent to the information displayed during an ILS approach. LCCS continuously provides the necessary information, on the basis of which the onboard equipment calculates the deviation from the given track line and in automatic (under control by the aircrew) or manual mode this deviation is compensated.

GBAS ground station failure information display and alarms are similar to ILS approach information. Along with signaling about failure, the LCCS constantly provides information on the integrity of the procedure in automatic mode. If the onboard equipment provides information about the impossibility of continuing the operation due to the LCCS failure or the impossibility of performing the intended operation using GNSS, the crew must immediately stop the operation and, together with the ATC controller, take measures to safely continue the flight.

To date, five wide-area differential correction systems have been presented:

- WAAS—wide-area system of functional additions, USA;
- WDCMS—wide-area differential correction and monitoring system, Russia;
- EGNOS—European Geostationary Navigation Service, EU;
- GAGAS—geostationary navigation supplement of the GPS, India;
- MSAS—Space Augmentation System, Japan.

There are proprietary global differential correction systems such as the StarFire navigation system (a commercial John Deere company system); Starfix DGPS and OmniSTAR (commercial system of Dutch Fugro N.V. company).

A separate place among the global differential correction systems is for the "PPP service" (Precise Point Positioning—high-precision absolute positioning). PPP technology is capable of providing positioning accuracy from decimeter to centimeter or more (for static mode) when combining accurate satellite orbits and clocks with a dual-frequency GNSS receiver (by taking into account the effect of the first-order ionosphere).

The main advantages of PPP technology over other differential positioning methods include the fact that only one receiver is necessary for PPP realization and no special base stations are required in the close vicinity of the user.

It is necessary to specially highlight functional supplement that differ from those traditionally used by users—pseudo-satellite and assisting functional augmentations.

Pseudo-satellite functional augmentations are one or more pseudo-satellites (navigation satellites located on the ground) that generate navigation signals in the GNSS format. They complement the GNSS global radio navigation field in a given area and usually have a local operating zone. Its size is determined by the power of the transmitter of the pseudo-satellites and the line-of-sight range.

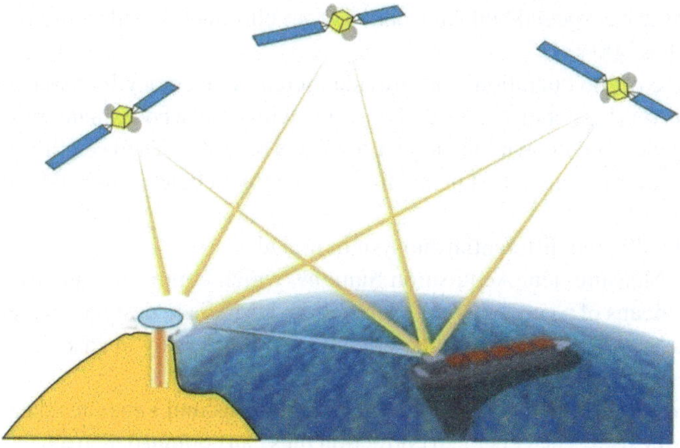

Fig. 14.16 Functioning principle of maritime DGPS

Assisting functional augmentations—systems that implement the "assisted GNSS" mode and form not corrective corrections, but additional auxiliary information to accelerate the entry into communication with the navigation space vehicle and increase the reliability of user's locations.

Marine differential subsystem (maritime DGPS).

In seaports and on approaches to them, in vessel traffic management centers (VTMS), at communication facilities and other points of coastal infrastructure, for the purpose of ensuring high-precision navigation, an information is transmitted on differential corrections of the control and correction station of the GLONASS/GPS global navigation satellite systems (Fig. 14.16).

A GNSS reference station (Unified Measurement Acquisition Station) is installed in the port area. The station complements the Global Positioning Systems and calculates the GLONASS/GPS signal correction for each satellite, providing localized pseudo-range corrections and support information. Corrections are sent along with information on the reliability of the station, the quality of the correction, and a notification whether a specific satellite should be used in the 285–325 kHz range. The transmission is carried out according to ITU-RM.823. The message types used correspond to RTCM SC-104, numbers 3, 6, 7, 9 and 16 using minimum shift modulation (MSK).

All Unified Measurement Acquisition Stations (UMAS) have an individual identification number transmitted in the DGPS signal. The data transfer rate ranges from 25 to 200 bps.

Corrections can be sent for a maximum of 12 satellites with elevation angles greater than 7 degrees. For maritime DGPS receivers of civil ships, the positioning accuracy is 10 m or better (in case of a successful constellation of satellites for the user equipment, i.e., DOP < 2 or 3). The interference immunity is increased if the antenna of the correcting receiver is referred to the H-field type (folded antenna) and if the receiver has special built-in technology to eliminate impulse interference. The range reaches 500 km.

The logic of the operation is to provide increased accuracy by using a reference GNSS receiver (base station), located at a point with known coordinates by comparing the coordinates of a known location with what is received. Then satellite corrections are calculated and transmitted in real time on the frequencies assigned by the UMAS to nearby users.

Traditionally, the differential subsystem includes:

Unified Measurement Acquisition Stations, which control the quality of relayed signals by means of a geodetic reference station. UMAS are designed with redundant configuration, which guarantees its reliability and autonomy in case of failures and disturbance.

A processor that calculates differential corrections and generates data for transmission to the user. The generated correction files can contain data from the receiving station and frequency and time standards.

Equipment for transmission of differential corrections (transmission is most often carried out "directly" from the UMAS).

Receiving equipment of users, providing reception and accounting of differential corrections (as a rule, combined with ship's GNSS equipment).

The UMAS includes the following components:

- two reference stations [main and backup sets (OS)] for determining differential corrections and generating corrective information;
- computer for remote control and operational monitoring of the UMAS state;
- two integrity monitors (main and backup sets of integral control station (ICSt);
- access control equipment;
- differential amendments transmitter;
- communication system (Ethernet, RDSI, GSM or Inmarsat) and uninterruptible power supply;
- checkpoint (CP).

The main task of the control point (CP) is to monitor the operation (maintenance) of Unified Measurement Acquisition Stations, communication lines (RDSI, GSM or Inmarsat) between them and the CP and a special data transmission channel (GIC—GPS integrity channel). As well as ensuring the integrity of observations of satellite radio navigation systems, the formation of integrity data for transmission to users.

Through a combination of satellite positioning systems, electronic cartography, communications systems and open information systems architecture, the maritime electronics industry has developed a navigation device called the Universal Automatic Identification System (UAIS) or simply Automatic Identification System (AIS).

An automatic identification system is mandatory for ships over 300 tons, including passenger ships, regardless of tonnage.

AIS stations exchange data with each other on two VHF channels. All AIS stations separate transmissions by time. AIS stations use internal GNSS receivers as a source of synchronization and uniform time.

The AIS station is connected to external ship equipment. Typically, the following systems are used as external equipment:

- sources of navigation data (receiver of the global navigation satellite system (GNSS), log, compass or its integrating device);
- electronic navigation devices capable of displaying and processing data issued by the AIS station: ECDIS, ARPS, pilot AIS.

Ships equipped with AIS equipment, while on the high seas or in coastal areas, regularly transmit standard messages in the VHF (ultra-short waves) range of the maritime mobile radio service containing static, dynamic and voyage information on the vessel. Simultaneously, each vessel equipped with AIS receives similar information from other vessels within the range limited by the propagation of VHF radio waves (20–30 miles). The received information is automatically processed and displayed on one of the ship's navigation displays. Synchronization of operation of all AIS stations (ship and coastal) is provided by GNSS, which is also a source of transmitted information on coordinates and velocity vector. In coastal areas where AIS base stations are installed, information transmitted by ships is received by base stations and enters the disposal of coastal services (VTS and ship reporting systems, search and rescue services, environmental control and pollution elimination services, border and customs authorities, various port services). Usually, to obtain a complete picture of shipping in a controlled area, AIS base stations are combined into networks that allow integrating information from remote base stations with each other, as well as with information received in the VTS and in mandatory ship reporting systems. In coastal areas, the accuracy of GNSS positioning of vessels, and therefore the effectiveness of AIS, can be improved by coastal reference stations and radio beacons that transmit differential corrections to vessels. AIS base (coastal) stations can operate in an active mode, controlling the mode of operation of ship stations and transmitting information related to the safety of navigation (local navigation warnings, differential corrections for GNSS, data on vessels maintained by a VTS).

The display of the received and processed information is carried out on the screen of the ship navigation graphic display (ARPS, electronic cartographic system, integrated navigation system). The result is a visual representation of the situation, much like an ARPS screen.

Despite the fact that one of the main sources of navigational information, both for ships and for coastal services, is a radar locator, under certain conditions of use it has a number of disadvantages:

- short detection range of targets, especially small ones in sizes;
- strong influence of interference from the sea or atmospheric phenomena;
- insufficient resolution capability;
- limitations due to the peculiarities of the beams passage (the target is not visible behind obstacles);
- the use of filtering algorithms makes it difficult to discriminate target maneuvers;
- transfer of markers of tracked target vessels (swopping) when they approach, during ARPS operation.

One of the ways to eliminate these shortcomings is the use of AIS.

AIS performs the following functions:

- automatic identification of ships;
- self-organization of the system and control of access to radio channels;
- reception of data via radio channel from other ships and coastal centers;
- transfer of own data to the radio channel for use by other ships and coastal centers;
- saving static data intended for automatic transmission to the radio channel;
- delivery of data received from the radio channel from other AIS objects for display in the data presentation device in AIS;
- determination of coordinates and parameters of vessel movement using the built-in GNSS receiver;
- receiving differential GNSS corrections from a GNSS control and correcting station or from a receiver of differential corrections of radio beacon channel and transmitting it via the AIS channel (base station);
- reception and transmission of data on the vessel's coordinates and parameters of its movement (and pitching angles, if available) from an external source;
- receiving static, additional dynamic data and binary messages for transmission to the radio channel from the AIS data presentation (storage) device;
- assignment (by coastal AIS) of the corresponding operating modes to ship and coastal stations, including assignment of areas, frequencies, radiation power, slots, periods of reports and the number of its repetitions.
- switching on/off the reserve coastal stations (repeaters) of AIS;
- output of information on the AIS state (to the main, e.g., ECDIS, auxiliary, e.g., pilot AIS, displays of the Control and Display Unit (CDU));
- outputs on CDU displays, calculated by the coordinates of ships.

The list of information transmitted by AIS is presented in Table 14.10.

In addition to this information, the following can be transmitted or received via the AIS channel: short text messages related to the navigational safety of navigation; ship collision prevention; meteorological conditions; interaction of ships and helicopters performing an operation to search and rescue a distressed ship; the ship's route, and can also be transmitted differential corrections of GLONASS and GPS, increasing the accuracy of determining the coordinates of the vessel to units of meters.

The use of this information on board the vessel should significantly increase the safety of navigation and maximally exclude ship collisions.

Table 14.10 Content of information transmitted by AIS

Type and content of information	Information particularities
Static	
IMO number	Introduced when installing the equipment. Not subject to change
MMSI	Maritime mobile radio service identifier. Introduced when installing the equipment. May change only when the vessel is re-registered
Name and call sign of the vessel	Introduced when installing the equipment. May change only when the vessel is re-registered
Ship type	Introduced when installing the equipment. Not subject to change
Length and beam of the vessel	Introduced in conjunction with the position of the GNSS antenna
Antenna location of position sensor	May vary if multiple receiving antennas are present
Vessel position sensor type	Introduced when installing the equipment, depending on the interfaced navigation equipment
Height above keel level	Additional information on the height of masts or other structures. Transmitted only at the initiative of the vessel or at the request of the coast station
Dynamic	
Vessel coordinates	Automatically read out from a position sensor connected to AIS equipment
Coordinate accuracy sign	Characterizes the accuracy of determining coordinates—worse than 10 m or better than 10 m (when using the DGNSS mode)
Coordinate determination time	UTC time. Automatically read out from a position sensor connected to AIS equipment
Course angle, ground speed (relative to the ground)	Automatically read out from a position sensor connected to the AIS equipment. (COG—course over ground / SOG—speed over ground)
Course	Automatically read out from the ship's heading indicator (gyrocompass) connected to the AIS equipment
Navigational status of the vessel	It is entered manually with a choice from the list (on the move, at anchor, uncontrolled). Changes are recommended to be made at the same time as the lights are turned on or when the COLREG signs are raised
Angular velocity	Rate of turn (course change). Automatically read out from the corresponding sensor, if present on the vessel
Steering angle	Automatically read out from the corresponding sensor, if available on the vessel and interfaced with AIS

(continued)

Table 14.10 (continued)

Type and content of information	Information particularities
Roll and pitching angles	Automatically read out from the corresponding sensor, if available on the vessel and interfaced with AIS
Voyage data	
Draft	It is entered manually and changed if necessary (e.g., when pumping out ballast before entering the port)
Dangerous cargo	If there is a dangerous cargo, it is entered manually before the start of the voyage
Destination, ETA	Entered manually before the start of the voyage, changed if necessary
Voyage plan	Defined by the coordinates of waypoints. It is entered manually before the start of the voyage, changed if necessary
Vessel POB	Additional Information. Transmitted only at the initiative of the vessel or at the request of the coast station

Chapter 15
Goniometric Radio Navigation Systems

Goniometric radio-navigational systems (GRNS) are intended for detection of directions from an object to navigational guide point (e.g., radio beacon or radar station) or from navigational guide point to an object.

GRNS can be classified by purpose into radio beacon, direction-finding and radio-compass systems.

15.1 Radio Beacon Goniometric Radio Navigation Systems

On air transport, the largest spread was gained by VOR (very high-frequency omnidirectional radio range) system. For the years of operation of this system, its construction has been repeatedly upgraded and it can have completely different configuration.

Radio beacons of VOR system are classified depending on volume of air space, in which its application is proposed. Since beacons operate in VHF band, then its maximum operational range is determined by direct visibility (line-f-sight) range and depends on AV flight altitude.

Radio beacons of T (terminal) class are intended for navigation in terminal area (approach navigation) and should provide an acquisition of navigational data at altitudes of not less than 300 up to approximate 4000 m at a distance of not less than 25 nautical mile (about 46 km).

Radio beacons of L (low altitude) class should provide a signal receiving at altitudes of not less than 300 m up to 18,000 ft (about 5500 km) at a distance up to 40 nautical miles (74 km).

Radio beacons of H (high altitude) class should provide a signal receiving (Fig. 5.1):

D. A. Akmaykin et al., *Theoretical Foundations of Radar Location and Radio Navigation*, Springer Aerospace Technology,
https://doi.org/10.1007/978-981-33-6514-8_15

- at altitudes from 1000 ft (300 m) up to 14,500 ft (about 4400 m) at a distance up to 40 nautical miles (74 km);
- at altitudes from 14,500 ft up to 60,000 ft (about 18,300 m)—at a distance up to 100 nautical miles (185 km);
- at altitudes from 18,000 ft up to 45,000 ft (about 13,700 m) at a distance up to 130 nautical miles (240 km).

It may cause consternation the fact that in accordance with mentioned above figures, the distance at altitudes of more than 45,000 ft is less than this altitude (100 nautical miles instead of 130). In fact, it should seem that the more altitude the more should be a distance. However, the mentioned distances are far from maximum distances at which a signal receiving is possible. As a rule, a signal can be received at large distances. These distances, besides ensuring of signal receiving, additionally guarantee that, being within its limits, AV will not be in operation zone of another RB, operating at the same or at a close frequency. Thus, a determined distance in 100 miles (for high altitudes) just guarantees that at smaller distances this will not happen.

Nowadays, the three main types of azimuthal RB of VOR system are used in civil aviation:

- RB with envelope phase measuring of amplitude-modulated (AM) oscillations (VOR);
- RB with two-stage phase measurement (PVOR);
- RB with use of Doppler effect (DVOR).

VOR radio beacons have two transmitting antennas: omnidirectional (non-directional) A_1 with directional pattern in horizontal plane $F_1(\alpha) = 1$; directional (beamed) antenna A_2 with directional pattern in horizontal plane $F_2(\alpha) = \cos\alpha$. In any azimuthal direction θ, a directional pattern magnitude A_2 is characterized by a value $F_2(\alpha - \theta) = \cos(\alpha - \theta)$.

Antenna A_1 sets a field with strength (density):

$$e_1(t) = E_{1M}\cos(\omega_0 t), \tag{15.1}$$

with amplitude E_{1M}.

Antenna A_2 sets a field in any azimuthal direction

$$e_2(t) = E_{2M}\cos(\alpha - \theta)\cos(\omega_0 t), \tag{15.2}$$

with amplitude $E_{2M}\cos(\alpha - \theta)$.

Usually for VOR RB, the following condition is fulfilled: $E_{1M} > E_{2M}$.

Antennas' radiation patterns of VOR radio beacon are shown on Fig. 15.1a. High-frequency signals are formed by one transmitter and radiated by antennas, having a common phase (electric) center. During field summation in air space, a resultant field of all-directional azimuthal VOR RB is formed $e_\Sigma(t) = e_1(t) + e_2(t)$ (Fig. 15.1b).

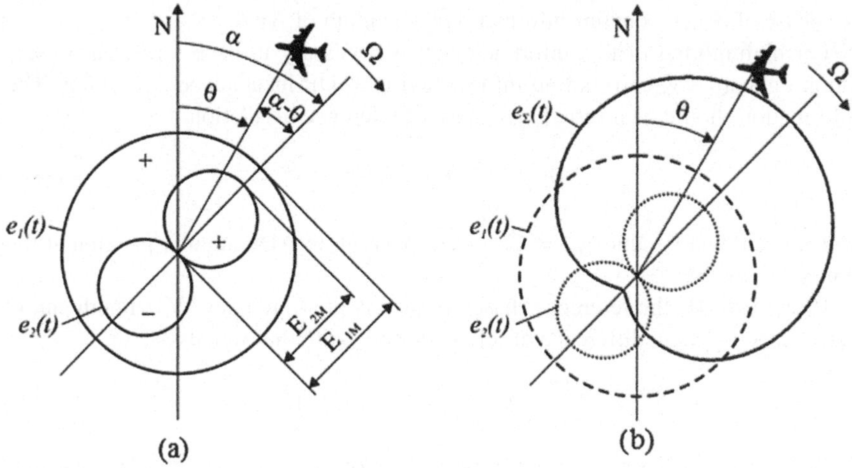

Fig. 15.1 Antennas radiation patterns of VOR RB

Considering (15.1) and (15.2) expressions, a resultant field value can be expressed as follows:

$$e_\Sigma(t) = E_{1M}\left[\frac{1 + E_{2M}\cos(\alpha - \theta)}{E_{1M}}\right]\cos(\omega_0 t), \tag{15.3}$$

Directional pattern A_2 rotates in horizontal plane with angular velocity $\Omega = \frac{2\pi n}{60}$ $\Omega = \frac{2\pi n}{60}$, where n—revolution rate of antenna directional pattern (ADP) per minute. ADP rotation (frequency) rate A_2 in VOR comprises $F = \frac{n}{60} = \frac{1800}{60} = 30\,\text{Hz}$. Directional pattern position A_2 (position of its maximums)—time function $\theta = \Omega t$. Antenna rotation will initiate a periodic variation (change) of resultant field.

Designate amplitudes relation $m_C = \frac{E_{2M}}{E_{1M}}$ and, by substituting in (15.3) of m_C and θ values, we obtain:

$$e_\Sigma(t) = E_{1M}[1 + m_C\cos(\Omega t - \theta)]\cos(\omega_0 t), \tag{15.4}$$

As a result, a field with depth amplitude modulation m_C is formed, by modulation frequency $\Omega = 2\pi F$ and envelope phase, depending on azimuth θ. Oscillations, received by onboard signals receiver of VOR RB, can be represented as follows:

$$U_{\text{rcv}}(t) = KU_m[1 + m_C\cos(\Omega t - \theta)]\cos(\omega_0 t). \tag{15.5}$$

where K—a coefficient, considering an attenuation.

After amplification and detecting, a low-frequency voltage can be selected:

$$u(t) = U_M\cos(\Omega t - \theta). \tag{15.6}$$

the phase of which contains information on azimuth of AV θ.

For highlighting of this information, it is necessary to have a reference oscillation onboard AV, carrying an information on ADP instantaneous position. This information should be inserted in a phase of reference oscillation

$$U_{\text{ref}}(t) = U_{\text{refm}} \cos(\Omega t), \tag{15.7}$$

with current phase value $\varphi_{\text{ref}} = \Omega t$, corresponding to ADP angular position at this time (at a time) t.

If onboard AV the reference (base) voltage is presented, an AV azimuth can be determined as phase difference of reference and azimuthal signals

$$\theta = \Delta\varphi = \varphi_{\text{ref}}(t) - \varphi_{\text{az}}(t). \tag{15.8}$$

Reference signal is necessary for operation of onboard equipment, where it should be equal for all AV. This signal is transmitted over communication channel on a carrier frequency ω_0, in the same way as an azimuthal. Separation of azimuthal and reference signals in channels is carried out at receiving part via frequency selection method of detected in amplitude combined signal. Such opportunity rises at use of reference signal transmission of double amplitude-frequency modulation.

Simplified functional chart of azimuthal channel of VOR system is depicted on Fig. 15.2. In transmitter (Trm), high-frequency oscillations ω_0 are formed. In power divider (PD) high-frequency signal is divided into two channels. Part of power goes to rotating antenna A_2. Antenna rotation rate is defined by antenna control unit (ACU) and equals to 30 Hz. Different techniques of antenna rotation have been used in RB. In the first VOR RB, an antenna rotation was performed mechanically using electric motor. In the modern VOR RB, an ADP electronic rotation technique is used

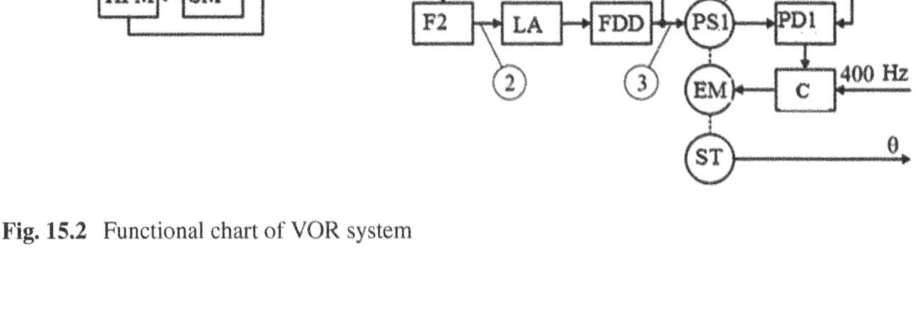

Fig. 15.2 Functional chart of VOR system

(electronic goniometer method), at which a rotation effect of ADP is achieved by supplying of two mutually perpendicular beamed antennas with patterns in shape of eight. Antenna supply is provided via balance-modulated oscillations with dephased modulation at 90°.

For shaping of a signal with double amplitude-frequency modulation, the oscillations are selected, the frequency of which is considerably more than ADP rotation rate (frequency), but considerably lower than carrier wave frequency, and these oscillations are used as an auxiliary. Auxiliary oscillations are called subcarrier, for which a condition $\Omega \ll \Omega_{\text{subc}} \ll \omega_0$ should be fulfilled, where Ω_{subc}—frequency of subcarrier oscillations. For VOR system, a subcarrier frequency equals to 9960 Hz.

In subcarrier modulator (SM), a frequency modulation of subcarrier is provided, by reference oscillations of 30 Hz with frequency deviation of 480 Hz at modulation index—16. In modulator (HFM), the high-frequency oscillations are modulated in amplitude by subcarrier voltage with modulation depth of 30%.

Antenna A_1 set up a field with density

$$e_1(t) = E_{1M}\left[1 + m_{\text{am}} \cos\left(\Omega_{\text{subc}} t + \frac{\Delta\Omega_{\text{subc}}}{\Omega} \cos \Omega t\right)\right] \cos \omega_0 t, \tag{15.9}$$

where m_{am}—amplitude modulation coefficient (factor); $\frac{\Delta\Omega_{\text{subc}}}{\Omega}$—frequency modulation coefficient; $\Delta\Omega_{\text{subc}}$—subcarrier frequency deviation.

The summary effects on antenna of onboard equipment A_0. We have a summary oscillation at the output of antenna:

$$u_\Sigma(t) = KU_M Z \cos \omega_0 t,$$
$$Z = 1 + m_c \cos(\Omega t - \theta) + m_{\text{am}} \cos\left(\Omega_{\text{subc}} t + \frac{\Delta\Omega_{\text{subc}}}{\Omega} \cos \Omega t\right). \tag{15.10}$$

Amplitude-frequency spectrum of summary oscillation is shown in Fig. 15.3.

It is necessary to extract from a summary for onboard equipment the azimuthal and reference signals and implement its comparison in phase.

After resultant (sum) signal conversion in receiving unit (Rcv), its amplifying and detection by amplitude detector (demodulator), an envelope is produced containing an azimuthal and reference signals of the following form:

$$u_{\text{env}}(t) = U_{2M} \cos(\Omega t - \theta) + U_{1M} \cos\left(\Omega_{\text{subc}} t + \frac{\Delta\Omega_{\text{subc}}}{\Omega} \cos \Omega t\right) \tag{15.11}$$

where U_{1M} and U_{2M}—amplitudes of composite signal components.

From Fig. 15.3, we can see that azimuthal and reference signals can be extracted via frequency discrimination (selection). For this purpose, from Rcv output the signal is delivered to two filters F1 and F2. In F1 filter, adjusted on 30 Hz frequency, azimuthal signal is extracted or variable phase signal, and in F2 filter, adjusted on 9960 Hz subcarrier frequency, the frequency-modulated under carrying oscillation

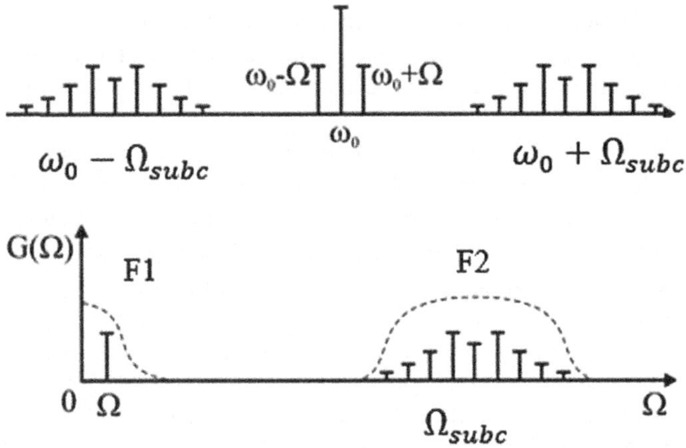

Fig. 15.3 Amplitude-frequency spectrum of received signal and envelope of received signal

is extracted. After symmetrical limiting in limiting amplifier (LA), the reference oscillation is extracted in frequency difference detector (FDD).

As a result of transformations, we obtain:

- azimuthal signal $u_{az}(t) = U_{2M}\cos(\Omega t - \theta)$;
- reference signal $u_{ref}(t) = U_{refM}\cos(\cos\ \Omega t)$.

Reference voltage delivers to phase shifters PS1 and PS2. In initial position, PS1 axis is turned at a random angle β, that causes an additional phase shift of a reference voltage on a value β

$$u_{ref}(t) = U_{refM}\cos(\Omega t - \beta),\ u_{ref}(t) = \Omega t - \beta. \tag{15.12}$$

Azimuthal and reference voltage is delivered to phase detector PD1 (PD). Phases difference between voltages at input $\Delta\varphi(t) = \theta - \beta$. Voltage at output of phase detector PD1:

$$U_{pd}(t) = U_{max}\sin\Delta\varphi. \tag{15.13}$$

This constant voltage is converted (into C) unbalance (error) signal with frequency of 400 Hz and delivers to control winding of electric motor (EM), which turns rotor axis of phase shifter PS1 as long as phase difference $\Delta\varphi$ becomes equal to zero. Herewith, $U_{pd} = 0$ and $\beta = \theta$. By this means, rotation angle (slewing angle) of phase shifter rotor FS1 becomes equal to AV azimuth. FS1 axis is connected with selsyn transmitter (ST) axis, which transmits measurement results to azimuth indicators.

In VOR system, an aircraft flight possibility in specified azimuth θ_{sp} is stipulated. For this purpose, the PD2 and FS2 are inserted into a circuit. FS2 axis is turned manually and is set at specified angle θ_{sp}. Herewith, a phase of reference voltage U_{ref} is additionally shifted to θ_{sp} value and becomes $\varphi_{ref}(t) = \Omega t - \theta_{sp}$.

This voltage delivers to PD2 input. Azimuthal voltage (pattern) u_{az} is delivered to the second input with phase $\varphi_{az}(t) = \Omega t - \theta$. Phases difference of azimuthal and reference voltages at PD2 input $\triangle\ \varphi = \theta - \theta_{sp}$. After phase detection at detector output—$U_{pd}(t) = U_{\max} \sin\ \triangle\ \theta$. When $U_{pd} = 0,\ \ \theta = \theta_{sp}$ AV azimuth coincides with specified direction. This problem is solved at AV flight to direction to VOR radio beacon or from it. For indicating flight towards a radio beacon or backward, the FD3 is introduced into a circuit, to which:

- azimuthal voltage $U_{az}(t) = U_{2M} \cos(\Omega t - \theta)$;
- reference voltage $U_{ref}(t) = U_{refM} \cos\bigl(\cos\ \Omega t - \theta_{sp} + 90°\bigr)$ are applied.

Phase difference of these voltages $\triangle\ \varphi = \theta - \theta_{sp} + 90°$. At flight on VOR RB, when $\theta = \theta_{sp}$, at PD3 output $U_{pd} = +U_{\max}$. "TO" placard on crew indicator corresponds to positive voltage presence. At RB flight, the current AV azimuth is changed at 180°, then $\theta = \theta_{sp} \pm 180°$. Azimuth change induces a voltage polarity change at PD3 output $U_{pd} = -U_{\max}$, herewith, "FROM" placard is indicated.

Voltage current waveforms (stress diagram) in representative points of onboard equipment circuit are depicted in Fig. 15.4.

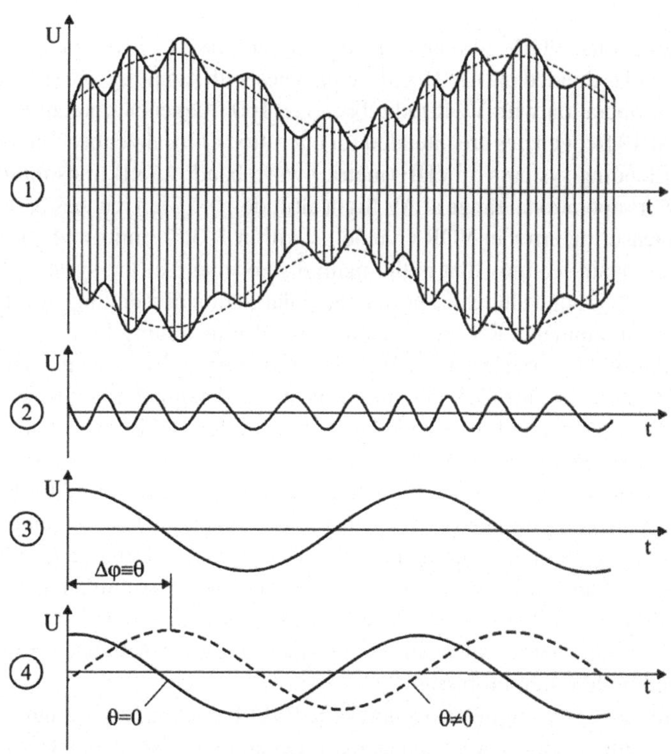

Fig. 15.4 Voltage waveforms in VOR onboard equipment

Fig. 15.5 ARB-90 and DME-90 radio beacons on site

Omnidirectional VOR radio beacons operate at 108–118 MHz band. At present time, this band is divided into 200 fixed frequencies with sampling rate (frequency) of 50 kHz. For operation with VOR radio beacons, 160 frequencies are allocated from 200, the rest 40 frequencies are intended for operation of localizer beacon (on-course beacon) of landing systems of VHF (metric) wave band. At RB transmitters power up to 50 W, its operational range at AV flight altitude 10–12 km reaches 250–370 km. Azimuth measuring error in VOR channel is within 1–3.5° limits and considerably depends on terrain features that is the main disadvantage of the system. To ease a countering with ground clutters, a horizontal polarization of radiating signals is used.

The system stipulates RB identification signal transmitting by Morse code via tone modulation of carrier waves at 1020 Hz frequency or by voice message.

As an example, in Fig. 15.5 a picture of positioned azimuth radio beacon (ARB-90) of VOR system combined with DME transmitter (DME-90) of the Russian manufacturing is represented.

PVOR (VHF omnidirectional radio range) system is a result of further development of VOR system, designed for increase of azimuthal measurements accuracy. Accuracy of phase measurement can be increased due to frequency increase of comparing oscillations. However, a frequency increase of azimuthal and reference signals leads to necessity to increase of rotation rate of directional (beamed) antenna, otherwise, azimuth measurement ambiguity arises. Increase of antennas rotation rate is connected with design problems.

To increase a measurement accuracy in PVOR system, two-channel azimuth measuring method is used. Two channels, rough and fine (precise), are intended for single-valuedness problems solution and phase precise measurement correspondingly.

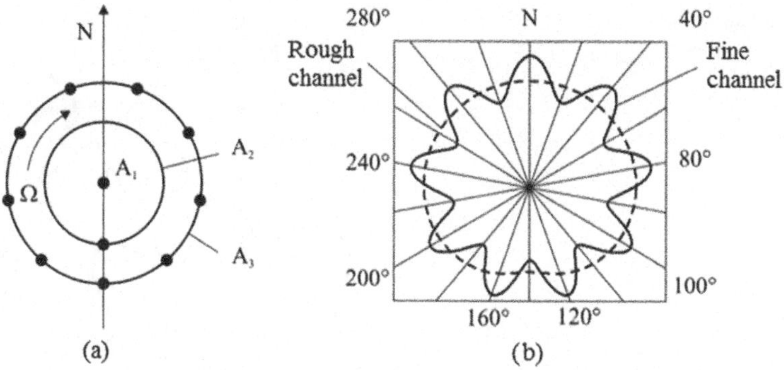

Fig. 15.6 RB PVOR antenna system: **a** antenna structure; **b** ADP in horizontal plane

RB PVOR system (Fig. 15.6a) consists of central antenna A_1, representing a vertical dipole, and two rotating on its axis coaxial cylinders A_2 and A_3.

Cylinders are made of radar-transparent material and are synchronously rotated at 15 Hz frequency. Passive reflecting element (reflector) is located along one of inner cylinder A_2 generatrix elements. The central antenna is powered by pulse-modulated waves of carrier frequency with constant filling coefficient (duty factor). The system, consisting of central dipole A_1 and cylinder A_2, has a directional pattern of "cardioid" type, in which a maximum of radiation is directed to north at that moment of time, when reflector is positioned in south direction. Increase of azimuth measuring accuracy is achieved by application of multi-beam ADP. To obtain such DP, nine (9) reflectors are mounted at outer rotating cylinder, which positioned along elements at equal space from each other (in a 40°).

Application of multi-beam directional patterns should in theory lead to errors decreasing, stipulated by re-reflected signals, by K times, where K—number of lobes. However, at large number of lobes a problem arises with ambiguity resolution of azimuth determination. With regard to all factors, effecting on azimuth measuring accuracy, an error in rough channel in tough terrain conditions can achieve 20° value. In connection with mentioned arguments, a number of lobes are selected as equal to 9.

At rotation of outer cylinder A_3 synchronously with an inner cylinder A_2, a ninth harmonic of oscillations (at 135 Hz frequency) is imposed upon main modulation (15 Hz). At 135 Hz frequency, an azimuth refining (precise channel) is carried out.

The system directional pattern (Fig. 15.6b) represents a cardioid (ADP of rough channel), at which a periodic function of azimuthal angle is superimposed, which is having 9 periods, each of which equals to 40° (ADP of fine channel). Pattern rotates in horizontal plane at 15 Hz frequency.

Distribution of radiation intensity in azimuthal plane in PVOR RB:

$$F(\alpha) = 1 + m_D \cos\alpha + m_T \cos K\alpha \tag{15.14}$$

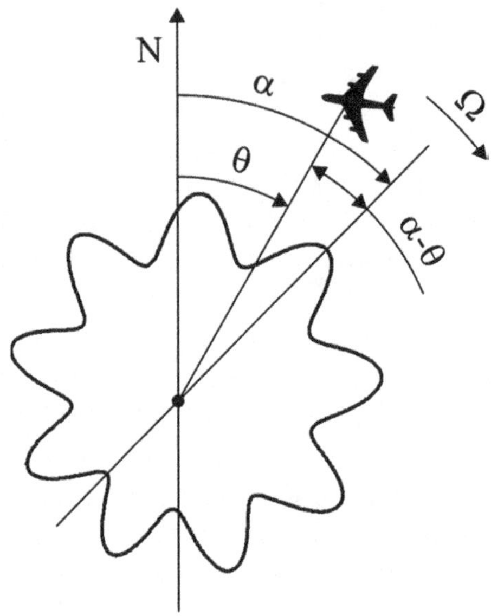

Fig. 15.7 Azimuth measuring in PVOR system

where m_D, m_T—are numerical coefficients, characterizing AM depth at corresponding modulation frequencies; K—a number of maximums of used directivity function.

At such directional pattern, radiation level (Fig. 15.7) in any azimuthal direction θ is characterized by a value:

$$F(\alpha - \theta) = 1 + m_D \cos(\alpha - \theta) + m_T \cos(K(\alpha - \theta)) \tag{15.15}$$

Radiated pattern is rotated in horizontal plane at angular velocity Ω and has an angular position $\alpha = \Omega t$. Rotation presence of antenna permits to form an azimuthal signal:

$$e_{\text{azs}}(t) = E_M F(\alpha - \theta) \cos(\omega_0 t). \tag{15.16}$$

By carrying on conversions, we obtain:

$$e_{\text{azs}}(t) = E_M[1 + m_D \cos(\Omega t - \theta) + m_T(K(\Omega t - \theta))] \cos(\omega_0 t). \tag{15.17}$$

After receiving, amplifying and detection of azimuthal signal in onboard equipment, a low-frequency voltage can be allocated:

$$u_{\text{az}}(t) = U_{\text{rchM}} \cos \varphi_{\text{rch}} + U_{\text{pchM}} \cos \varphi_{\text{pch}}, \tag{15.18}$$

where $\varphi_{\text{rch}}(t) = \Omega t - \theta$—azimuthal signal phase of rough channel; $\varphi_{\text{pch}}(t) = K(\Omega t - \theta)$—azimuthal signal phase of precise channel.

This voltage contains an information on ADP rotation rate (rough channel frequency) $\Omega = \frac{2\pi n}{60} = 2\pi F_r$, where from at $n = 900$ rpm, $F_r = 15$ Hz.

The voltage also contains an information on precise channel frequency $K\Omega = K\frac{2\pi n}{60} = 2\pi F_p$, where $F_p = 135$ Hz.

Signal, containing reference oscillations, is radiated through the central antenna A_1. Reference signals of rough and precise channels are transmitted via pulse-code modulation (PCM).

Transmitter of PVOR radio beacon operates in pulse mode. Each signal of a transmitter represents a group of two pulses of 3,2 μs duration with fixed interval between them equals to 12 μs. Radio beacon signals, excepting signals of reference voltage, have a random distribution in time. Number of chaotically running in time signals equals to 2700 pulses per second or 180 pulses per one antenna revolution.

Reference signals of rough measuring are transmitted once per one antenna revolution, when a maximum of radiation comes through north direction. North reference signal represents a sequence of twelve (12) pairs of pulses, following one by one with a fixed interval of 30 μs. Consequently, for rough measuring, $12 \times 15 = 180°$ pairs of pulses are transmitted.

Reference signals of precise measuring are transmitted each time when consecutive maximum of nine-lobed characteristic goes through north direction. Reference signal of precise measuring represents a sequence of six (6) pairs of pulses, following one by one with a fixed interval of 24 μs. Totally, $6 \times 8 = 75°$ pairs of pulses are transmitted per second.

Total amount of pulse pairs, radiated per second by PVOR RB, reaches 3600.

Low-frequency voltage of reference signal is produced onboard an AV.

$$u_{\text{ref}}(t) = U_{\text{rrefM}} \cos \varphi_{\text{rref}} + U_{\text{prefM}} \cos \varphi_{\text{pref}}, \tag{15.19}$$

where $\varphi_{\text{rref}}(t) = \Omega t$—reference signal phase of rough channel; $\varphi_{\text{pref}}(t) = K\Omega t$—reference signal phase of precise channel.

Availability of two azimuthal and two reference signals permits to conduct two steps of measurement of phase difference: rough at 15 Hz frequency, precise at 135 Hz. At rough measurements $\Delta\ \varphi_r(t) = \theta_r$. Azimuth is univalently determined, but with low accuracy. At precise measurements $\Delta\ \varphi_p(t) = K\theta_p$, azimuth is determined precisely, but not univalently (single-valued). Two-step measurements permit to determine azimuth univalently and with high accuracy. Two-step measurement technique is shown in Fig. 15.8.

Rough measurement of phase difference $\Delta\ \varphi_r$ permits to define single-valued read-out zone of $360°/K$ width, within limit of which an AV azimuth is located and to obtain a number of such k zones, including in azimuth. Precise measurement of phase difference $\Delta\ \varphi_p$ permits to define a precise AV position inside of this zone θ_p.

AV azimuth is a sum of measurement results: $\theta = k \cdot 40° + \theta_p$, $0 \le \theta_p \le 40°$, $k = 0, 1, 2, \ldots, 8$.

If both measurement steps are performed by the same units (phase meters) and in equal conditions, then measurement errors of phase difference can be considered as equal. Due to oscillation frequency increase by K times, an accuracy should be

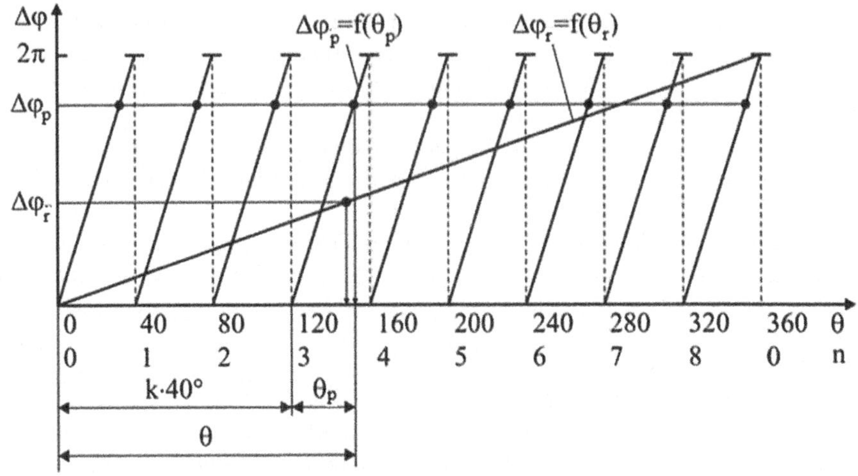

Fig. 15.8 Two-step azimuth determination of PM PVOR

increased by K times. True measurement accuracy in PVOR system is approximately 4…5 times more than in VOR system with the same conditions.

Functional chart of PVOR system azimuth meter is depicted on Fig. 15.9. Single pulses are coming to azimuthal channel input from a receiver output after decoder. Onboard equipment decoder passes (gates through) pulsed signals with code spacing of pulses 12 μs. Voltage of variable phase and reference voltage are produced in individual channels.

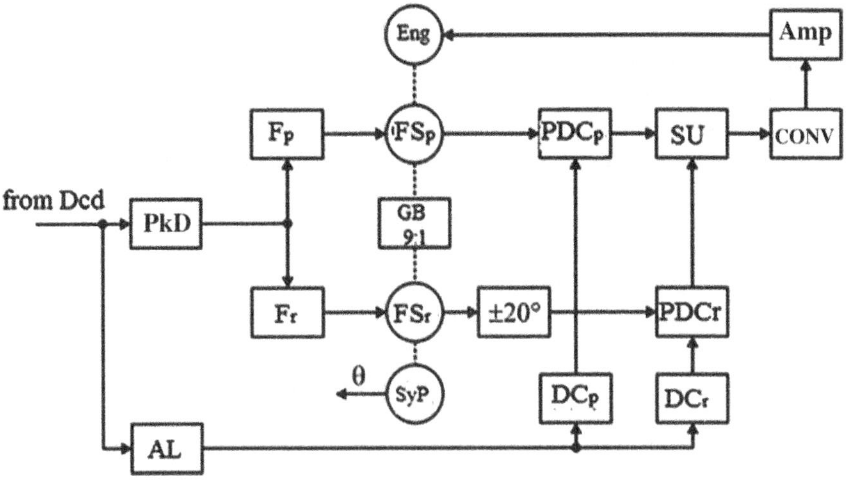

Fig. 15.9 Functional chart of PVOR azimuth meter

To get reference voltages, the pulses from decoder output are amplified and limited in amplifier-limiter. Pulses of constant amplitude come to decoding cascades. Decoding cascade of rough signal (DC_r) extracts from chaotic sequence of pulses the signals, corresponding to groups of 12 pulses, following with interval of 30 μs. These signals are converted into reference voltage of 15 Hz. Decoding cascade of precise signal (DC_p) extracts pulsed sequences, consisting of six elements with interval of 24 μs, and forms a frequency reference voltage of 135 Hz. Reference voltages of 15 and 135 frequencies come to phase detectors of rough (PDC_r) and precise (PDC_p) channels, correspondingly.

Voltage is produced in channel of variable phase, presenting signals amplitude modulation law of radio beacon. Envelope extracting is carried out by peak detector. Detector output voltage is delivered to filters of rough (F_r) and precise (F_p) channels, which extract frequency oscillations of 15 and 135 Hz, correspondingly.

Onboard equipment includes two automatic follow-up (servo) phase meters, one of which operates at 15 Hz frequency and is intended for multiple-valuedness elimination, another one—at 135 Hz frequency and is intended for precise measuring of azimuth. Phase meter operation principle is analogue to a phase meter, examined at an example of VOR system meter. Corresponding rotors of phase shifters of precise (PS_p) and rough (PS_r) channels are connected between each other mechanically through gearbox with 9:1 transmission ratio (GB 9:1).

In "Search" mode, an engine (Eng) rotates phase shifters rotors till the moment of azimuth measuring in rough channel. For measuring in channel, a pulse gate ±20° of 15 Hz with phase, depending on AV azimuth, is shaped in channel. Pulse gate comes to input of phase detector (PD_r). After rough measuring of azimuth, the system switches to "Tracking" mode. In this mode an output voltage of phase detector (PD_p) via closed-loop switching unit K comes C to converter. Engine rotates till the moment of voltage outage from PD_p output, i.e., till the moment of azimuth measuring in precise channel.

Hence, an angular position of phase shifters rotors corresponds to azimuth value. Azimuth value is transmitted to indicator via synchro pickup (SyP).

Bandwidth, used in PVOR, ranges from 962...1213 MHz. As compared to VOR, the two-step azimuth measuring system has a higher accuracy (0, 75...1°), relatively small sizes and weight. Application of two-step method stipulates the use of special onboard equipment. Two-step azimuth determination method is realized in tactical air navigation system (TACAN) system.

In PVOR radio beacons the special mode of call signal sending is stipulated. At sending of call signals the transmitter generates signals with constant repetition rate of 2700 Hz. Herewith, at output of receiving unit, the tone alternating voltage of 2700 Hz is produced. By manipulating of these signals transmission time, it is possible to transmit signals via the Morse code.

To increase azimuth measurement accuracy by VOR onboard equipment the radio beacons were developed, the operation principle of which is based on use of Doppler effect.

DVOR system (Doppler VOR). DVOR RB antenna system consists of central and side antennas (Fig. 15.10). The central antenna A_{cnt} is located in the origin of

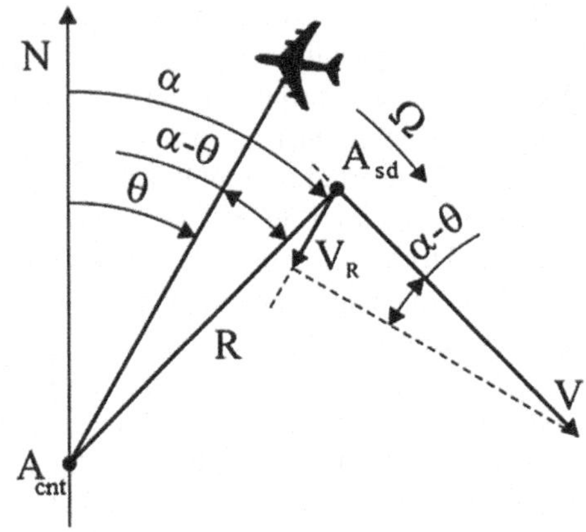

Fig. 15.10 Application of Doppler effect in DVOR

coordinates, the side antenna A_{sd}—at R distance from the central antenna at α angle to base reference line (direction to north).

Signals from antennas A_{cnt} and A_{sd} are received by onboard equipment in remote point with θ azimuth. Side antenna is rotated in radius circle R with angular velocity $\Omega = \frac{2\pi n}{60}$. Linear rotation rate of antenna is $V = \Omega R$. Then radial component of velocity in θ direction:

$$V_R = V\ \sin(\alpha - \theta). \tag{15.20}$$

Current angular position of side antenna $\alpha = \Omega t$. When $\alpha = 0, t = 0$, the antenna is located in base reference line. Radial velocity:

$$V_R = \Omega R\ \sin(\Omega t - \theta). \tag{15.21}$$

At receiving of oscillations from rotating side antenna A_{sd}, the Doppler frequency shift arises at receiving point $F_D = \frac{V_R}{\lambda}$.

Considering an expression (15.21)

$$F_D = \frac{\Omega R\ \sin(\Omega \mathrm{t} - \theta)}{\lambda}. \tag{15.22}$$

and denoting $\frac{\Omega R}{\lambda} = F_{\mathrm{Dm}}$, we obtain

$$F_D(t) = F_{\mathrm{Dm}} \sin(\Omega t - \theta) \tag{15.23}$$

From (14.24) expression, we can see that Doppler frequency shift, resulting from side antenna rotation, is connected with AV azimuth.

By receiving such signal on AV, it is possible to extract operating voltage of the form $u_{\mathrm{azs}}(t) = U_{\mathrm{SM}} \cos(\Omega t - \theta)$, the phase of which depends on azimuth.

The reference voltage of the form $u_{\text{ref}}(t) = U_{\text{refM}} \cos \Omega t$ comes onboard AV through the central antenna A_{cnt}, the phase of which does not depend on azimuth. By comparing of reference and variable signals phase, an AV azimuth is determined in onboard equipment.

The main advantage of Doppler VOR RB comparing with standard VOR RB—is a high suppression efficiency of clutter influence on operation accuracy. For an effective suppression, the antennas' rotation radius R should be relatively large and comprises $\frac{2R}{\lambda} = 4 \ldots 6$, and direction finding rate should be maintained high that demands a high rotation rate. For this reason, instead of rotating antennas the fixed antenna arrays are installed in modern DVOR systems, comprising a large amount of circularly positioned antennas, and electronic commutation (switching) of antennas is used. Herewith, signal formats of Doppler VOR radio beacons are to be selected as equal with VOR RB in order to have a possibility of its receiving using a single-type onboard equipment without any upgrading or replacement.

Antenna system of DVOR radio beacon (Fig. 15.11) consists of a large number, e.g., fifty, of dipoles $D_1 \ldots D_{50,}$ circular spaced of R radius. Opposite dipoles, e.g., $D_1 \ldots D_{26}$, powers a current with $f_{1,26} = f_o \pm f_{\text{so}}$ frequencies, where f_0—carrier frequency, f_{so}—frequency, equals to subcarrier frequency of VOR system 9960 Hz. Alternate switching of dipole pairs to HF sources imitates its circular rotation at rotation frequency of 30 Hz.

Oscillation, received onboard AV, due to presence of Doppler shift, has $f_{1,26} = f_o \pm f_{\text{so}} \pm F_D$ frequencies, i.e., oscillations are modulated in frequency with frequency deviation $\Delta F_D = \frac{\Omega R}{\lambda}$.

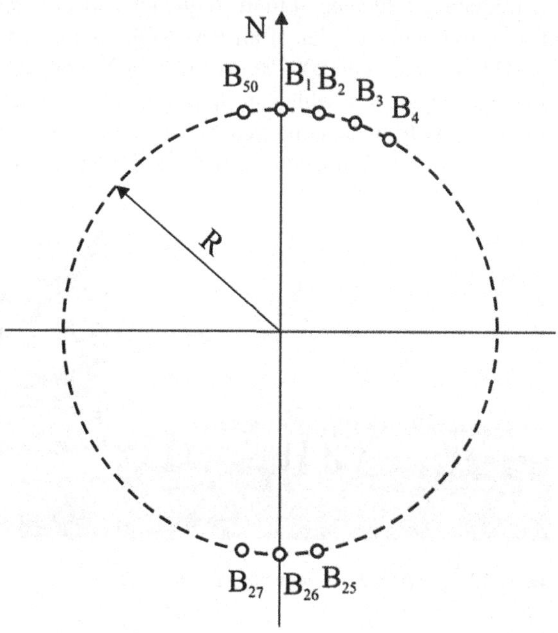

Fig. 15.11 Antenna system of DVOR radio beacon

Reference signal is radiated through the central antenna, representing an amplitude-modulated oscillations:

$$e_{\text{cnt}}(t) = E_{\text{cntM}}(1 + m_{\text{ref}} \cos \Omega t) \cos \omega_0 t.$$

A signal is generated at receiving point as a result of fields summing of central antenna and side dipoles:

$$\begin{aligned} e_{\Sigma}(t) &= E_{\text{cntM}} Z \cos \omega_0 t, \\ Z &= 1 + m_{\text{ref}} \cos \Omega t + m_{\text{subc}} \cos(\Omega_{\text{subc}} t + m_{\text{FM}} \cos(\Omega \text{t} - \theta)), \end{aligned} \quad (15.24)$$

where $m_{\text{FM}} = \frac{2\pi R}{\lambda}$—frequency modulation index.

From (15.24) expression, we can see that this signal in its structure is identical to signal of standard VOR. The difference of signal processing of DVOR radio beacon consists in that azimuthal signal (variable phase signal) is transmitted via FM channel and extracted by F2 filter, and reference signal is transmitted via AM channel and is extracted in onboard equipment by F1 filter.

DVOR radio beacon installed at airfield area position and combined with DME is depicted in Fig. 15.12.

VOR radio beacons operate at 111.975–117.975 MHz bandwidth, besides the fact that 108–111.975 MHz frequencies can be used in those cases when the use of such frequencies is an acceptable. The highest assigned frequency equals to 117.950 MHz. Channel separation is 50 kHz, beginning from the highest assigned frequency. In those areas, where a channel separation of 100 or 200 kHz is often used the frequency tolerance of high-frequency carrier is ±0.005%. Frequency tolerance of high frequency carrier of all new VOR in areas, where a channel separation of 50 kHz is used, is ±0.002%. In those areas, where new VOR are installed with assigned frequencies, which are spaced at 50 kHz relatively to frequencies, assigned by existing VOR in the same area, first of all the steps are taken to decrease a frequency tolerance of high-frequency carrier, using in the existing VORs, up to ±0.002%.

Fig. 15.12 DVOR radio beacon at position

Table 15.1 VOR radio beacon specifications

No.	RB type	DME-90	VOR-734	DVOR 2000
1	Bandwidth, MHz	108…118	108…118	108…118
2	Coverage zone in HP: deg.	0…360	0…360	0…360
	km	300	300	300
3	Coverage zone in VP, deg.	0…40	–	0…40
4	Azimuth measuring error, deg.	±1	–	±0.2
5	Operational environment: temperature, °C	−50…+50	−50…+70	−50…+50
	Wind load, m/s	50	55	50
	Precipitations, mm/min	–	–	up to 3

It is almost managed to exclude an influence of terrain features on azimuth channel accuracy at development of Precision Doppler VOR (PDVOR). In ground radio beacons of this system, a signal of reference phase is transmitted via frequency modulation of auxiliary subcarrier frequency of 6500 Hz. The advantages of PDVOR system can be realized only by using special onboard equipment. Standard VOR receiver operates with PDVOR radio beacons in the same way as with DVOR radio beacons. For this purpose, the components in PDVOR signal spectrum are remained, corresponding to DVOR reference signal.

Table 15.1 states specifications of modern VOR RB.

Localizer (azimuth transmitter) beacon (LOC) and glide-path (elevation transmitter) beacon (GP) of instrument landing systems type are referred to goniometric RNS of radio beacon type. They have the same principle of operation and are functioning as equisignal beacons. Regarding this, the system operating principle will be examined at an example of localizers of ILS landing system.

Radio beacon (LOC) produces a HF electromagnetic field in air space, simultaneously amplitude-modulated by two different frequencies. As assigned by radio beacon direction (course for landing), the modulation coefficients for these frequencies are equal (Fig. 15.13). At deviation from this direction, the modulation coefficients are happen to be unequal, where the more deviation the more difference of modulation coefficients, and at deviation into different directions, the relation of modulation coefficients is changed to an opposite value. An aerial vehicle position

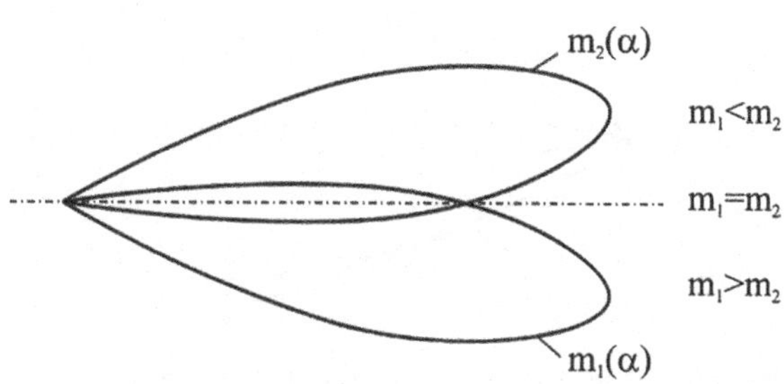

Fig. 15.13 Dependence of modulation coefficient from direction

relative to assigned direction can be judged by relation of modulation coefficients, i.e., via envelope amplitudes comparison of two modulation frequencies.

Localizer beacon antennas have combined electrical centers, and currents, supplying these antennas are in phase, and therefore in air space, surrounding a radio beacon, two in-phase fields are formed.

In air space, electromagnetic fields, excited by each antenna, are composed and as a result a combined field is produced, an expression for strength of which has the following form:

$$e_{\Sigma}(t) = E_M F_3(\alpha)\left[1 + \frac{F_1(\alpha)}{F_3(\alpha)} m \cos \Omega_1 t + \frac{F_2(\alpha)}{F_3(\alpha)} m \cos \Omega_2 t\right] \cos \omega t, \quad (15.25)$$

or

$$e_{\Sigma}(t) = E_M F_3(\alpha)[1 + m_1(\alpha) \cos \Omega_1 t + m_2(\alpha) \cos \Omega_2 t] \cos \omega t,$$

where $F_3(\alpha) = F_1(\alpha) + F_2(\alpha)$; E_M—amplitude in peak of directional pattern; $m_1(\alpha)$—modulation coefficient for frequency Ω_1; $m_2(\alpha)$—modulation coefficient for frequency Ω_2.

Expression (15.25) shows that combined field of a beacon is simultaneously modulated by two Ω_1 and Ω_2 frequencies. It formed from three components: combined field of carrier frequency ω and two fields of side frequencies of modulation $\omega \pm \Omega_1$ and $\omega \pm \Omega_2$. Field of a carrier frequency has a directivity $F_3(\alpha)$, a field of side modulation frequencies $\omega \pm \Omega_1$—directivity $F_1(\alpha)$ and a field of side modulation frequencies $\omega \pm \Omega_2$—directivity $F_2(\alpha)$. Due to the fact that intensities of carrier frequency field and fields of side modulation frequencies are changed depending on direction based on different laws (Fig. 15.14), then modulation coefficients for $m_1(\alpha)$ and $m_2(\alpha)$ depend on direction.

According to (15.25):

$$m_1(\alpha) = \frac{F_1(\alpha)}{F_3(\alpha)} m, m_1(\alpha) = \frac{F_2(\alpha)}{F_3(\alpha)} m. \quad (15.26)$$

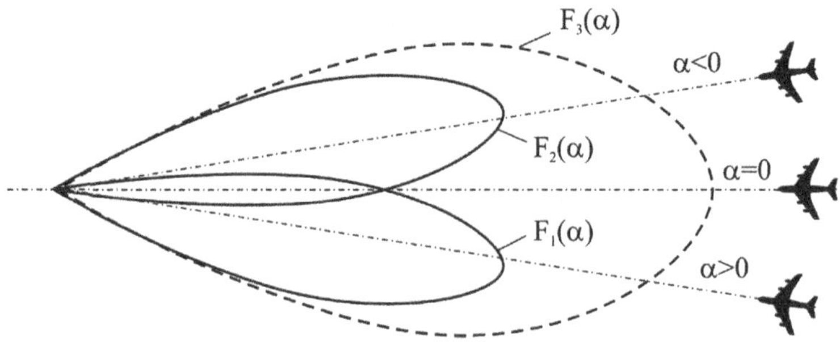

Fig. 15.14 LOC beacon directional pattern of ILS type

From (15.26) expressions, it is followed that in that direction where $F_1(\alpha) = F_2(\alpha)$, modulation coefficients for Ω_1 and Ω_2 frequencies are equal. Hence, a radio beacon produces an equisignal radiation of modulation side frequencies in direction of directional patterns intersection $F_1(\alpha)$ and $F_2(\alpha)$ of beacon antennas. This equisignal direction serves for assignment of landing flight path (trajectory). At deviation from equisignal direction the modulation coefficients are changed.

Onboard an aerial vehicle, it is possible to determine its position relative to equisignal direction, assigned by radio beacon, via comparison of envelope amplitudes of Ω_1 and Ω_2 frequencies. Herewith, signal division, transmitted by different antennas, is possible, since spectrums of these signals differ from each other (Fig. 15.15).

In radio beacon of ILS system type the modulation frequencies of 90 and 150 Hz are used, herewith in glide-path beacons a lower antenna is supplied by signals with 90 Hz modulation frequency, and upper antenna—by signals with 150 Hz modulation frequency.

LOC and GP beacons of III and II categories landing systems should provide a high stability of assigned directions in air space and other output characteristics. Such requirements can't be provided by equisignal radio beacons used in I category of ILS systems. Hence in landing systems of high accuracy (III category, and even II category), the so-called radio beacons with reference null are used as LOC and GP beacons. Examine in brief an operation of such radio beacons (Fig. 15.16).

Radio beacon with reference zero operation principle concludes in the following: Two antenna systems A_1 and A_2 are used in radio beacon. A_2 antenna system has two-lobed directional pattern $F_2(\alpha)$ (Fig. 15.17), the direction minimum of which coincides with assigned direction, and at transition through a direction minimum the ADP sign changes to an opposite. A_1 antenna system has a single-lobed directional pattern $F_1(\alpha)$, the direction maximum of which coincides with assigned direction.

A_2 antenna is feed by difference of two balanced-modulated oscillations $\cos\,\Omega_1 t\,\cos\,\omega t$ and $\cos\,\Omega_2 t\,\cos\,\omega t$ with different modulation frequencies, which are formed by two balanced modulators BM1, BM2 and high-frequency oscillator (HFO). In ILS system, the modulation frequencies $\Omega_1 = 90\,\text{Hz}$, $\Omega_2 = 150\,\text{Hz}$ are

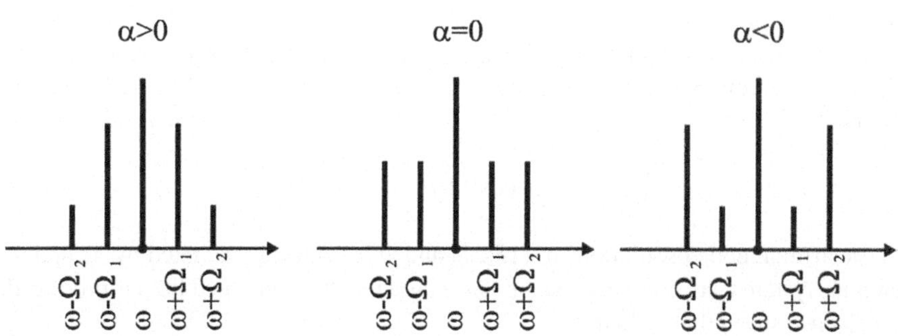

Fig. 15.15 LOC beacon signal spectrums

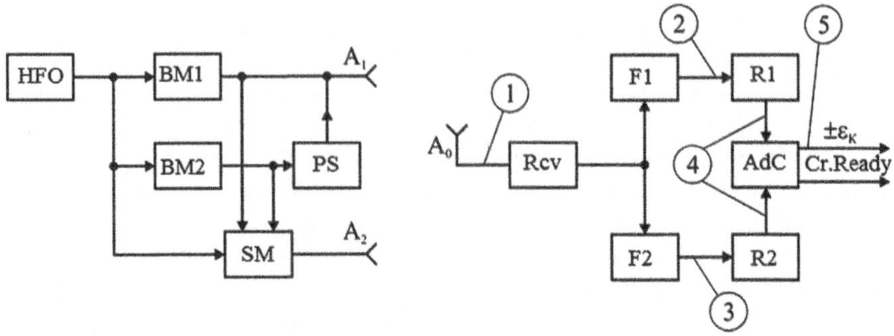

Fig. 15.16 Functional chart of ILS course channel

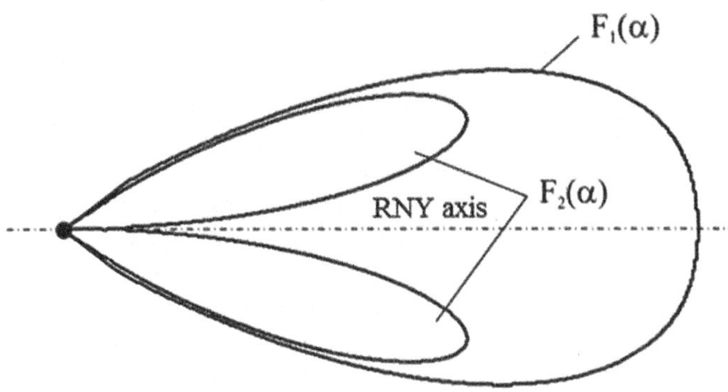

Fig. 15.17 Antenna system directional pattern of radio beacon with reference null

used. Oscillation difference is produced as result of its antiphase summing, and phase shift at 180° is provided by phase shifter (PS).

Electric field strength, radiated by A_2 antenna:

$$e_2(t) = E_{M2}F_2(\alpha)[\cos\,\Omega_1 t - \cos\,\Omega_2 t]\cos\,\omega t. \tag{15.27}$$

A_1 antenna is feed by a sum of balanced-modulated oscillations and oscillations of a carrier frequency, which is formed in summator (SM). A field is radiated into air space

$$e_1(t) = E_{M1}F_1(\alpha)[1 + m\,\cos\,\Omega_1 t + m\,\cos\,\Omega_2 t]\cos\,\omega t. \tag{15.28}$$

Electromagnetic oscillations of (15.27) and (15.28) forms, radiated by A_1 and A_2 antenna systems, are composed in air space, as a result a combined field is produced, the electric strength of which:

$$e_\Sigma(t) = E_{M1}F_1(\alpha)Z\,\cos\,\omega t,$$

$$Z = 1 + \left[m + \frac{E_{M2}F_2(\alpha)}{E_{M1}F_1(\alpha)}\right]\cos\,\Omega_1 t + \left[m - \frac{E_{M2}F_2(\alpha)}{E_{M1}F_1(\alpha)}\right]\cos\,\Omega_2 t. \quad (15.29)$$

The structure of combined field (15.29) of RB shows that it is a HF oscillation, simultaneously amplitude-modulated by oscillations of two modulation frequencies, the modulation coefficients of which depend on direction:

$$m_1(\alpha) = m + \alpha\frac{F_2(\alpha)}{F_1(\alpha)},\; m_2(\alpha) = m - \alpha\frac{F_2(\alpha)}{F_1(\alpha)},\; \alpha = \frac{E_{M2}}{E_{M1}} \quad (15.30)$$

Then, (15.29) can be written as:

$$e_\Sigma(t) = E_{M1}F_1(\alpha)[1 + m_1(\alpha)\cos\,\Omega_1 t + m_2(\alpha)\cos\,\Omega_2 t]\cos\,\omega t. \quad (15.31)$$

It is evident that structure of combined signal of equisignal radio beacon is the same as radio beacon with reference null has. Consequently, the main properties of these signals (shape and spectral content, changing character depending on direction etc.) are equal, though forming ways of electromagnetic field in these beacons are different.

Modulation coefficients $m_1(\alpha)$ and $m_2(\alpha)$ are changed from α angle according to different laws (Fig. 15.18). At assigned direction (landing course), where $\alpha = 0$, modulation coefficients are equal $m_1(\alpha) = m_2(\alpha) = m$. At deviation from assigned direction in one way ($\alpha > 0$) $m_1(\alpha) > m_2(\alpha)$, and at deviation into another way ($\alpha < 0$) the modulation coefficients relation is changed into opposite, i.e., $m_1(\alpha) < m_2(\alpha)$. Difference in depth modulation $m_1(\alpha) - m_2(\alpha)$ is changed from a direction in the same way as in equisignal radio beacon.

Convert (15.31) expression into form:

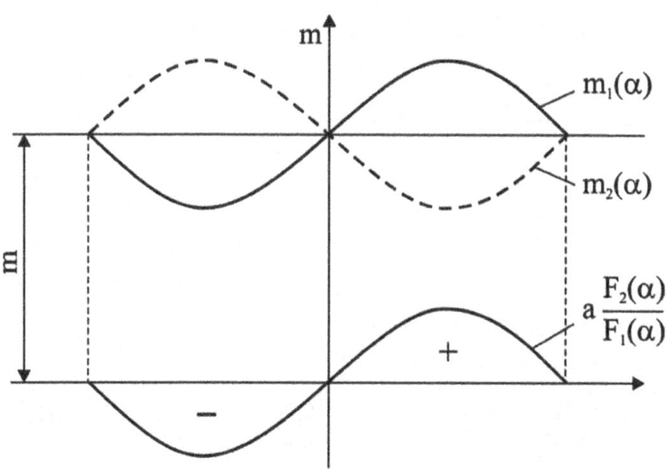

Fig. 15.18 Dependence of modulation coefficient from direction

$$e_{\Sigma}(t) = E_{M1}F_1(\alpha)\cos\omega t + E_{M1}F_1(\alpha)m_1\cos\Omega_1 t\,\cos\omega t + E_{M1}F_1(\alpha)m_2\cos\Omega_2 t\,\cos\omega t. \quad (15.32)$$

It is obviously that combined signal of radio beacon consists of three components: oscillations of carrier frequency, oscillations of modulation side frequencies $\omega \pm \Omega_1$ and oscillations of modulation side frequencies $\omega \pm \Omega_2$. Amplitudes of the mentioned components depend on aircraft position relatively to course line and are defined correspondingly by functions $F_1(\alpha)$, $F_1(\alpha)m_1(\alpha)$ and $F_1(\alpha)m_2(\alpha)$. In direction $\alpha = 0$, the radiated oscillations (waves) of modulation side frequencies $\omega \pm \Omega_1$ and $\omega \pm \Omega_2$ are equal. Hence, a radio beacon with reference null produces an equisignal waves radiation of modulation side frequencies into direction of directional pattern minimum $F_1(\alpha)$. This equisignal direction serves for assignment of landing course (heading) in LOC beacons or glide path in GP beacons.

Timing waveform diagram in certain points of onboard equipment is represented in Fig. 15.19. Combined signal, which represents a HF signal, simultaneously amplitude-modulated by two $\Omega_1 = 90$ Hz and $\Omega_2 = 150$ Hz frequencies, is received by LOC beacon antenna system. After amplification and detection of combined signal, the sum of two low-frequency voltages with 90 and 150 Hz frequencies is extracted, the amplitudes of which are proportional to modulation coefficients m_1 and m_2, correspondingly. These voltages are extracted by two filters F1 (2) and F2

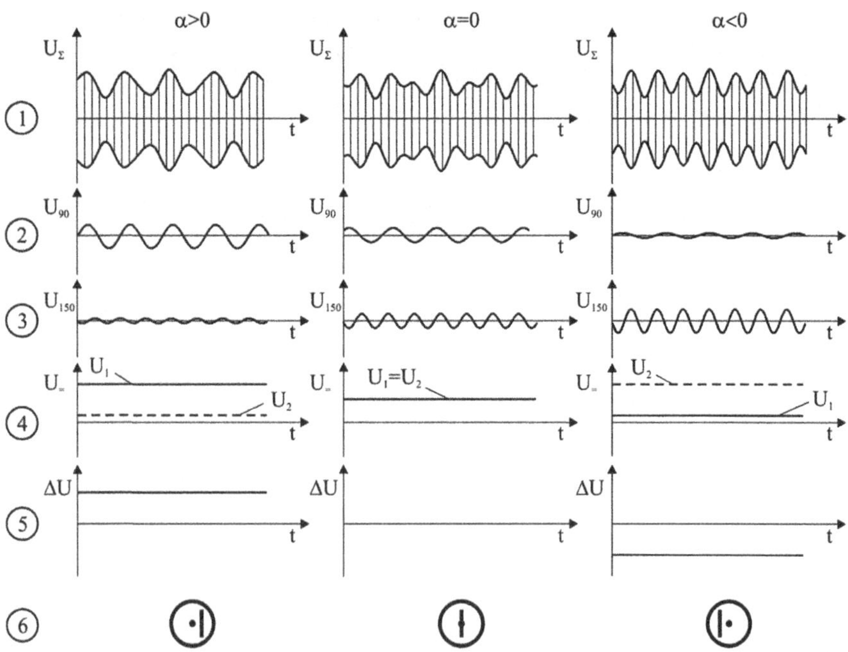

Fig. 15.19 Timing waveform diagram in certain points of ILS system onboard equipment

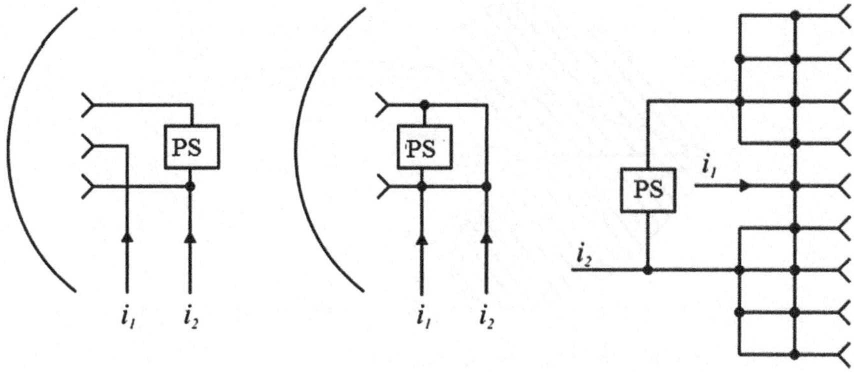

Fig. 15.20 LOC beacon antennas with reference null

(3), and adjusted at 50 and 150 Hz frequencies and rectified by R1 and R2 rectifiers (4).

Constant voltages are induced after rectifiers, proportional to modulation coefficients m_1 and m_2. Constant voltages come to adding circuit (AdC) where its sum and difference are formed. Difference of these voltages (5) delivers to needle of indicating unit of landing system. The unit needle deflection is directly proportional to angular deviation from specified direction at low deviations (6).

Sum of constant voltages serves for generation in AdC of "Course readiness" voltage (Cr. Ready), coming to (indicating) displaying units and onboard systems.

In LOC beacons with reference null the antenna systems of different type are used. So that, parabolic antennas with three or two exciters can be used (Fig. 15.20). If three exciters are used, then a central exciter combined with reflector (scatterer) serves as A_1 antenna and forms a directional pattern $F_1(\alpha)$. Side exciters are fed in antiphase and are used in combination with reflector as A_2 antenna with directional pattern $F_2(\alpha)$. At use of two exciters, A_2 antenna is formed due to antiphase feed of exciters, and A_1 antenna—at in-phase feed. There are linear series type antennas that can be used as well. Such antenna consists of several dozens of (weakly) near-omnidirectional antennas (e.g., closed-loop-type or director-type (Yagi) antennas), positioned at one line perpendicular to runway (RNY), and at the same distance from each other and fed with certain amplitudes and phases to get a necessary ADP.

When use of linear series-type antennas, A_1 antenna is formed due to in-phase feed of exciters, and A_2 antenna system—at antiphase feed of two antenna groups, the centers of which are positioned at a certain space from each other, and at in-phase antenna feed, forming each group.

In GP beacon with reference null, two near-omnidirectional antennas are used, suspended over a ground surface at different heights, where suspension height of upper antenna is twice more than lower antenna height. Upper antenna combined with specular reflecting ground surface is used as A_2 antenna system. Lower antenna combined with ground surface operates as A_1 antenna system. For assignment of

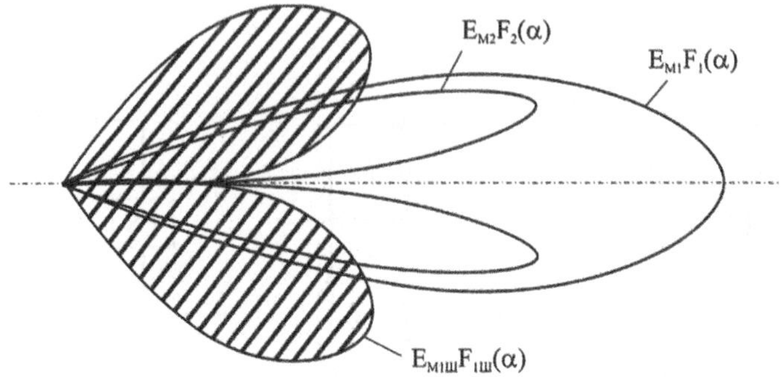

Fig. 15.21 Radiation diagram of two-channel LOC beacon

glide-path, the lower lobe of directional pattern $F_1(\beta)$ and two lower lobes of directional pattern $F_2(\beta)$ are used. Since antenna suspension heights differ in two times, the first minimum of pattern $F_2(\beta)$ coincides with a maximum of lower lobe of lower antenna directional pattern.

At increase of LOC and GP beacon operation accuracy, there is a necessity to increase not only the position stability in air space of course line and glide-path, but to decrease its course lines bend. Requirement to minimal bend of course line and glide-path is in collision with necessity to provide quite enough wide coverage zone in horizontal plane for LOC beacon and operation at low elevation angles for GP beacon. At extension of coverage zone, the more terrain features and irregularities participate in formation of reflected signals, in consequence of which the amplitude of bends increases. The mentioned contradiction is eliminated in radio beacons, using two-channel operation principle.

Two-channel radio beacons. Examine the operation principle of two-channel radio beacon as an example of LOC beacon. Two-channel LOC beacon includes two channels: narrow and wide, in each of which its own antenna system is used. Narrow channel antenna system forms quite narrow ADP of 6–12° in horizontal plane (Fig. 15.21). Wide channel antenna system forms wide ADP, ensuring the specified width of coverage zone (±35°).

Narrow channel forms almost straight course line, since coverage zone of this channel is free from reradiated terrain features and irregularities. This channel is used to control aerial vehicle at small deviations from landing heading plane. At large deviations, a wide channel is used, which is effected by reradiated signals; however, there are no strict requirements for coverage zone of this channel to precise characteristics due to large deviations from landing heading plane. In order to reduce signals influence of wide channel on operation of beacon narrow channel, the null in ADP of wide channel is envisaged in coverage zone of narrow channel.

Both radio beacon channels have the same operation principle, examined before. On aerial vehicle the signals of narrow and wide channels should be used separately, and with this purpose a carrier frequency, differing from narrow channel carrier

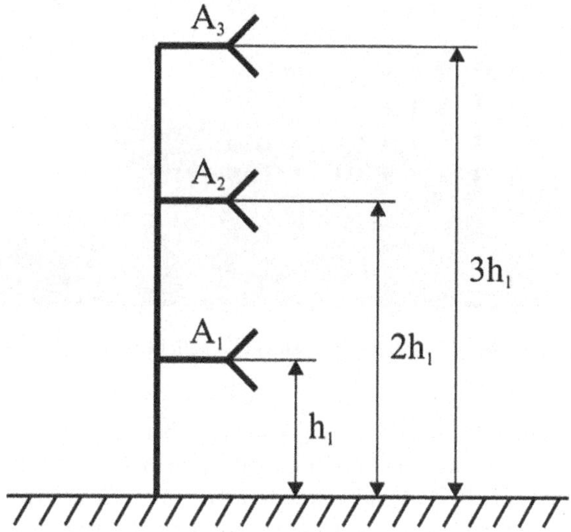

Fig. 15.22 Antenna system of two-channel GP beacon

frequency at 9–11 kHz, either is used in wide channel or equal carrier frequency with narrow channel is used, but channels signals have a phase shift at 90°.

In glide-path radio beacon, to reduce a glide-path bend a radiation is compensated at low elevation angles, and to receive information on AV position in this area an additional channel is used. For this purpose in two-channel GP beacon, besides two antennas (lower and upper), the third A_3 auxiliary antenna is used, located at height in three times higher than suspension height of lower antenna (Fig. 15.22). In main channel of GP beacon A_1 and A_2 antennas are used, and in auxiliary channel—A_1 and A_3 antennas. Phases and amplitudes of antennas feeding currents are selected in order to reduce a field level at low angles to horizon that leads to reducing of signals, reflected by terrain irregularities, and, consequently, to reducing of glide-path bend amplitude.

To interact with onboard equipment, the radio beacons of ILS landing system can be adjusted at 40 fixed frequencies, LOC beacon in 108–112 MHz band, GP beacon—329–335 MHz.

As an example, localizer (at the left) and glide-path (at the right) beacons of ILS-410 system are illustrated in Fig. 15.23.

15.1.1 Direction-Finding Goniometric Radio Navigation Systems

Ground-based radio direction-finders are intended for signals receiving and bearing measurement to onboard VHF radio transceivers. As it was mentioned in the first

Fig. 15.23 ILS-410 localizer and glide-path beacons

section, the radio bearing is an angle between magnetic north of an object or radio-navigational point (RNP) and direction to RNP or to an object. Radio bearing from an object to RNP is called true (magnetic) bearing of radio transceiver (TRB or MRB), from RNP to an object—true (magnetic) bearing of an object (TOB or MOB).

Nowadays, radio direction-finders, operating in VHF waveband, since a radio communication is carried out in this bandwidth, are used in civil aviation. They are called VHF Direction Finding Equipment, and in abbreviated form VDF. For such radio direction-finders, a 118–137 MHz bandwidth is allocated.

Doppler radio direction-finders are widely used in civil aviation. Doppler effect is manifested at moving of radiator element relatively to receiving point or at moving of receiving point relatively to radiator element. Hence, if in radiation field of such source an antenna is moved, then a component (F_D Doppler frequency) will be added to its frequency, which depends on velocity and direction of antenna movement. If this antenna circularly rotates, then added Doppler frequency component will be changed to sinusoidal.

Let's examine an example of antenna rotation relatively to rotation point with R radius for aircraft bearing $\alpha = 0$ (Fig. 15.24). We assume a radiation point (radio transceiver of aircraft) as stationary. We call a reference point as antenna position in point 1. At antenna moving from point 1 into point 2 with rotation rate Ω_{rt} (V_{ant}—velocity vector), the radial (R-component) component of its velocity V_r is changed from zero to V_{r2}, depicted on a figure. For further antenna positions in points 2…12, the R-component of velocity changes by achieving a maximum value in points 4 and 10 ($\varphi = 0°$ and $\varphi = 270°$) and changes its direction (sign) from point 7 (at $180° < \varphi < 360°$).

Graph of velocity R-component variance V_r is shown in Fig. 15.24b. Dependence of velocity R-component from antenna slewing angle is defined by the following expression:

$$V_r = V_{\text{ant}} \sin \varphi \tag{15.33}$$

Correspondingly, a frequency of received signal is also changed by sine law: is reduced at radiator moving within limits $0° < \varphi < 180°$ and is increased at $180° < \varphi$

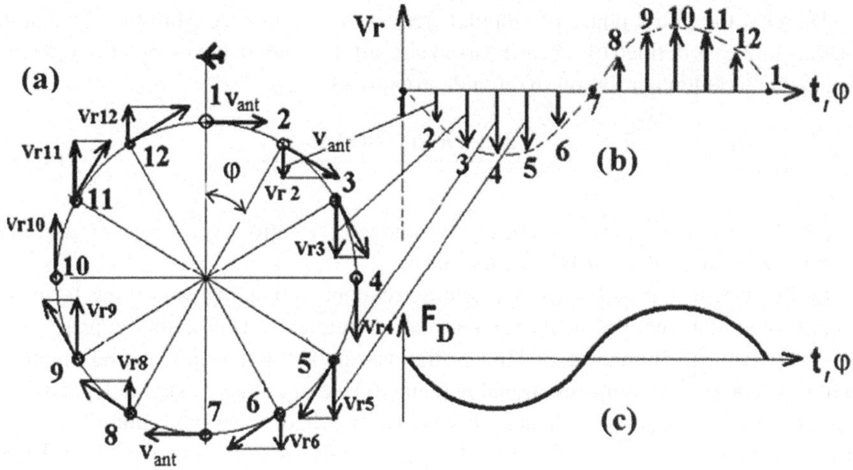

Fig. 15.24 Antenna rotation in radiation field of aircraft radio transceiver in circular (aircraft bearing 0°)

< 360°. Maximum frequency change will be at antenna slewing angles within $\varphi = 90°$ and $\varphi = 270°$. Frequency change of received signal relatively to radiator frequency is Doppler frequency. Doppler frequency change graph from antenna slewing angle is depicted in Fig. 15.24c. As we can see, the initial phase of Doppler frequency dependence from time (radiator slewing angle) comprises 0°.

A case of *R*-component change of antenna movement velocity for aircraft position in west (aircraft bearing $\alpha = 270°$) is examined in Fig. 15.25. Antenna position in point 1 is also assumed as reference point. As we can see in this case, the initial phase of Doppler frequency dependence from time (radiator slewing angle) already comprises 270°.

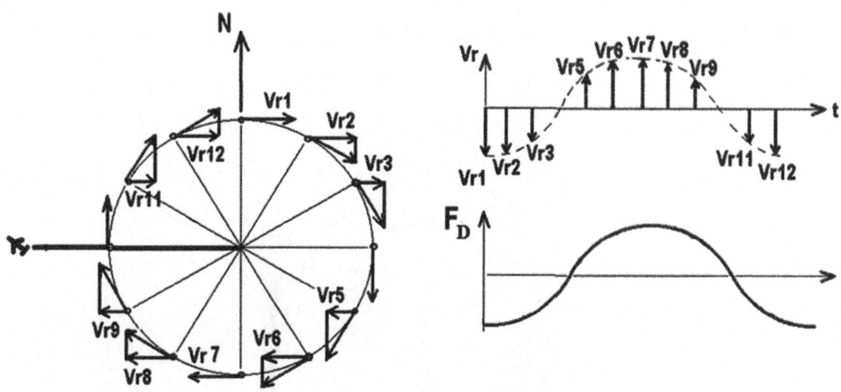

Fig. 15.25 Antenna rotation in radiation field of aircraft radio transceiver in circular (aircraft bearing 270°)

Thereby, the initial phase of Doppler frequency dependence from time (radiator slewing angle) is defined by a bearing to an aircraft. In general, the Doppler frequency value, due to antenna rotation in circular, comprises:

$$F_D(t) = \frac{2\pi\Omega_r R \sin(2\pi\Omega_r t - \alpha)}{\lambda}, \tag{15.34}$$

where R—radiator rotation radius; Ω_r—antenna rotation velocity; α—aircraft bearing; λ—wavelength of HF oscillations.

In receiver, it is possible to extract this frequency changing law in the form of voltage or signal, changes with antenna rotation frequency. Extraction is carried out by comparison in frequency of HF oscillations, picked out from rotating antenna and oscillations from non-directional antenna. Phase of extracted signal depends on direction to an aircraft, and hence, this signal is called a "variable phase" signal. It is possible to measure a phase of this signal only by comparison with a phase of another reference signal, the phase of which will not depend on aircraft bearing. Consequently, a signal is called "uniform (constant) phase" signal. "Uniform phase" signal in radio direction-finder is formed by reference oscillator.

Radio direction-finder is adjusted in that way that phases of signals will coincide in direction to north. In any other direction, a phase of "variable phase" signal falls back from a phase of "uniform phase" signal at an angle between this direction and direction to north. Consequently, by measuring a value of such backward it is possible to determine an object bearing.

Radio direction-finder antenna system consists of vertically symmetrical half-wave dipoles, equally spaced in circular, forming a ring array (Fig. 15.26), in the center of which the central (non-directional) antenna is positioned—vertical asymmetrical dipole.

Due to sequential connection of dipoles to general load (switching frequency 525 Hz), a rotation imitation of receiving point is conducted relatively to rotation

Fig. 15.26 Radio direction-finder antenna system

Table 15.2 ADF technical specifications

Parameter description	ADF-75	ADF-80	ADF-95	DF-2000
Bandwidth, MHz	118–136	118–136	118–136.975	118–137
Operational range, km				
$H = 1000$ m	–	–	100	120
$H = 3000$ m	–	–	180	200
$H = 10{,}000$ m	300	200	300	360
Bearing error, deg.	1	1.5	1	1
Channels no	1–8	1–2	2–16	2–16

center that leads to, due to Doppler effect, frequency change of receiving signal. In this case, pseudo-Doppler effect concludes in that at switching moment a phase of receiving signal is jump-like changed, and joint curve of phase change, consisting of discrete points, repeats a Doppler envelope. All modern airdrome radio direction-finders are constructed on the base of such quasi-Doppler principle and hence are related to Doppler phase direction-finders.

Re-reflections from ground features effect in general on bearing accuracy of Doppler automatic direction-finders (ADF), and including signals, which are close in frequency to ADF operating frequency. Hence, direction-finders are installed in airdrome area, on the terrain away from reradiators (power transmission lines, high-rise structures, trees, etc.) with regard to requirements for ensuring of electromagnetic compatibility.

Table 15.2 states technical specifications of some ADF types of the Russian manufacturing.

15.2 Radio Direction-Finding Goniometric Radio Navigation Systems

Onboard automatic radio compasses (ARC) are intended for determination of heading angle HA (relative bearing) of ground locator beacons (LB) or broadcasting radio stations (BCRS). Relative bearing measuring permits to solve onboard an aircraft the following tasks of aerial navigation:

- flight toward a radio station (and from it);
- determination of current coordinates of AV;
- fly-over LB;
- plotting and control of maneuvering during approach for landing, etc.

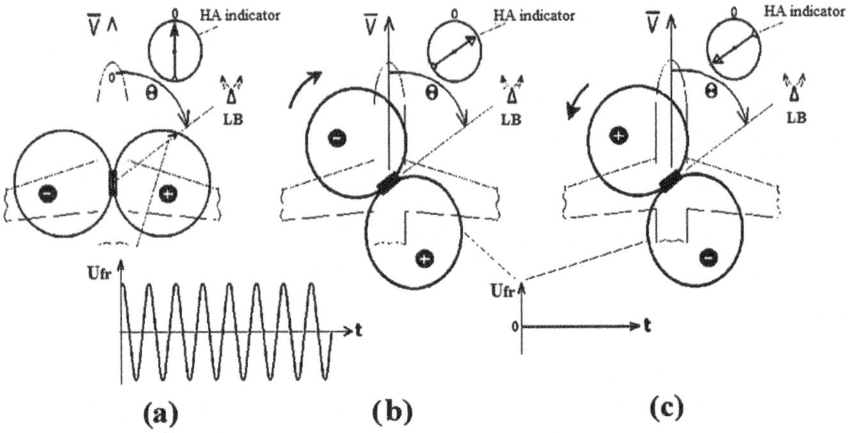

Fig. 15.27 ARC operation principle

For HA determination in ARC, the antennas with clear-cut directional properties in horizontal plane are used, which have amplitude of output signal depends on its orientation to LB. Such antennas are closed (frame) and goniometric (Bellini–Tosi) antennas.

Directional pattern of closed and goniometric antennas (further will be denoted as frame) has two zero receptions that stipulates two-valuedness measurement of HA (either correct or with error at 180°). Clarify this assertion using Fig. 15.27. We orient a frame with position of zero reception along aircraft longitudinal axis and let the readings of HA indicator here equal to zero (Fig. 15.27a). Assume that LB positioned in direction relatively to longitudinal axis at some angle $\Theta \neq 0$. Then at frame output HF oscillations are produced of corresponding amplitude and initial phase. If we rotate a frame clockwise at this angle Θ, then a signal at its output becomes equal to zero (Fig. 15.27b). As an angle of turn of a frame is transmitted for turn of indicator needle, then it is possible to take a reading of HA by its scale.

But, if we turn a frame counterclockwise till a position, at which a signal at its output becomes equal to zero, then indicator readings will differ from actual value of HA at 180° (Fig. 15.27c). Hence, depending on rotation direction of a frame and indicator needle, the measured HA value either corresponds to the actual one or differs from it at 180°.

Let's relate rotation direction with initial phase of the frame HF oscillations, e.g. at initial phase $\varphi_0 = 0°$—to the right, and at initial phase $\varphi_0 = 180°$—to the left. In this case, the measured heading angle HA values will correspond to an actual. Consequently, a rotation direction of a frame and indicator needle should be determined by the initial phase of the frame HF oscillations. But, oscillations phase can be defined only in comparison with oscillations phase, received from another source. The non-directional antenna is used as such source, the voltage of which:

$$U_v = U_v \sin \omega t.$$

If oscillation phases of frame and non-directional antenna are coincided (co-phased), then rotation direction of frame should be to the right, if in antiphase—to the left. As HF oscillations of frame are cosinusoidal, and non-directional antenna—sinusoidal, then in order to compare them (in-phase or antiphase), the oscillation phase of frame is turned further at 90° using capacitor.

Hence, the ARC principle of action in automatic measuring of HA is based on use of directional properties of frame antenna and signals phase comparison, received by frame and non-directional antennas. Because of phase comparison (in-phase of antiphase), the frame rotation direction and HA indicator needle are determined, and its turn is carried out until the moment, when a signal at frame output becomes equal to zero. Since "zero" readings of RV indicator needle correspond to frame position along longitudinal axis of an aircraft, then its turn from reference position will point a heading angle (relative bearing).

Nowadays, **ARC with inner phase modulation** are widespread used since they are more stable against different interference influence, rather than ARC with inner amplitude modulation.

The time diagrams illustrating the ARC with inner phase modulation operation principle are shown in Fig. 15.28. At output of phase switch (Fig. 15.28a), the HF balanced-modulated oscillations is formed, which have the initial phase switched through half cycles of low-frequency signal from 90° into 270° or vice versa. After summation of HF oscillations of non-directional antenna (Fig. 15.28b) and balanced-modulated, the phase-modulated oscillations are formed (Fig. 15.28c).

For time intervals t_1, t_2 and t_3 (Fig. 15.28d), vector diagrams of combined signals are shown: U_{non}—signal vector of non-directional antenna; U_{mBM}—vector of balanced-modulated (BM) signal, the direction of which differs from a direction of a previous vector at 90° or 270°. Deviation angle of combined signal vector from direction of U_{non} vector corresponds to a phase shift $\Delta\varphi$. As we can see from Fig. 15.28, a phase shift value depends on amplitude value of BM oscillation, and a sign—from relation of phases of comparing signals.

As the amplitude U_{mBM} is changed according to a law sin Ωt, where $\Omega = 2\pi F$, F—a frequency of low-frequency signal, then $\Delta\varphi$ is change according to the same law.

From Fig. 15.28e, we can see:

$$tg\Delta\varphi = \frac{U_{mBM}}{U_{non}} = \frac{U_{mBM}\max}{U_{non}} \sin \text{HA} \sin \Omega t, \tag{15.35}$$

where $U_{mBM}\max$—maximum value of BM oscillation at HA = 90° or 270°.

Phase modulation index (maximum phase deviation):

$$\Delta\varphi = arctg\left(\frac{U_{BM}\max}{U_{non}} \sin \text{HA} \sin \Omega t\right) \tag{15.35}$$

As you can see, phase modulation index depends on HA.

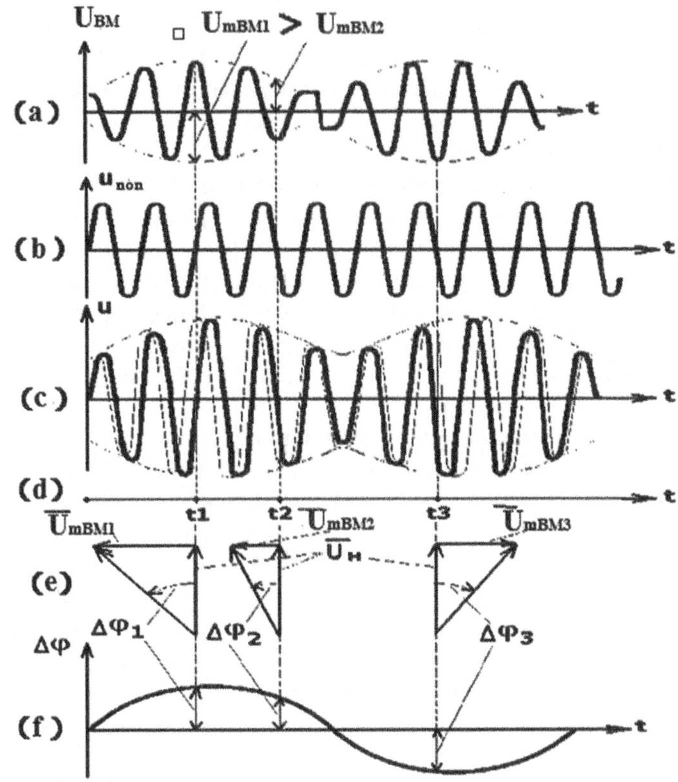

Fig. 15.28 Waveform diagrams in ARC comparison circuit with inner phase modulation

Hence, information on direction of radio waves arrival, concluding in amplitude and initial phase of a signal of frame antenna, converted, correspondingly, into index and sign of phase modulation of combined signal. At frame turn by zero reception into direction to LB, we obtain a minimum of phase modulation depth. At summation of oscillations, the resulting signal amplitude is also changed, i.e., there is also amplitude modulation with 2F frequency. This amplitude modulation is considered as parasitic, i.e., interfering, and therefore, in the further ARC cascades should be eliminated. Elimination is conducted using an amplitude limiter. In this case, we get an extra bonus, since at amplitude limiting the influence of LB call signals on operation of ARC compass part is eliminated. Using the following ARC circuits, we can extract a voltage, an amplitude and phase of which corresponds to phase change law of phase-modulated oscillation, and use it as a signal, feeding the control winding of motor.

Let's examine some facts and phenomena effecting on HA calculation error.

"**Coastal effect**" is exerted at flights near sea shore. Radio waves, radiated by locator beacons, located on surface, at intersection of two mediums borders with different electrical properties, change a propagation direction (kind of deflected).

Additional measuring error of HA, stipulated by this effect, can be 5°. Minimal aircraft altitude (height) $H_{\min}$, at which this error is minimum or even is missing.

$$H_{\min} \geq \frac{900{,}000}{f_L} \tag{15.37}$$

where f_L—LB operating frequency.

This error also decreases at LB bearing, located as close as possible to perpendicular line from aircraft to medium coastal line.

Polarization error. Mid-band radio waves can propagate along ground surface by enveloping it, and also (especially at night time) via reflection from ionosphere. At reflection from ionosphere, an electromagnetic field effects on a frame antenna, the direction vector of which comprises some angle with horizontal plane. Horizontal component of this vector induces in horizontal arms of a frame. As a result, an additional directional pattern is formed, depicted in Fig. 15.29, the sizes of which are proportional to elevation angle of spatial radio wave arrival. Direction of "zero reception" is correspondingly shifted. In Fig. 15.29, a direction of longitudinal axis of an aircraft is coincided with direction to locator beacon, i.e., HA = 0. But, due to additional directional pattern, the indicator needle will point a value of $0 + \Delta\Psi_1$ in case (a) $0 + \Delta\Psi_2$ and in case (b).

Besides, a field of reflected from ionosphere signal has an elliptic polarization, at which a vector of field electrical component rotates in air space with a frequency of a signal. This leads to minimum bluntness of a frame directional pattern that is followed with additional decreasing of accuracy.

Propagation conditions of spatial waves at presence and absence of illumination of lower ionosphere layer are different and are changed drastically in morning and

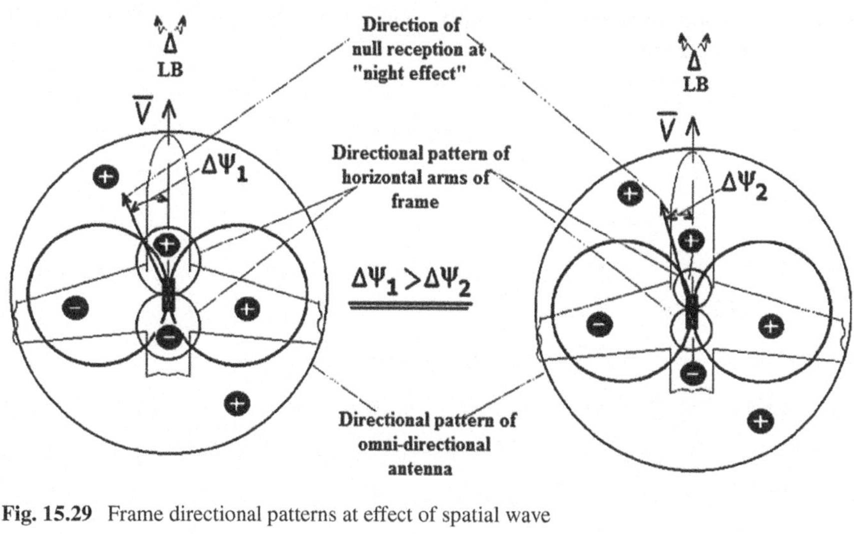

Fig. 15.29 Frame directional patterns at effect of spatial wave

evening time. Ionosphere instability is especially observable before two hours of sunset (sunrise) and during two hours after sunset (sunrise).

Herewith, not only parameters of polarization ellipse are considerably changed, but incidence angle of radio wave arrival to a frame. As a result, sizes of additional directional pattern will be changed, and HA indicator needle will swing.

At adverse conditions, the pointer swinging can be a dozen of degrees. The phenomenon was named as "**night effect**."

To decrease a bearing error, stipulated by "night effect," it is necessary to keep up with the following:

- average out the measured values of HA;
- If there is a possibility to select LB, it is better to use for bearing radio stations located at close range from an aircraft, since at small distances (dozens of kilometers) the measurements are performed at surface wave;
- If there is a possibility to select LB, it is necessary to use for bearing radio stations operating at longer wave (low-frequency waves).

Radio bearing deviation error Δ is stipulated by effect of reradiators, i.e., by aircraft construction components, which are excited under action of incoming signal and radiate its own electromagnetic field. Radio bearing deviation Δ depends on installation place of a frame and on HA. At ARC installation on an aircraft, the deviation Δ dependence from direction of radio waves arrival (radio deviation chart). Then using a deviation compensator, installed in HA data transmission system to indicator, a deviation is eliminated (deducted) until the certain minimum. Residual deviation (error) value is usually not exceeding 1…2°.

"**High-terrain (mountain) effect**" is observed in mountain terrain during flights at relatively low altitudes. In these conditions, the radio waves are coming to ARC frame directly from LB and reflected from mountains. During an aircraft, moving a direction of resulting signal is constantly changed, and indicator needle position is correspondingly changed.

Chapter 16
Hyperbolic Radio Navigation Systems

Long-range radio technical navigation systems (LRRNS) implementing the differential-range-measurement principle of an object position finding are almost not used. These systems, after successful introduction of satellite radio navigation systems SRNS and its functional updates, will not be used in near future on maritime and air transport. Even now modern aerial vehicles are not equipped with LRRNS systems, however its use is still provided on some aircraft of old types. There are no new development works on LRRNS systems.

Alongside with that, it's necessary to mention that there are developments on integrated equipment of LRRNS and SRNS, and scientific research works on deeper integrating of these systems in order to provide noise immunity increase and user equipment operation in high latitudes are carried out.

In this regard, we'll concentrate only on differential-ranging principle of an object position finding.

The existing differential-ranging LRRNS operate in low-frequency (kilometric) and myriametric waves (long and very long waves) to provide the required operation range based on use of surface radio wave propagation properties. Hyperbolic (differential-ranging) radio navigation systems (HRNS) can be related to range-measuring system without responder.

HRNS systems create a grid (lattice) of position lines representing hyperbolas with focal points in position dots of a pair of ground-based reference stations (Fig. 16.1). Distance difference ΔD, which is the main radio-navigational parameter in these systems, is defined by time interval Δt between signals receiving from the corresponding pair of ground-based reference stations (A and B):

$$\Delta D = \sqrt{\left(x_{\mathrm{gsA}} - x_{\mathrm{ob}}\right)^2 + \left(y_{\mathrm{gsA}} - y_{\mathrm{ob}}\right)^2 + \left(z_{\mathrm{gsA}} - z_{\mathrm{ob}}\right)^2} - \sqrt{\left(x_{\mathrm{gsB}} - x_{\mathrm{ob}}\right)^2 + \left(y_{\mathrm{gsB}} - y_{\mathrm{ob}}\right)^2 + \left(z_{\mathrm{gsB}} - z_{\mathrm{ob}}\right)^2} = c\Delta t, \quad (16.1)$$

D. A. Akmaykin et al., *Theoretical Foundations of Radar Location and Radio Navigation*, Springer Aerospace Technology,
https://doi.org/10.1007/978-981-33-6514-8_16

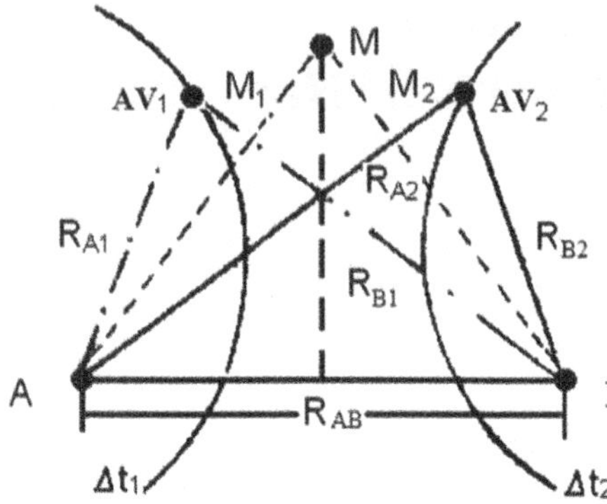

Fig. 16.1 Operation principle of pulsed HRNS

where x_{gsA}, y_{gsA}, z_{gsA}, x_{gsB}, y_{gsB}, z_{gsB}—coordinates of A and B ground reference stations.

The basic classification feature of HRNS is an informative signal parameter, i.e., a parameter of received signal (amplitude, frequency, phase, delay time), functionally related with a radio navigational parameter.

Due to informative parameter the phase, frequency, time (pulsed), and pulse-phase HRNS systems are distinguished.

The difference of radio pulses travelling time of A and B stations up to user is measured in (time-difference) **pulsed HRNS systems**. If transmitters of A and B ground-based stations are operating synchronously, then a time interval Δt between pulses at receiver output of onboard receiver-indicator will be proportional to difference of distances from M point (object position) up to ground stations. (A and B) (Fig. 16.1).

Particularly, to receiver input of user onboard equipment, positioned in M point, at a time $t_A = \frac{D_A}{s}$, where D_A—distance from M point up to A ground station, a pulse (signal) will arrive, radiated by its transmitter of A point, and at a time $t_B = \frac{D_B}{s}$, where D_B—distance from M point up to B ground station, a pulse comes of this station transmitter. A point transmitter pulse from receiver output comes to measuring circuit and is used in it as a reference. Then arrival moment of B point transmitter pulse to measuring circuit defines a current value of distances difference.

$$\Delta t = t_B - t_A = \frac{D_B}{s} - \frac{D_A}{s} = \frac{\Delta D}{s}. \tag{16.2}$$

At simultaneous arriving of pulses of A and B stations transmitters, an ambiguity arises in determination of time interval Δt, and consequently of an object position line as well.

Let's examine points of possible position of M1 and M2 objects, located symmetrically to M point, lying on perpendicular, rebuilt unto AB base center. Then $D_{A1} = D_{B2}$ and $D_{A2} = D_{B1}$, and time intervals Δt_1 and Δt_2 will be defined by as follows:

Fig. 16.2 Pulsed HRNS position lines

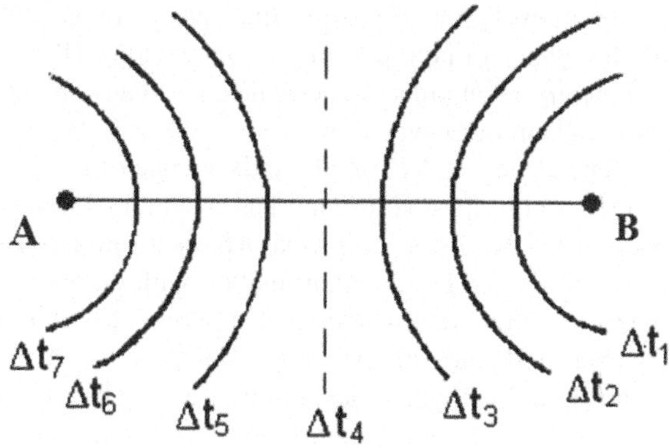

$$\Delta t_1 = t_{B1} - t_{A1} = \frac{D_{B1}}{s} - \frac{D_{A1}}{cs},$$
$$\Delta t_2 = t_{B2} - t_{A2} = \frac{D_{B2}}{c} - \frac{D_{A2}}{c} \quad (16.3)$$

and turned to be equal to absolute value $|\Delta t_1| = |\Delta t_2|$. So far as $D_{B1} > D_{A1}$, and $D_{B2} < D_{A2}$, then $\Delta t_1 > 0$ and $\Delta t_2 < 0$. Since indicator doesn't respond to a sign of time interval Δt, then an ambiguity arises of measurements results.

To exclude the read-out ambiguity, it is necessary that a condition $\Delta t > 0$ is always fulfilled, independently on what point of HRNS operating area an object is positioned. It is clear that if a radiation moment of radio pulses by B station will be delayed relatively a radiation moment of A station radio pulse at time $\Delta t_{AB} = \frac{D_{AB}}{s}$, then $\Delta t > 0$ condition will be always fulfilled.

Delay, equals to Δt_{AB}, will be observed in the case, when B station will be triggered off by pulses, radiated by A station. Such operation principle is implemented in the current-pulsed HRNS systems; herewith, A ground-based station is called a master (primary) station and B station—a slave (secondary) station. Pulses, radiated by A master station will reach M point in $\Delta t_A = \frac{D_A}{s}$ after a moment of its radiation by A master station. Pulses of B slave station will reach the same M point in $\Delta t_B + \Delta t_{AB} = \frac{D_B}{s} + \frac{D_{AB}}{s}$.

From AMB triangle (Fig. 16.1), we can see that wherever a point M (vertex of a triangle) is positioned, a value of one of its side |AM| will be always less then a sum of its two others sides |AB| + |BM|, i.e., a condition will be always fulfilled.

$$\Delta t = (t_{AB} + t_B) - t_A > 0. \quad (16.4)$$

The navigational experience shows that for an object (marine vessel, aircraft) positioning (coordinates determination) finding the special-purpose maps with marked position lines, corresponding to certain Δt (Fig. 16.2) value, was used. Position lines were plotted typographically with interval of each 50–100 μs. If it was necessary to find out not-mapped intermediate position lines, an interpolation between neighboring digitized position lines has to be used. This required a certain skill. It's necessary to note that one of the existing HRNS disadvantages is conversion complexity of an object hyperbolic coordinates in geographical ones, used often in ship- and aircraft navigation.

Dependence of ΔD on phase difference of two coherent oscillations, created at receiving point by two ground-based stations located in positions with the known coordinates, has been accepted as a basis for **phase HRNS** operation. The measured phase difference is in certain relation with difference of distances from a moving object to ground-based stations that permits to determine an object position line by the measured phase difference.

Position line (hyperbolas) family is set up by two ground-based stations forming a system base. To determine an object position, as for pulsed HRNS, it is necessary as a minimum two pairs of stations, creating position lines. An object position is determined as an intersection point of two position lines, created by different pairs of stations.

All discussions relatively to pulsed HRNS are correct for phased HRNS as well, excluding a method accepted as a basis for measuring of distance difference, i.e., phase difference measurements of two oscillations at receiving point (onboard an object).

Phase difference measurement of two oscillations can be carried out either on high (carrier) frequency or on beat (difference) frequency. Grids of position lines can be determined by a frequency, different from a frequency on which the measurements are performed. Correspondingly, phased HRNS are distinguished with grids set up of position lines: on carrier frequency; on reduced pattern frequency; beat frequency; on combination frequencies.

As an example, let's examine HRNS operation principle with phase difference measurement at carrier frequency.

For simplicity of arguments let's consider that a pair of ground-based stations radiates oscillations of the same frequency ω_0 and that operation of these stations is coincided in the way that difference of initial phases $\Delta\varphi_0$ of radiated oscillations remains unchangeable and can be admitted as equal to zero. Then current phases φ_{A} and φ_{B} of oscillations radiated by both stations will be written as follows:

$$\varphi_{\text{A}} = \varphi_{\text{B}} = \omega_0 t + \varphi_0. \tag{16.5}$$

Oscillations of A station in M point are received with a phase

$$\varphi_{\text{AM}} = \omega_0 t + \varphi_0 - \frac{\omega_0}{s} D_{\text{A}}, \tag{16.6}$$

and oscillations of B station—with a phase

$$\varphi_{\text{BM}} = \omega_0 t + \varphi_0 - \frac{\omega_0}{s} D_{\text{B}}, \tag{16.7}$$

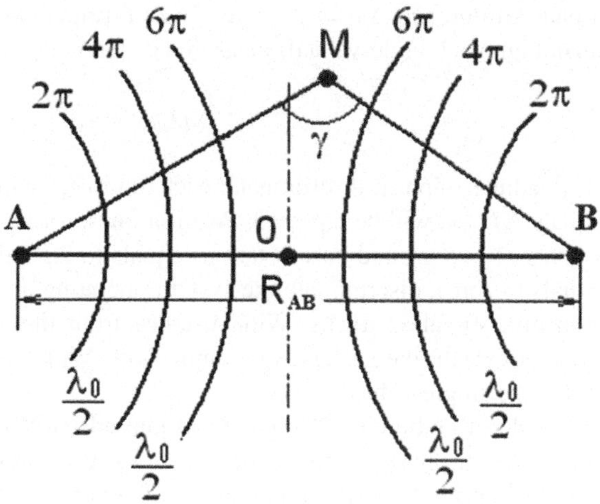

Fig. 16.3. Position lines of phased HRNS

In these formulas, D_A and D_B are distances to an object and A and B stations, correspondingly. We consider that oscillations of both stations can be received separately onboard an object by receiver-indicator, amplified and delivered to phase discriminator, measuring a phase difference:

$$\Delta\varphi = \varphi_{BM} - \varphi_{AM} = \omega_0 t + \varphi_0 - \frac{\omega_0}{s} D_B - \left(\omega_0 t + \varphi_0 - \frac{\omega_0}{s} D_A\right)$$
$$= \frac{\omega_0}{s}(D_B - D_A) = \frac{\omega_0}{s}\Delta D. \tag{16.8}$$

The measured phase difference corresponds to difference of distances:

$$\Delta D = D_B - D_A = \frac{s}{\omega_0}\Delta\varphi = \frac{\lambda_0}{2\pi}\Delta\varphi. \tag{16.9}$$

Position lines also plotted on a map and digitized from 0 up to 360°. Navigating officer, based on measured by HRNS equipment values of phase difference $\Delta\varphi$, defined a corresponding position line on a map, and when appropriate resort to interpolation (Fig. 16.3).

Phased HRNS have a high accuracy of measurement ΔD as comparing with pulsed HRNS. At measurement errors of phase difference at 1°, a linear error in determination of position line in RNS, operating in long-wave band, will not exceed (without considering conditions of radio wave propagation) values of several hundred meters.

Read-out ambiguity is inherent to phased range-difference (hyperbolic) RNS. This is explained by the fact that phase discriminator is capable to measure phase difference only within limits of full phase cycle from 0 up to 2π. By substituting a phase difference $\Delta\varphi = 2\pi$ into (16.8) formula, we obtain a maximum value of unambiguous (single-valued) measured difference of distances:

$$\Delta D_{\text{amb}} = \lambda_0. \tag{16.10}$$

Readings of phase discriminator each time after achieving of distances difference, equals ΔD_{amb}, will be repeated. Position lines, for which phase difference between signals of two ground-based stations equals to 2π value (or, put it in another way, equals to zero), discriminate the system operating area into zones of single-valued read out, so-called tracks. With distance from the system base, the track width increases. At the base, it has a minimum width. Within one track, phase measurements turn to be single-valued.

Read-out ambiguity elimination in phased HRNS is carried out via reading of full cycles of phase difference increment of received oscillations, or by creating of additional grids of position lines at use of auxiliary pattern (comparison) frequency. Stations operate at different frequencies, and phase difference measurement is carried out at reduced comparison frequency. This is caused by the fact that at HRNS operation on one frequency, the interference of fields takes place, creating a disturbed sum signal at receiving point.

Use of advantages of pulsed and phased HRNS has lead to development of pulse-phase HRNS. The relative radio signal delay in pulse-phase HRNS is defined based on information on simultaneously measured navigational parameters. Δt and $\Delta\varphi$.

To this date, several types of HRNS, relating to phased and pulse-phase HRNS, are still in service.

The phased HRNS are introduced by the Russian "Marshrut" HRNS-20 system. "Marshrut" phased HRNS (another system designation is "Alfa") has been developed simultaneously with the US Omega-phased HRNS (withdrawn from service in 1997) and brought into service in 1972. The system operational range is up to 10,000 km from a master station. Position finding accuracy is 2.5…7 km. The system has been updated by 1999 at position finding accuracy level of not worse than 1.2…1.5 km.

Nowadays, the pulse-phase HRNS are represented by the Russian "Chaika" and the US "Loran-C."

HRNS "Chaika" (another designation is "Tropik") includes several chains of ground-based station of HRNS-3, HRNS-4, HRNS-5, HRNS-10 types (mobile version). Total area of operating zones of all "Chaika" HRNS chains is about 20 million of square kilometers. The system operational range is up to 20,000 km, determination accuracy of an object hyperbolic coordinates comprises about 700 m. Stations' operating frequency of the system is 100 kHz. Nowadays, "Chaika" pulse-phase HRNS is considered as an effective supplement for GLONASS SRNS, ensuring an increase of access and integrity of the system in favor of commercial customers.

HRNS "Loran-C" (LOng RAnge Navigation)—is the US-developed ground-based radio navigation system introduced into service by the end of 1950. Loran-C system operates at 100 kHz frequency. Estimated accuracy of the system comprises 150–200 km at distance up to 1500 km. The system operational range is up to

20,000 km. There were 34 chains of ground-based stations around the USA, North Europe territory and neighboring sea areas of the Northern hemisphere.

The Loran-C system finished its operation from 2010 on the US territory and the US Loran-C stations in composition of Russian–American chain Chaika/Loran and within American–Canadian chain have terminated its operation as well. As a result, nowadays the Loran-C HRNS operation on the US territory is completely finished. It is recommended to use GPS systems for users.

Chapter 17
Radio-Navigational Speed and Drift Angle Meters

Doppler speed and drift angle meters (DSDM) are intended for determination of AV velocity vector components relatively to underlying surface and information output for aircrew and to data navigational complex, necessary for tasks performing of aircraft navigation.

DSDM are classified into airborne and helicopter-borne by the purpose. Airborne DSDM are designed for measuring of actual (ground) speed and drift angle. Nowadays, DSDM gauges are not installed on civil aircraft. An exception is only for aircraft of old types. Helicopter-borne DSDM are intended for determination of vector components of full longitudinal, transverse (lateral) and sometimes vertical velocity. The main distinction of helicopter-borne DSDM from airborne DSDM concludes that they should measure a Doppler frequency shift almost from a zero and consider its sign, and airborne, since an aircraft flies only forward with certain speed, measure only absolute magnitude (modulus) of Doppler frequency shift and with minimum values at that, differing from a zero. Helicopter-borne DSDM are more complicated in terms of design structure.

In practice, we found DSDM application with three- and four-beam antenna system. The most frequently DSDM beams are directed as ribs of regular pyramid with a base located in horizontal plane. DSDM antenna systems as a rule are rigidly body-mounted on AV, however can be installed on rotating platforms.

Let's examine DSDM operation principle in terms of three-beam system. Let's consider that DSDM antenna coordinate system is coincided with aircraft in reference system XYZ (Fig. 17.1). Determine a relation between measured values of Doppler frequencies and AV flight-path velocity projection in horizontal earth-referenced coordinate system.

For simplicity, we consider that air-velocity vector is directed along OX-axis. Flight-path velocity projections of AV on XYZ coordinate system denote as W_x, W_y, W_z correspondingly, and through $\vec{k}$, $\vec{l}$, $\vec{m}$, $\vec{n}$—unit vectors along conjugated beams.

Doppler frequency shift value for k beam is defined as follows:

D. A. Akmaykin et al., *Theoretical Foundations of Radar Location and Radio Navigation*, Springer Aerospace Technology,
https://doi.org/10.1007/978-981-33-6514-8_17

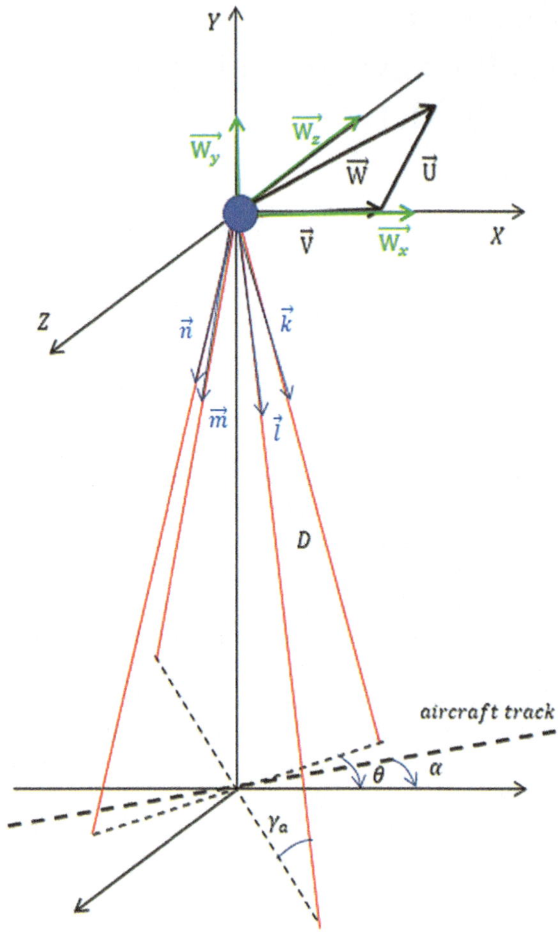

Fig. 17.1 Doppler frequencies' calculation principle of multiple-beam DSDM

$$f_D = \frac{2}{\lambda_0} W \cos \eta_k, \quad (17.1)$$

where λ_0—wavelength of emitted oscillation, η—angle between flight-path velocity vector and direction of sighting beam.

Express (17.1) via projections of flight-path velocity vector:

$$f_{Dk} = \frac{2}{\lambda_0}\left(W_x k_x + W_y k_y + W_z k_z\right), \quad (17.2)$$

where k_x, k_y, k_z—projections of $\vec{k}$ vector into corresponding axes.

By analogy with (17.2) for other beams $\vec{l}$, $\vec{m}$, $\vec{n}$ can be written as follows:

$$f_{Dl} = \frac{2}{\lambda_0}\left(W_x l_x + W_y l_y + W_z l_z\right), \tag{17.3}$$

$$f_{Dm} = \frac{2}{\lambda_0}\left(W_x m_x + W_y m_y + W_z m_z\right), \tag{17.4}$$

$$f_{Dn} = \frac{2}{\lambda_0}\left(W_x n_x + W_y n_y + W_z n_z\right). \tag{17.5}$$

Projections k_x, k_y, k_z values equal:

$$\begin{aligned} k_x &= \cos\gamma_a \cos\theta, \\ k_y &= -\sin\gamma_a, \\ k_x &= -\cos\gamma_a \cos\theta, \end{aligned} \tag{17.6}$$

where γ_a—elevation angle of beam alignment, θ—azimuth of beam boresight.

Similarly, projections of $\vec{l}, \vec{m}, \vec{n}$ unit vectors are defined. For equally spaced beams location of DSDM antenna, the Doppler frequencies are defined by the following expressions:

$$f_{Dk} = \frac{2}{\lambda_0}\left(W_x \cos\theta \cos\gamma_a - W_y \sin\gamma_a - W_z \cos\gamma_a \sin\theta\right), \tag{17.7}$$

$$f_{Dl} = -\frac{2}{\lambda_0}\left(W_x \cos\theta \cos\gamma_a + W_y \sin\gamma_a + W_z \cos\gamma_a \sin\theta\right), \tag{17.8}$$

$$f_{Dm} = -\frac{2}{\lambda_0}\left(W_x \cos\theta \cos\gamma_a + W_y \sin\gamma_a - W_z \cos\gamma_a \sin\theta\right), \tag{17.9}$$

$$f_{Dm} = \frac{2}{\lambda_0}\left(W_x \cos\theta \cos\gamma_a - W_y \sin\gamma_a + W_z \cos\gamma_a \sin\theta\right). \tag{17.10}$$

Equations (17.7)–(17.10) permit to determine vector components of AV flight-path velocity. Methods of pairwise and separate signals processing are applied in up-to-date DSDM.

At pairwise signals processing in four-beam DSDM, the radiation and receiving of signals are conducted simultaneously for two beams, e.g., *k*, *m* and *l*, *n*. Each pair of beams is turned on alternately (one by one), and signals receiving for each pair are carried out alternately. Lobing frequency (beams switching frequency) as a rule comprises 5 Hz.

At horizontal flight, the Doppler signal spectrum, simultaneously received, is equal and medium Doppler frequencies are equal $f_{Dk} = |f_{Dm}|$, $f_{Dl} = |f_{Dn}|$ (Fig. 17.2a). If AV performs horizontal flight $\left(W_y \neq 0\right)$, then medium Doppler

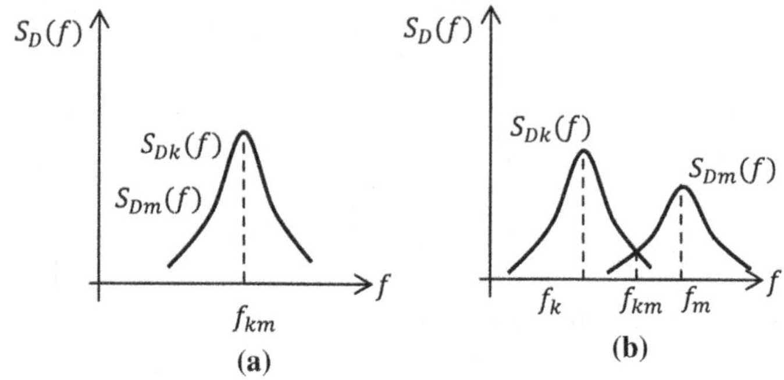

Fig. 17.2 Doppler spectra of signals

frequencies in each pair of beams become diverse, i.e., $f_{Dk} \neq |f_{Dm}|$, $f_{Dl} \neq |f_{Dn}|$. In this case, spectrums' medium frequencies of each pair of beams are calculated (Fig. 17.2b):

$$f_{Dkm} = \frac{1}{2}(f_{Dk} + |f_{Dm}|), \tag{17.11}$$

$$f_{Dln} = \frac{1}{2}(f_{Dl} + |f_{Dn}|). \tag{17.12}$$

Considering (17.7)–(17.10), we can write down the following:

$$f_{Dkm} = \frac{2}{\lambda_0}(W_x \cos\theta \cos\gamma_a - W_z \cos\gamma_a \sin\theta), \tag{17.13}$$

$$f_{D\,\mathrm{ln}} = \frac{2}{\lambda_0}(W_x \cos\theta \cos\gamma_a + W_z \cos\gamma_a \sin\theta). \tag{17.14}$$

As can see, at pairwise signals processing, an information on W_y is faded, and determination algorithm of angle drift and ground speed vector is reduced to the following expressions:

$$tg\alpha = \frac{f_{Dln} - f_{Dkm}}{f_{Dln} + f_{Dkm}} ctg\theta, \tag{17.15}$$

$$W = \lambda_0 \frac{f_{Dln} + f_{Dkm}}{4\cos\theta \cos\gamma_a \cos\alpha}. \tag{17.16}$$

At separate signals processing, an information is processed on each beam. In this case, expressions for calculation of ground speed vector of (18.7)–(18.10) are excessive and we can step to three-beam systems, for which values of ground speed vector components are calculated in accordance with following formulas:

Fig. 17.3 Dependence of specific RCS from wave incident angle

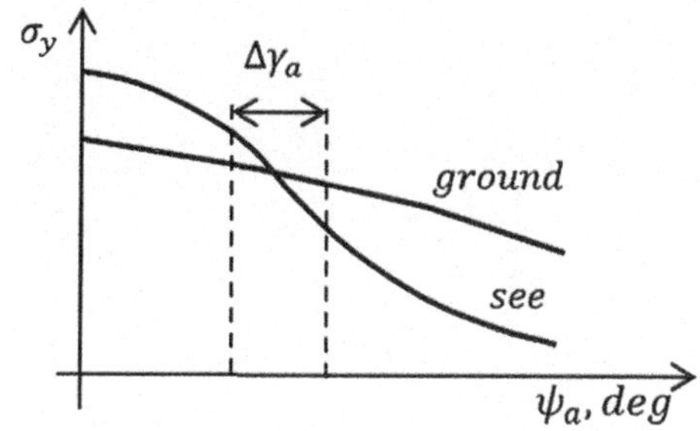

$$W_x = \lambda_0 \frac{|f_{Dl}| + f_{Dk}}{4\cos\theta\cos\gamma_a}, \tag{17.17}$$

$$W_y = \lambda_0 \frac{|f_{Dm}| - f_{Dk}}{4\sin\gamma_a}, \tag{17.18}$$

$$W_z = \lambda_0 \frac{|f_{Dl}| - f_{Dm}}{4\sin\theta\cos\gamma_a}. \tag{17.19}$$

Drift angle is defined from the following relation:

$$tg\alpha = \frac{W_z}{W_x} \tag{17.20}$$

It is obvious that separate signal processing is preferable for helicopter-borne DSDM.

Modern DSDM operate at continuous non-modulated of frequency-modulated oscillations mode in frequency range 8.8–9.8 GHz and 13.25–13.4 GHz.

DSDM operation error depends largely on trend of reflecting surface. At flight performance above sea surface, radio wave reflection coefficient is changed abruptly within a directional pattern of DSDM antenna. As a result, Doppler signal spectrum experiences a distortion. The dependence of specific radar cross section from wave incident angle is depicted in Fig. 17.3. Evidently, that within limits of a beam width ($\Delta\gamma_a$) the value of specific RCS is slightly changed, and for sea surface, these changes are more significant.

As a result of influence of so-called sea effect, DSDM will give out underestimated ground speed and probably a value of drift angle. In up-to-date DSDM gauges, this influence is eliminated by corrections (adjustments) introducing when reflecting surface trend is changed.

Bibliography

1. Bakulev PA Radar systems: textbook for universities. In: Bakulev PA (ed) 2nd, reprint
2. Botov MI, Vyakhirev VA (2013) Fundamentals of the theory of radar systems and complexes. Textbook. SFU, Krasnoyarsk
3. Bots MI (2012) An introduction to the theory of radar systems: monograph. In: Bots IM, Vyakhirev AV, Devotchka V, Bots MI (eds) Siberian Federal University, Krasnoyarsk
4. Brylev GB, Gashina SB, Nezdoiminogo GL (1986) Radar characteristics of clouds and precipitation. Gidrometeoizdat
5. Demidenko PP (2003) Ship radar and radio navigation systems. Phoenix, Odessa
6. Dudko BP (2003) Radionavigation: textbook. TUSUR, Tomsk
7. Shirokov Yu F (2012) Fundamentals of the theory of radar systems. SSAU, Samara
8. Theoretical foundations of radar. In: Shirman JD (ed) Sov. Radio
9. Sosulin Yu G (1992) Theoretical foundations of radar and radio navigation. In: Radio and communications
10. Finkelstein MI (1984) Fundamentals of radar. In: Radio and communications
11. Sosnovsky AA, Khaimovich IA (1982) Aviation radar. Guide. In: Davydov PS (ed) Transport
12. Dulevich VE (1978) Theoretical foundations of radar. Training manual for universities, 2nd edn. In: Korostelev AA, Klyuev NF et al. (eds) reprint and add. Sov. Radio
13. Karutin SN (2014) Differential correction and monitoring of global navigation satellite systems (GNSS). MSU Publishing House, Moscow
14. Kondratenkov GS, Yu FA (2005) Radio vision. Earth-sensing radars. Textbook for higher educational institutions under. In: Kondratenkov GS (ed) Radio engineering
15. Sokolov AV (ed) (2003) Questions of perspective radar: monograph. In: Radio engineering
16. Akhmedov RM et al (2004) Automated air traffic control systems: new information technologies in aviation: textbook. Manual. In: Pyatko SG, Krasov AI (eds) Polytechnic, Petersburg
17. Kozlov AI, Logvin AI, Sarychev VA (2005) Polarization of radio waves. KN. 1. Polarization structure of radar signals. In: Radio engineering
18. Kozlov AI, Logvin AI, Sarychev VA (2007) Polarization of radio waves. KN.2. Radar polarimetry. In: Radio engineering, Moscow
19. Kozlov AI, Logvin AI, Sarychev VA (2008) Polarization of radio waves 3. Radiopolarimetry of complex signal structures. In: Radio engineering
20. Solovyov Yu A (2003) Satellite navigation and its applications. In: Eco-Trends
21. Labels MS (1985) Statistical theory of radio navigation. In: Radio and communications
22. Labelkov MS (2017) Meander noise-like signals (BOC-signals) and their varieties in satellite radio navigation systems. In: Radio engineering

D. A. Akmaykin et al., *Theoretical Foundations of Radar Location and Radio Navigation*, Springer Aerospace Technology,
https://doi.org/10.1007/978-981-33-6514-8

The manufacturer's authorised representative in the EU is Springer Nature Customer Service Centre GmbH, Europaplatz 3, 69115 Heidelberg, Germany. If you have any concerns regarding our products, please contact ProductSafety@springernature.com

Printed and bound by CPI Group (UK) Ltd, Croydon, CR0 4YY
30/09/2026
02225251-0009